DO 0330488 4

AF616288

BOURNEMOUTH
UNIVERSITY
LIBRARY
WITHDRAWN

HARDWARE DESCRIPTION LANGUAGES

IEEE Press
445 Hoes Lane, P.O. Box 1331
Piscataway, NJ 08855-1331

IEEE Solid-State Circuits Society, *Sponsor*
SSC-S Liaison to IEEE Press, Stuart K. Tewksbury

IEEE Circuits and Systems Society, *Sponsor*
CAS-S Liaison to IEEE Press, Gordon Roberts

Cover design: William T. Donnelly, *WT Design*

Technical Reviewer

Frederick J. Hill, *University of Arizona*

HARDWARE DESCRIPTION LANGUAGES

Concepts and Principles

Sumit Ghosh
Arizona State University

IEEE Circuits and Systems Society, *Sponsor*

IEEE Solid-State Circuits Society, *Sponsor*

Stuart K. Tewksbury, Joe E. Brewer, *Series Editors*

The Institute of Electrical and Electronics Engineers, Inc., New York

This book and other books may be purchased at a discount
from the publisher when ordered in bulk quantities. Contact:

IEEE Press Marketing
Attn: Special Sales
445 Hoes Lane
P.O. Box 1331
Piscataway, NJ 08855-1331
Fax: (732) 981-9334

For more information about IEEE Press products, visit the
IEEE Press Home Page: http://www.ieee.org/organizations/pubs/press

Printed in the United States of America

10 9 8 7 6 5 4 3 2 1

ISBN 0-7803-4744-7
IEEE Order Number PC5791

Library of Congress Cataloging-in-Publication Data
Ghosh, Sumit, 1958–
Hardware description languages : concepts and principles / Sumit Ghosh.
p. cm. – (IEEE Press series on microelectronic systems)
ISBN 0-7803-4744-7
1. Computer hardware description languages. I. Title.
II. Series.
TK7885.7.G46 1999 99-27111
621.39′2—dc 21 CIP

In memory of my loving parents

Contents

Preface

My association with hardware description (design) languages for digital systems dates back to 1981 when, as a graduate student at Stanford University, Professor Willem vanCleemput taught us EE481: Computer-aided design (CAD) of digital systems. Prior to that, Professsor R. M. K. Sinha at IIT Kanpur had thoroughly enthused us by teaching computer architecture design, utilizing AHPL, the APL-based hardware programming language, proposed by Professsors Hill and Peterson in their book on a hardware programming language (AHPL). My own Ph.D. dissertation on an Ada-based distributed hardware design language and simulation environment with Professsor W. vanCleemput, my investigations at Fairchild Advanced Research & Development Laboratory (Palo Alto) on a behavior-level fault simulator in ADLIB-SABLE, my research at Bell Labs Research with Dr Erik Debenedictis, Dr. Meng-Lin Yu, and Dr. P. A. Subrahmanyam, my collaborations with my Ph.D. advisees—Peter Walker at Brown University and Jerry Schumacher at Arizona State University—and my interactions with Professsor Norbert Giambiasi of the University of Marseilles in France, have imparted to me a deep appreciation and understanding of the fundamentals of HDLs. This, I wish to share with my readers.

My motivation for writing the book is three-fold. First, it is an expression of my gratitude and it is my humble effort to reciprocate the kind teachings of all of my teachers and the extraordinary books that had educated me in the art of thinking from first principles. Second, in the course of my research with my Ph.D. advisees and while teaching CSE 517: Hardware Design Languages (HDL), to computer science and electrical engineering graduate students at Arizona State University and the technical staff from Motorola Inc., Intel Corp., VTI, etc., in the Phoenix valley in Fall 1996–7, I realized that for many of the key HDL concepts, the reasonings and fundamental principles were either scattered in journal or conference papers or were simply unavailable. Following a week-long take-home exam, many of the students had remarked to me that, despite reading most of the available books on VHDL and a thorough search of the literature utilizing DIALOG, INSPEC, UNCOVER, and WWW, they could not obtain the relevant material. Professsor Giambiasi had remarked to me that, while teaching VHDL in France, his students found many of the language constructs as unnatural, deviant from the underlying hardware design principles, and difficult. Third, throughout my education, my teachers in high school and my professors at Indian Institute of Technology, Kanpur, and at Stanford University had always stressed the importance and long-term value of understanding the basics. When one understands the underlying principles and can relate them to one's own basic understanding, learning becomes fun and enjoyable. It becomes a captivating and intoxicating exercise in creativity. One acquires a fine, subtle, sense of the whys underlying the whats and one can "see" where the future is headed. The existing knowledge no longer appears to be shrouded in mystery but is seen as natural. The need for memorization followed by regurgitation disappears. Most important of all, the understanding and the knowledge become firmly integrated with one's own thought processes. I have made every effort and it is my earnest hope that, when the reader has reached the end of this book, the evolution of the HDLs, the complexity of the language constructs, and the seemingly unending possibilities of complex interactions between the different language constructs, will evoke a single thought, "Yes, it is what it should be." At the same time, the reader will "see" where the HDLs must be headed in the future.

The field of HDL is rich and fascinating for it represents an unique interaction of a number of disciplines from electrical engineering and computer science—namely, language design, hardware design and computer architecture, compilers, language environments, simulation, distributed algorithms, parallel processing, and yes, philosophy. An effort to understand the basics, starting from the first principles, promises a number of additional advantages. First, it reveals the reason for each specific language construct, leading to a better appreciation of how to use it to realize the maximum benefit. As a result, one evolves into a superior decision maker and develops superb digital designs. Second, one develops a deeper understanding of the whats and whys of the underlying limitations. As circuits become faster and more complex in the future, one is better equipped to address the problems and may even initiate changes to update the HDLs. Third, knowledge of the fundamentals provides a continuity of understanding through the hundreds of publications in the literature on HDLs. Fourth, fundamentals are very convenient since they are usually few in number and they reflect a highly condensed and crystallized form of knowledge. The fundamentals are like a tiny seed that holds the many details of the huge banyan tree of the future. Fifth, virtually all of the HDLs to-date including Verilog HDL and VHDL execute only on uniprocessors, implying excruciatingly long simulation times for large hardware systems. The two key reasons include the lack of an intricate and correct weaving of concurrency into the HDLs and the absence of a successful asynchronous, distributed event driven simulation algorithm that detects and discards inconsistent events while being free of deadlocks. This book addresses both limitations, starting from the first principles. Last, should a new HDL come around in the future, one that is based on new principles superseding today's time-based and event driven simulation principles, to one conversant in the basics it offers little resistance to understanding and mastering it.

My own experience with CSE517 at Arizona State University has been particularly revealing. At the onset, I had stated to the class that I would start from first principles and proceed from one HDL to the subsequent one, starting with the first HDL and working our way towards the state-of-the-art in hardware design language—VHDL. In the process, we will examine each evolutionary development, reasoning about it from the fundamentals and analyzing its purpose and utility from the underlying principles of hardware design and the primary intent of HDLs. A number of students, however, were driven by their desire to quickly learn VHDL to either complete their assignments at work or to secure jobs in the semiconductor industry. Their impatience persisted through the first half of the semester until we started to critically analyze the VHDL constructs. Suddenly, as it were, the entire class stumbled upon the realization that the basic principles of HDL design form a continuous thread through all HDLs starting with the first one and that the elements of VHDL fall straight out from these basic principles. They could literally see through the past 30 years of HDL efforts and, if asked, they could even undertake the design of the principal elements of VHDL from scratch.

This book presents HDLs as a science, not a mysterious elite art. It starts with clear and unambiguous logical principles that are grounded in physics, reality, and the first principles. The book argues and reasons about these principles and from it develops practical mechanisms for describing hardware, constructing simulators, synthesizing executable hardware descriptions, and executing them on a computer. Not only will these basic principles provide the key to understanding the secrets of all HDLs ever invented, they will continue to remain in effect even if the current HDLs are replaced with new ones in the future. The book has been written for practicing electronic CAD engineers, researchers in simulation and verification of electronic CAD, graduate and doctoral students in computer design, and undergraduates specializing in electronic hardware design.

The book starts from the very basics and assumes only that the reader is familiar with the rudiments of digital design and is willing to reason, think, and introspect. My conviction in this Socratic style of education is based on my sincere belief that all knowledge is inherent in the human mind and it needs to be brought forth by the right stimulus. The book has been so structured that anyone with a minimal background can reason from the fundamentals and discover the principles for oneself. For one adept in HDL usage, knowledge of the enormous possibilities and fundamental limitations will foster greater creativity in achieving higher quality hardware designs.

Chapter 1 introduces HDLs, traces the origins of the early HDLs, why they were invented, and what is the fundamental definition of an HDL. Chapter 2 follows the evolutionary development of the early HDLs, revealing every major improvement achieved by the successive HDLs. It also analyzes the fundamental differences between HDLs and programming languages to show how the HDL effort was able to maintain its unique identity as distinct from the general-purpose programming languages. The analysis presented in Chapter 2 leads directly into the issue of behavior-level HDLs. Chapter 3 presents the fundamentals of any behavior-level HDL, arguing from a combination of first principles, physics, and the underlying philosophy of this universe. It is probably the most basic and revealing of all the chapters in the book and, upon completion, it is my hope that the reader will have acquired a vision that will allow him or her to comprehend the past 30 years of HDL effort in a nutshell. Chapter 4 examines the first behavior-level HDL very carefully and critically, and aims to strengthen the principles of Chapter 3 through an example. Chapter 5 analyzes the Verilog hardware description language proposed by Cadence, Inc. in the light of the fundamentals. Chapter 6 presents an exercise towards a simple and effective HDL design. Chapter 7 analyzes the massive VHDL effort in the light of the fundamentals established in Chapter 3. It analyzes both its achievements and weaknesses, reasons the causes for the limitations, and presents mechanisms to address the limitations. Chapter 8 illustrates the development of hardware descriptions from first principles through a series of case studies—three real-world complex digital systems. A thorough understanding of these representative example systems will provide adequate knowledge to the reader to undertake the writing of accurate hardware descriptions of just about any complex digital system. Chapter 9 presents the principles underlying the concurrent simulation of the HDLs in general and VHDL in particular, and explains why this effort is unique in the discipline of Computer Science and

Engineering. Chapter 10 addresses the issue of transport delay that is gaining increasing importance in the current era of higher VLSI densities, higher clock speeds, and newer bus design techniques. This chapter is unique in that it shows how two apparently diverse branches of knowledge—languages and grammars from computer science and the classical transmission line theory in electrical engineering—must come together to solve a problem. Finally, Chapter 11 summarizes the book and presents some parting thoughts on where HDLs are headed in the future and relevant philosophical reflections.

Sumit Ghosh
Arizona State University

Acknowledgments

I am indebted to the U.S. Army Research Office and the BMDO Office for their continued encouragement and support. Sincere thanks are due to Dr. Peter Walker, H. J. (Jerry) Schumacher, Professor Norbert Giambiasi, Dr. Willem vanCleemput, the entire CSE 517 class (Fall 1996–7), Qutaiba Razouqi, Ricardo Citro, and Dr. S. S. Joo. I am indebted to the anonymous referees for their highly constructive suggestions. My deepest gratitude is for Professsor Stuart Tewksbury, invisible waves of hope and encouragement from whom inspire me continuously.

To the readers, I enthusiastically welcome your thoughts, comments, and feedback.

Sumit Ghosh
Arizona State University

Acknowledgments

List of Figures

List of Tables

1

The Origin of HDLs

1.1 INTRODUCTION

A key characteristic shared by all HDLs is that they form a subset of computer programming language designs. The dictionary definition [1] of a language is any method of communicating ideas, as by a system of signs, symbols, gestures, or the like. While natural languages such as English satisfy the above definition, they lack the precision and the absence of ambiguity that constitute the key requirements of every computer programming language intended for execution on precise computers. The imprecision and ambiguity in natural languages is a direct consequence of their broad scope and the desire for creative expression. In contrast, the desire to execute programs, written in the language, on computers to obtain accurate and unambiguous results inevitably limits the scope of the programming language. The character of computer programming languages is defined by the syntactical rules that constitute a manifestation of the grammar and the semantics of the language constructs that define the meaning or effect following execution. The general-purpose programming language community believes [2, 3] that the semantics of a language embodies its fundamentals and that a study of several programming languages, with the semantics of each language exhibiting distinct features, is likely to illuminate the issues involved in selecting a language for a specific application. While this is true and is corroborated by our ordinary experience that a person conversant in multiple languages exhibits superior mastery over languages, there is another, alternate point of view.

If the intent of a computer programming language is known, i.e. one has the knowledge of the characteristics of the underlying process(es) that need to be expressed in the language, and the expectations following execution of a program on a computer(s) are specified, one may develop a basic set of (not necessarily canonical) language constructs and associated semantics that satisfy the intent. Clearly, the semantics will be a function of the underlying process(es) that are modeled in the language and the fundamental characteristics of the host computer(s) on which the executable programs are executed. The desirable properties of such a language would include accuracy of representation, precision of the execution results, faithfulness to the intent, simplicity, fast execution, and "naturalness" [3], i.e. the appropriateness of its elements relative to the underlying process(es). Given that the characteristics of the digital hardware processes (designs) are known, there is strong hope that this approach may successfully apply to HDLs.

There is another very important requirement of all computer programming languages. A language encapsulates the complete knowledge of all processes within its scope of modeling. That is, it must permit a programmer to express, accurately and precisely, every possible executable program that the programmer may design without violating any of the syntactic and semantic rules laid down by the language. Clearly, the syntax and semantics together must encapsulate the total character of the underlying execution. Thus, the language designer is required to carefully synthesize the grammar and the semantics of each of the language constructs, taking into consideration every possible and legitimate interaction between the individual constructs, and guaranteeing that accurate and unambiguous results are generated following the execution of any legitimate program. Any failure in this guarantee may seriously jeopardize the language design effort. In addition, for greater acceptance by the community and effectiveness of usage, the language designer is required to make every effort to realize straightforward and simple-to-use language constructs. The guideline to the language designer includes the precise purpose of the language which, in turn, imparts to it a mandate, a well defined set of characteristics, a clear scope, and a precise set of limitations. In our case, the intent is to capture the hardware design process faithfully and accurately and to yield fast and accurate results following the execution of the programs written in HDL.

A key characteristic of digital hardware designs is timing, especially relative timing, which underlies the occurrence of all hardware activities or events. In essence, timing constitutes the external and observable manifestation of the fundamental, causal relationship between the activities. Timing determines the correctness of and directly affects the performance of digital designs. While the functional correctness is of primary concern with the simpler, combinatorial designs, for synchronous sequential designs, timing is extremely important and for asynchronous sequential designs, it is absolutely critical. One cannot but marvel at the subtle intricacies of the timing of information flow among tens of thousands of tiny wires in a tiny integrated circuit chip that is probably flying a huge airplane or controlling a massive electric power plant. Clearly, the issue of timing is integral to HDLs and the principal question is: What is the nature of timing from the perspective of HDLs?

In our reality, i.e. at the level of human comprehension, time is synonymous with the wall clock which ticks every second. At best, our senses and faculties can grasp time ranging from a fraction of a second to possibly tens of years. Although many digital designs may operate in the time units of seconds, the subcomponents generally operate at nanoseconds to picoseconds, i.e. at rates a billion to a trillion times faster than our reality. Last, when a digital hardware is modeled by a program written in an HDL and executed on a computer(s), the occurrences of the hardware events are labeled by time instants within the execution. The nature of the modeling permits one to control the time during the execution at will—step through linearly, non-linearly, compressed, or dilated. For instance, the passage of an electron through a P–N junction, actually requiring only 1 ns, may be modeled and its execution may require an hour to complete on a computer. Conversely, the cycle of birth and death of stars spanning billions of human years may be modeled and executed in 5 minutes on a computer. Thus, the question is; What is the common thread, if any, that binds these different notions of time? Fundamentally, this universe is composed of space, time, and causation [4]. Everything that we know or can possibly know must be subject to causation. According to the principle of causality, for every cause there is an effect and for every effect realized there must have been a cause. Thus, causality is the primary, fundamental truth and it constitutes the common thread that binds the different notions of time. In turn, time serves as an external manifestation of the causal relationship between two or more events. Given that cause C bears an effect E, the time value of E must be greater than that of C, at least by an infinitesimal amount. In contrast, however, given two events at two distinct time values, a causal relationship is not necessarily implied. This fundamental nature of causality is further amplified by the following scenario. Consider a design with two unconnected components that are obviously located at two geographically dispersed points in space. For each individual component, two successive hardware events may be labeled by two distinct values of its own notion of time. However, for any two events, one in each component, respectively, no remarks can be made with certainty about their simultaneity in time since, according to Feynman [5], it follows from the theory of relativity that we can never determine their simultaneity in an absolute sense. Simultaneity of events is strictly local. In contrast, where the two components are connected together, the principle of causality is capable of characterizing precisely the presence of any causal dependence. According to Feynman [5], although the law of conservation of energy continues to hold true, nature is tricky in that energy is continuously being transformed into a large number of forms that appear at unlikely places. In theory, nature can play similar tricks with causality, making the search for the cause and effect most elusive. Much to our relief, however, a basic assumption has held true in hardware designs in that, if two components are causally related, there must exist an electrical connection through which an electrical signal propagates from one component to the other.

For digital hardware designs, the external stimulus at the primary inputs constitute the primary cause. In general, one or more of the stimuli may be asynchronous, i.e. occur irregularly in time. For instance, to a digital controller that manipulates the signals at an intersection, the arrival of vehicles is generally asynchronous. That is, relative to its understanding of the progress of time which is encapsulated by its clock, the controller may never know precisely when a vehicle will arrive at the intersection. However, for the controller to make a decision about whether to allow a vehicle to pass or stop, the knowledge of the time value of the vehicle's arrival must first be understood in terms of its own clock. This process is termed synchronization and it is generally achieved by the use of synchronizers. While the most fundamental synchronizer is the flip-flop, sophisticated synchronizers utilized in the communication between two or more systems operating asynchronously are designed around the basic flip-flop. Under these circumstances, in theory [6–8], it is not possible to guarantee that the device specifications—setup and hold—will be satisfied by the input signals. When setup and hold time constraints are violated, the result is malfunctions that are transient, hard to correct, and occasionally devastating. In general, most processes in nature appear independent and asynchronous to us humans. Examples include the emission of alpha particles in radioactive decay, the appearance of earthquakes, and the occurrence of accidents. One could argue that it is due to our ignorance of the precise causal relationship and its characteristics, as is stated in the Law of Karma and hypothesized by Penrose [9] in the book, "*The Emperor's New Mind Concerning Computers, Minds, and the Laws of Physics*." Synchronous processes, on the other hand, are generally man-made and they reflect the character of our comprehension. Examples include the meeting of the students and the teacher at 8:40 A.M. in Room 100 for the Math 101 class or the US citizens voting for their President on November 4. As a result of the ubiquity of asynchronous inputs, virtually all hardware and software systems, including distributed networked systems, are prone to failure at any time. To date, we have not acquired the knowledge with which to eliminate this uncertainty completely. Clearly, both digital hardware designs and HDLs inherit this fundamental limitation.

1.2 ON THE NATURE OF HARDWARE VERSUS SOFTWARE

Since HDLs relate to hardware, we will first focus on the definition of hardware and understand its basic characteristics. Generally speaking, the notion of hardware creates in our mind an image of transistors, gates, flip-flops, and CPU boards. Software, on the other hand, evokes the image of programs and code. However, it is not uncommon to find a computer in which the entire operating system, traditionally a piece of software, is stored in a ROM, a traditional piece of hardware. Conversely, one can write program fragments to emulate just about every major piece of hardware, including flip-flops and CPUs. In fact, that is precisely what HDLs are designed to accomplish. Even if we were to admit that the distinction between hardware and software is increasingly blurred, we would still cling to these terms in an intuitive way, in both theory and experimental circles. To examine whether there is a real distinction between them, consider the principal theme of the stored program control concept. Imagine that we have an objective—to compute the sum of the first N natural integers.

Clearly, we need a piece of hardware, i.e. a collection of gates, that can perform actual addition. At this point, we have two choices. First, we can either design the hardware such that it reads the first N natural numbers and computes their sum. Such a hardware design will be inflexible and potentially uneconomical. Second, we can incorporate flexibility in the hardware design so that it may be utilized to perform a number of different functions at different times. The stored program control [10] concept, attributed to von Neumann, is the most basic concept underlying computers and it supports the second choice. It argues that the flexibility may be manifested through a software "program." That is, the software possesses the ability to realize the execution of different functions with the same hardware. This is termed "control." Clearly, the program is a physical entity and the need to "store" it gives rise to the notion of memory or storage. It is critical to note that, while it encapsulates flexibility, the program or software is inert, i.e. it cannot execute on its own. Only the hardware can execute it. The hardware is labeled the central processing unit. In summary, therefore, under von Neumann's stored program control, the program encapsulates the control and while it is executed by the hardware, it, in turn, directs the hardware to perform a specific execution. The program is inert and it cannot control the hardware all by itself. It needs the active cooperation of the hardware. Thus, the relationship between the hardware and the software is symbiotic. The definition of control [11] is derived from a strict function, i.e. an entity or a signal is determined to be control when its state must be known in detail to compute the overall system behavior. Under specific circumstances, such as the execution of the "jump" statement, the otherwise sequential execution of the program is interrupted by the hardware generating a negative value, say, and the hardware could appear to exercise control. However, only the program software qualifies as control since its state must be known in detail to compute the overall system behavior at any instant of time. Thus, while the separation between the hardware and the program is increasingly blurred, the issue of control will distinguish the entity representing the stored program from that corresponding to hardware.

Fundamentally therefore, software is the inert flexibility and hardware is the rigid vehicle that can execute operations and the software. Their relationship is such that one without the other is useless. A natural question is which of them holds the intelligence.

Although software appears to be the most likely candidate of the two, it is inert and cannot manifest its intelligence on its own. Therefore, neither the hardware nor the software contains the intelligence. In truth, the intelligence lies in the human designers, the ones who conceive of them and put them to use through programs and hardware modules.

1.3 THE ORIGIN AND DEFINITION OF HARDWARE DESCRIPTION LANGUAGES

Given an input–output function, an architect may succeed in designing a combinatorial hardware if the function is simple or may need to resort to a sequential design where the function is complex. In the event that a sequential design is necessary, generally the designer's first preference is to synthesize a synchronous design utilizing the well developed theory of synchronous finite state machines. Where an asynchronous design is implied, while the individual synchronous modules are synthesized utilizing the synchronous finite state machine theory, the asynchronous interactions between the modules require elaborate analysis. While manual design may be adequate for relatively simple, moderate-sized input–output functions, for today's complex input–output functions, manual design techniques are overwhelming and impractical. At first, to relieve the designer of the clerical and administrative burden of design, computer program aids were developed, one of the first examples being IBM's software for wire list preparation [12]. Then, tools to capture the interconnection of high-level computer system building blocks such as the PMS [13] were developed. Gradually, hardware design languages emerged as a practical vehicle, as early as the 1970s, to facilitate the design of complex hardware in the same way that for developing large complex programs, assembly language programming must yield to high-level general-purpose programming languages.

The immediate question is: What are hardware description languages? Given that the hardware design process is complex and evolving, the literature lists four key definitions of HDLs, in chronological order. According to Su [14], in 1974, HDLs are languages that are descriptive yet efficient for simulation and automatic hardware implementation. Given that the digital engineering process views hardware at different levels of abstraction, conceivably the development of a number of appropriate and convenient languages for the different levels is warranted. vanCleemput [15] observes that the complexity in the hardware design process requires a successive or iterative refinement strategy which must be facilitated by HDLs. He enumerated the four major applications of HDLs: (1) documentation of the behavior specification to allow communication between designers and users and as a tool for teaching hardware design and computer architecture, (2) as an input to the system level simulator, (3) as an input to the automatic hardware compiler or synthesizer, and (4) as an input to the formal verification system. Subsequently, in 1984, vanCleemput and Ofek [16] revised the list of applications and dropped item (4). According to Baudet and colleagues [17], an HDL is a class of representations which, given a known sequence of operations, leads to a hardware realization. They clarify that the definition allows for several levels of abstraction or design hierarchy with behavior-level being at the top and structural at the bottom. Barbacci and Uehara [18] view HDLs as the bridge between software and hardware and define HDLs as high-level languages that permit designers to express asynchronous operations, parallel control, and the structure of a hardware system.

Analysis of the four definitions reveal that HDLs must describe hardware, support different levels of abstraction, permit the execution or simulation of hardware descriptions to allow for the successive refinement of the design, and facilitate the realization of hardware implementations. Barbacci and Uehara's claim that HDLs must describe asynchronous operations and parallel control is not accompanied by any supporting reasons. What is missing from the above definitions is a basic understanding of the purpose and role of HDLs.

Fundamentally, the goal of an HDL is to model in a host computer any hardware system, synchronous or asynchronous, at any given level of abstraction, as faithfully and accurately as possible. HDLs must facilitate hardware descriptions that are natural and easily comprehensible to the designer. Once developed, the model or the hardware description serves as an accurate replica of the original (or intended) hardware—whether already built or currently under development. Thus, the results obtained from executing the model on the host computer must match those of the actual hardware system. The model may then be either functionally simulated for design correctness and performance estimation, provided as an input to a tool for checking formal correctness, or used as an input to a hardware synthesis tool. Clearly, without a host computer, HDLs are of limited value in exactly the same manner that the ordinary natural languages would be meaningless if no human beings were around.

Thus, HDLs allow a hardware designer to describe accurately the target hardware design or an existing hardware system. The primary goal is to execute the description on a host computer so as to detect design flaws and errors and yield performance estimates. The execution of a hardware description on a host computer mimics the operation of the actual hardware and is termed functional simulation. Thus, the role of the host computer is to emulate the target hardware. The description may imply ease in the comprehension of an otherwise large and complex design, serve as documentation for the design, and facilitate teaching the hardware design process and computer architecture. Other uses of hardware descriptions in HDLs include the ability to perform fault simulation and test generation to aid in developing reliable designs, timing verification to quickly ensure the timing accuracy of the designs, formal verification of designs, and automatic synthesis of the target hardware. Conceivably, a description of a hardware at a given level of detail may not be ideally suited for all of the different uses. For example, while accurate delays are essential for functional simulation and timing verification, current synthesis tools are essentially limited to zero delay assignments. This underscores the need to know the basic philosophical aims of an HDL and all of its potential uses, prior to the design of the HDL.

To summarize, the objectives of HDLs are to

- Provide a means of describing a circuit either through a structural model, i.e. employing the physical elements utilized to implement the circuit, or a behavior model, i.e. the behaviors of the physical models are emulated in software.
- Permit the use of multiple models of the same circuit, to facilitate simulation, synthesis, etc.
- Provide a foundation for simulating circuits by encapsulating the concept of timing, including delays.
- Provide a foundation for defining a circuit completely, independent of a specific implementation. The definition will constitute a starting point for developing an implementation for a specific technology (hardware/software) today as well as future realizations in advanced technologies.

2

Evolutionary Development of the Early HDLs

Although HDLs have been in existence since the 1960s [15], they gained visibility in the 1970s following a concerted effort to increase their usability in the digital design process. The earliest predecessors of HDLs included software programs [12], developed at IBM, for automating logic diagrams where the designer was relieved of the clerical functions of drawing, saving, and reproducing interconnected networks of blocks representing Boolean functions. These programs motivated the development of the early HDLs at the logic design level, termed gate-level simulators, where the syntax of the language allowed the designer to specify instances of gates from a limited vocabulary of gate types, and the interconnection. The semantics of the language included the logical behavior of the gate types and was intimately linked with the operation of the underlying simulator. Examples include the LAMP simulator [19] (Logic Analyzer and Maintenance Planning), SALOGS [20] (Sandia Logic Simulation System), and TEGAS [21] (TEst Generation And Simulation system). Although logic-level simulators were popular in the late 1970s and early 1980s, as the digital systems experienced growth, the large number of gates caused difficulty in comprehension and also slowed the simulation speed. To address these problems, register transfer-level languages such as CDL [22, 23] (Computer Design Language), DDL [23, 24] (Digital System Design Language), and AHPL [10, 23] (A Hardware Programming Language) and architectural simulators such as ISP [25] (Instruction Set Processor notation) were invented. As a result, designers focused on the higher levels of abstraction where the digital modules were complex and represented the equivalent of tens to hundreds of gates. These developments proved timely since the host computers had also become powerful and were capable of quickly executing the sophisticated descriptions of the complex digital modules. At the same time, sophisticated tools such as SDL [26] (Structural Description Language) were developed to capture the evolving multiple levels of abstraction in digital design and would later prove very useful in behavior-level HDLs.

A unique characteristic of the early HDLs was their ability to represent information including signal values in vector form [27]. This implied efficient usage of precious memory and faster processing and proved especially valuable in parallel fault simulation of a circuit under test. A number of machines could be simulated utilizing a single Boolean operation.

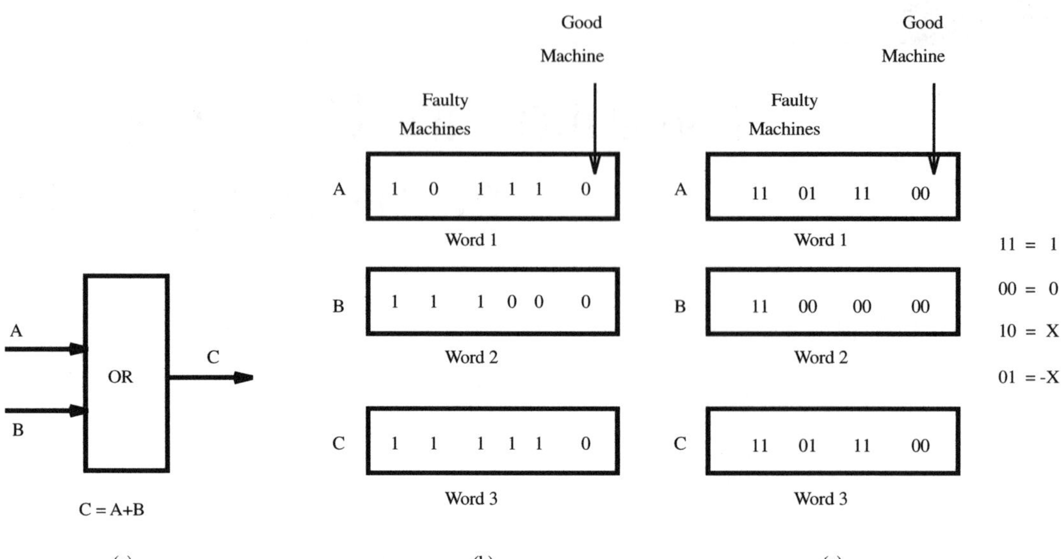

Figure 2.1 Representing signals in vector form in the early HDLs.

One of the machines is a good machine, i.e. the circuit with no faults, and each of the other machines represents a copy of the circuit in the presence of a specific fault.

Consider Figure 2.1, developed along the lines of an example in [28], which represents an OR gate (Figure 2.1(a)) under parallel fault simulation. The input pins are labeled A and B and the output pin is C. Assuming a 2-valued logic representation, i.e. {0,1}, Figure 2.1(b) describes computer words 1 and 2 which hold the signal values corresponding to pins A and B. The rightmost bit of words 1 and 2 contain the bit value for the good OR gate, i.e. with no fault, while other bits of the words 1 and 2 contain the bit values corresponding to the faulty machines, i.e. the OR gate in the presence of specific faults. A bit-wise Boolean OR operation between the words 1 and 2 in the computer yields word 3 which represents the output signal values of pin C for the good and faulty machines. This ability to simulate multiple machines with a single Boolean operation implies high efficiency. Where a 4-valued logic representation, $\{0, 1, X, -X\}$, is used, as shown in Figure 2.1(c), 2 bits of the computer word are required, at a time, to store each signal value. Clearly, this reduces the number of machines that may be simulated through one Boolean operation while it yields higher precision. As in Figure 2.1(b), word 3 is obtained in Figure 2.1(c) by applying the bit-wise Boolean OR operation to words 1 and 2. It represents the output signal C for the good and faulty machines.

2.1 LAMP (LSI-LOCAL)

The LAMP system [29] was developed at Bell Labs in the late 1960s for logic simulation, fault simulation, timing simulation, and automatic test pattern generation for circuits, expressed at the gate level. The HDL used to describe the circuit was termed LSI-LOCAL and its vocabulary included the basic logic elements: NANDs, NORs, ANDs, ORs, and NOTs. The language processor first performed syntax checks, i.e. for missing parameters and illegal characters, circuit connectivity, and consistency checks such as fan-in/fan-out limits. Then, the language processor compiled the description into tables to be used by the underlying simulators. Although LAMP has been extended to functional simulation, i.e. utilizing blocks such as encoders, flip-flops, multiplexors, etc., this chapter focuses solely on its gate-level structure.

Figure 2.2, copied from [29], presents a gate-level circuit of an exclusive-OR function, and the corresponding circuit description in LSI-LOCAL is shown in Figure 2.3. The LAMP description starts with a name of the circuit, under the label CKTNAME, and then lists the inputs and output. In this case, A and B are the inputs and X is the output. The circuit contains only two types of gates—NOT and NAND—and their numbers are 2 and 3, respectively. For each line, describing a specific gate instance, the four fields are the type of the gate, the name attributed to the gate, the list of inputs, and an optional output. Upon completion of the circuit description, the designer may specify the type of simulation desired through the command language structure. The LAMP system will execute the simulation for the user specified input stimulus and display the output results.

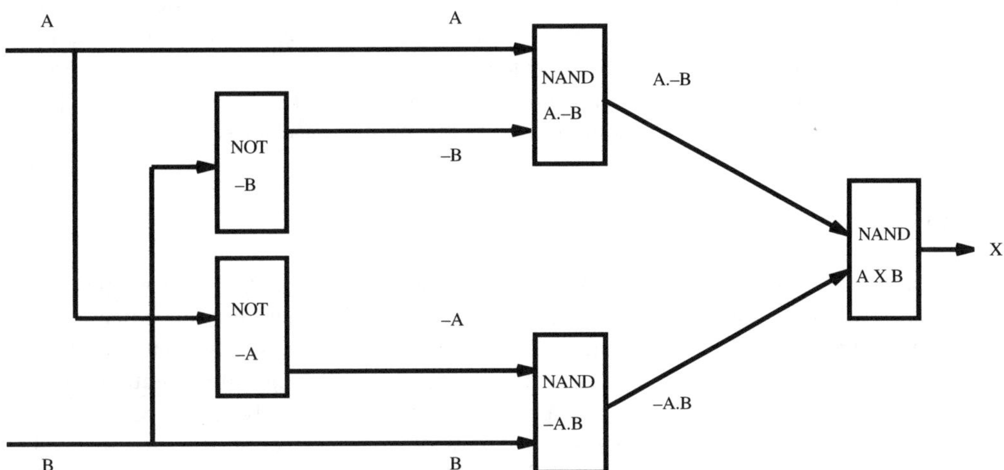

Figure 2.2 A gate-level exclusive-OR circuit.

The execution semantics may be described as follows. The 5 statements, L1 through L5, correspond to the two NOT and three NAND gates. The gates are concurrent entities and, therefore, the statements may be executed in any order. Furthermore, a gate may be executed as soon as a new stimulus is received at any of its input ports. The gates corresponding to L1 and L2 receive stimulus from the external world and since their outputs serve as inputs to the three NAND gates, none of the statements L3 through L5 may be executed prior to the completion of execution of L1 or L2. Consider that, at an instant of time, a new stimulus is received at input A. The NOT gate corresponding to the statement at label L1 executes, following which the statements at labels L3 and L4 execute and then the statement at label L5. Subsequently, a new stimulus is received at input B causing the NOT gate corresponding to the statement at label L2 to execute. Thereafter, the statements at labels L3 and L4, respectively, execute where the NAND gates use the current value of B and $-B$ and the

```
CKTNAME: XOR;
  INPUTS: A,B;
  OUTPUTS: X;
L1:NOT:   -A,          A;
L2:       -B,          B;
L3:NAND:  A.-B,        (A,-B);
L4:       -A.B,        (-A,B);
L5:       A X B,       (A.-B, -A.B) (X)
          (gate name)  (input list) (optional output)
```

Figure 2.3 A gate-level exclusive-OR circuit described in LSI-LOCAL.

previously received values of A and −A. Then, the gate corresponding to the statement at label L5 executes.

Thus, the timing model reflects concurrency relative to the execution of the gates. The issue of concurrency is, however, purely academic, since the underlying host computer is a uniprocessor. As described in [29], LSI-LOCAL, no delays are associated with any gate and, in theory, the execution times of the gates are determined solely by the arrival times of the external stimulus at inputs A and B. While this may not pose a problem for combinational circuits, as in Figure 2.2, serious inconsistencies may result during the simulation of sequential digital systems with zero delay gates [27]. The absence of the propagation delays also greatly diminishes the utility of the simulation results.

2.2 CDL

Computer Design Language (CDL) [22, 23] was proposed in 1965 to describe the computer elements and hardware algorithms at the register-transfer level, just above that of electronics, i.e. the gate level. The methodology consists of five basic steps: describing the digital design problem, choosing a storage structure, formulating an algorithm, setting up a control structure, and finally, describing the design in CDL. Clearly, CDL marks a significant departure from the gate-level HDLs in that it is no longer confined to simply hardware but aims to model control. Entertain the following scenario. A 5-bit register, A, contains a word that must be complemented, given that the operation supports complementing only the rightmost bit of the register.

Clearly, the problem must be solved in 5 steps, with one bit of the total of 5 bits complemented at each step. Furthermore, following complementing a bit, the content of A must be rotated, i.e. the least significant bit must be ejected and then stored at the most significant bit position, so that the subsequent bit is placed for complementing at the next step. To keep track of the 5 steps, a 3-bit counter is required which can count up to at least 5. To indicate the end of the operation, a FINI register may be used while a start signal may be used to start the operation. The synchronous finite state machine, by definition, implies the use of a clock which is represented by P. Figure 2.4(a) lists the necessary resources and Figure 2.4(b) represents the flowchart reflecting the algorithm. To realize the algorithm, a controller in the form of a 3-state finite state machine is designed and also shown in Figure 2.4(c). The three states of the finite state machine are labeled through $T = 100$, $T = 010$, and $T = 001$, respectively. The transition from one state to the subsequent state is synchronized through a clock pulse, P.

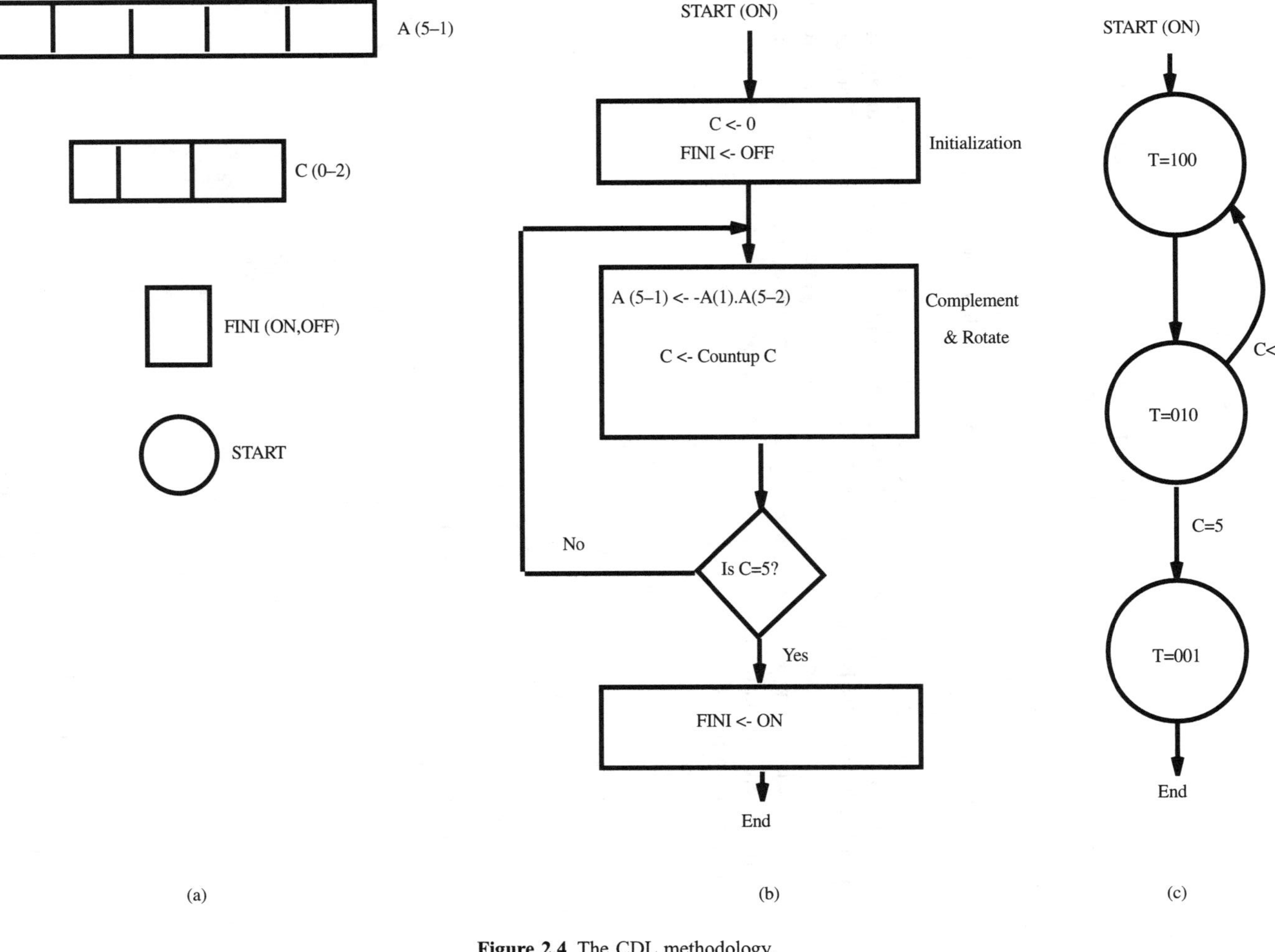

Figure 2.4 The CDL methodology.

```
/START(ON)/        T <- 100
/T (=100) * P/     T <- 010
/T (=010) * P/     IF (C=5) THEN T <- 001 ELSE T <- 100
/T (=001) * P/     T <- 0
```

Figure 2.5 The CDL control description.

The CDL control description, shown in Figure 2.5, utilizes guarded commands. That is, the statement on the right is effected only when the Boolean expression, enclosed within the slashes, evaluates to TRUE. Thus, the state is set to 100 when the START signal is TRUE. When the machine is in state given by T = 100 and the clock pulse, P, is TRUE, the system undergoes a state transition to T = 010. When the system is in state T = 010 and a clock pulse P is asserted, the subsequent state transition is dependent on whether the counter C is equal to 5 or not. That is, when the complement and rotate operations have not yet executed for 5 times, the system returns to the previous state, T = 100. Otherwise, the task is complete and the final state is given by T = 001. It may be pointed out that an advantage of using the guarded commands is that the individual lines in the CDL control description are rendered concurrent and execution of the control description will yield correct results regardless of the order in which they may be arranged. Figure 2.6 presents the complete CDL program. A complement operation is indicated by a dash preceding the variable and concatenation is denoted by a period.

```
Register A(5-1), C(0-2), T(1-3),
         FINI(ON,OFF)
Switch   START(ON)
Clock    P

/START(ON)/       T <- 100, C <-0, FINI <- OFF
/T (=100) * P/    T <- 010, A(5-1) <- -A(1).A(5-2), C <- Countup C
/T (=010) * P/    IF (C=5) THEN T <- 001 ELSE T <- 100
/T (=001) * P/    T <- 0, FINI <- ON

END
```

Figure 2.6 The complete CDL program.

2.3 AHPL

A Hardware Programming Language (AHPL) [10, 23] was introduced as an HDL at the register-transfer level. It employs the notational conventions of the APL programming language and utilizes only those operations that may be readily interpreted as hardware primitives. AHPL is based on the philosophy that most digital designs may be organized through a control section and a section containing data registers and logic. This philosophy underlies both CDL [22] and DDL [24]. However, AHPL differs from both CDL and DDL in that it lacks guarded commands and that the program sequence, similar to conventional programming languages, is designed for sequential execution. AHPL permits both clocked assignments to registers and asynchronous combinatorial assignments. The methodology in AHPL consists of first specifying the problem, then designing an algorithm, and finally synthesizing the AHPL code that reflects both the control flow and the data operations.

Consider the task of describing a digital system in AHPL that identifies the sequence 0011 from a string of 0s and 1s and generates an output 1. Although AHPL can accommodate both Mealy and Moore style synchronous finite state machines, this example will assume a Moore model. Figure 2.7 presents the synchronous, finite state machine with five states. Except for state 1, labeled initial, the four states are labeled as state 2 through state 5. The variable X represents the input string of 0s and 1s while Z represents the output. A special START signal is assumed to initiate the operation and bring the system to state 1, the initial state. Figure 2.8 presents the AHPL description.

In Figure 2.8, the five states are represented by the five labels or steps, 1 through 5. The output Z is set to 0 at every step except at step 5 when the sequence 0011 has been identified and, therefore, the output Z is 1. At step 1, the transfer statement implies the following. If the input signal X is 0, indicated by −X in the second half of line 1 under step 1, the subsequent state is 2. Otherwise there is no state transition. The "−" symbol implies a complement operation on the signal with which it is associated. Given that the flow of control is normally sequential, even if the transfer statement is modified to "`->(X x 1)`," the code would execute correctly. That is, as long as the input signal remains at 1 there is no state transition. In each of state 2 through state 4, the transitions corresponding to $X = 0$ and $X = 1$ are explicitly stated. For example, in step 3, there is a state transition to state 4 when the input, X, is 1. Otherwise, for the input, X, equal to 0, the system remains at state 3.

Although the underlying timing model of AHPL restricts it to synchronous digital designs, Hill and Peterson [10] present AHPL as capable of modeling asynchronous subsystems as well as a monostable multivibrator (one-shot). However, this capability is incomplete due to the following reasons. First, every step in AHPL is assumed to require one clock period. Second, the one-shot is included in the standard AHPL syntax as "ONE-SHOT; ysyn (6)," where the width, 6, of the one-shot refers to 6 clock periods. Both of the above requirements are incompatible with the asynchronous design philosophy wherein the clocks of every subsystem are unique, independent, and unrelated. Third, Figure 5.21 [10, page 152] models an asynchronous subsystem consisting of two modules A and B that interact with one another. AHPL requires the insertion of a synchronizer module, external to B, that accepts B's clock to synchronize the signal from A prior to using it within B. This implies AHPL's inability to realize an independent description of module B, which is critical to model asynchronous systems.

2.4 ISP

The primary goal of ISP [25] was to model computers at the instruction set level. ISP is termed an architectural HDL because of its ability to model computer architectures, i.e. processors, memories, I/O devices, and their interconnections. Through hiding the lower-level implementation details, ISP presents a high-level view of an instruction set computer and can aid in verifying the efficiency of the chosen instruction set, compute the traffic between registers, memory, and other computer elements, and estimate performance for representative programs. Fundamentally, however, it is very much a register-transfer level language since it

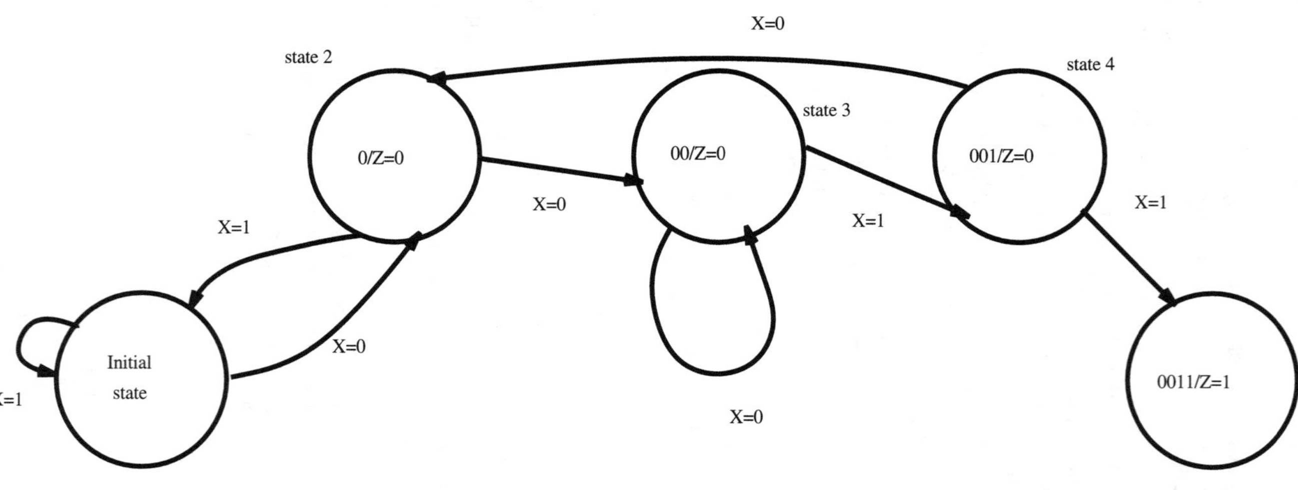

Figure 2.7 Modeling a synchronous, finite state machine in AHPL.

```
1. -> (X x 1 + -X x 2)
   Z=0
2. -> (X x 1 + -X x 3)
   Z=0
3. -> (X x 4 + -X x 3)
   Z=0
4. -> (X x 5 + -X x 2)
   Z=0
5. Z=1
```

Figure 2.8 The AHPL description.

assumes a synchronous computer design, i.e. the presence of a single clock, and it primarily models register transfer and memory operations. Thus, ISP cannot model asynchronous memory operations, a characteristic shared by many of the modern computers. ISP is also unable to model computers that are developed under the data-flow paradigm. The core ISP philosophy centers around the instruction set model of computers where the basic operations are instruction fetch, instruction decode, operand fetch, execution, and result store. An unique feature of ISP is the ability to describe concurrent and sequential statements through the use of the "next" clause. Two statements that are terminated by semicolons are viewed as concurrent while the presence of "next" forces them to execute sequentially. The concept is useful only as a teaching tool since ISP lacks provisions for any concurrent execution. Figure 2.9 presents a simplified ISP description of a IBM S/370 processor. At the top of the description, all of the registers and memory elements are declared. The main executable program fetches the instructions, decodes, and executes them, one at a time. Between two consecutive instructions, the main procedure executes the interrupt servicing program.

2.5 COMPARATIVE ANALYSIS OF HDLS

HDLs, by definition, must describe hardware. Thus, gate-level tools such as LSI-LOCAL qualify as true HDLs since they describe only hardware. In contrast, both register-transfer level and architecture-level tools expanded the role of HDLs with the aim of describing systems, i.e. hardware plus control. Although CDL, DDL, AHPL, and ISP are capable of describing control, they are limited to synchronous finite state machines and the instruction set processing paradigm of instruction fetch, decode, operand fetch, execution, and result store. While, in theory, CDL, DDL, AHPL, and ISP can describe concurrent operations, they are subject to strict synchronous control.

The evolution beyond CDL, DDL, AHPL, and ISP appears to be dictated by a key limitation that is shared by all of them. Analysis of the literature reveals that the term behavior-level HDL has been grossly over-used and has even been used synonymously with architecture-level HDLs. In truth, behavior-level HDLs differ uniquely from gate-level, register-transfer level, and architecture-level HDLs and mark a significant turning point in the evolution of HDLs. They possess the ability to model asynchronous systems. The motivation for the evolution to behavior-level HDLs had been presented in Chapter 1 and is briefly recapitulated here.

We human beings prefer synchronous systems for the sake of convenience, to invoke a sense of order, and for ease in debugging such systems. For instance, the Presidential election in the US (and most countries) is pre-arranged to take place on a specific date of the year and all US citizens from the East Coast to the West Coast are asked to cast their votes. Despite the

```
S370:=
 begin declare
       Memory[O:FFFFFF]<0:7>;                         Primary Memory
       R[0:15]<0:31>;                                 General Purpose
                                                      Registers
       PSW<0:63>                                      Processor Status Word
       ....                                           Other Registers
       eralced
Run:= begin                                           Main Executable Program
      IFetch:= begin                                  Instruction Fetch
               MAR <- PSW<40:63> next                 Initial Instruc.
                                                      Address
               Instr<0:15> <- Memory[Mar:Mar+1] next  Read First Half-Word
               PSW<32:33> <- Instr<0>+Instr<1>+1 next Instruction Length
               PSW<40:63> <- PSW<40:63> + PSW<32:33>*2 Program Counter
               next
               ....                                   Fetch Rest of Instruc.
               end;
     IExec:= begin                                    Instruction Execution
             decode Instr<0:1> =>                     Select Instruction Type
             RR:= begin                               RR Instr. Decode Table
                 (decode Instr<2:7> =>                Select RR Instructions
                 end;
             RX:= begin                               RX Instr. Decode Table
                 Mar <- Instr <20:31> next            Displacement
                 (if Instr<16:19> => Mar<-Mar + R[Instr])        next  Base
                 (if Instr<12:15> => Mar<-Mar + R[Instr<12:15>]) next  Index
                 (decode Instr<2:7> =>                Select RX Instructions
             end;
             RSS1:= begin                             RSS1 Instr. Decode Table
                   ...
                   end;
             SS:= begin                               SS Instr. Decode Table
                  ...
                  end;
             end;
     INT:= begin ... end next                         Interrupt Handling
     Run                                              Repeat Main Procedure
     end
end
```

Figure 2.9 ISP description of a simplified IBM S/360 processor.

fact that the date may be highly inconvenient to many of the individual citizens, the importance of the election exceeds the individual convenience, and the arrangement has stayed. Thus, the election constitutes a synchronous operation. In contrast, virtually all processes in nature are asynchronous, i.e. they occur irregularly in time. Since most real world systems interact with one or more of the natural processes, their behaviors are also rendered asynchronous. Consider the design of a digital traffic light controller. Although the controller may be developed as a synchronous design, the fact that cars are likely to arrive at the intersection irregularly in time renders the overall system asynchronous in nature. Thus, as systems grow in complexity, the overall system behavior tends to be asynchronous in nature even if the behaviors of its constituent individual modules continue to remain synchronous.

2.6 ON THE RELATIONSHIP BETWEEN HDLS AND GENERAL-PURPOSE PROGRAMMING LANGUAGES (PLs)

Thus far, we have traced the evolution of HDLs from gate level to register-transfer and architecture level and have examined the argument for the subsequent transition to the behavior level. A key question emerges: What are the similarities and fundamental differences, if any, between PLs and HDLs? The question presupposes the thought that PLs have been around longer, the development of HDLs is relatively recent, and that the language designers must justify their HDL efforts, as distinct from the PLs. In truth, however, HDLs and PLs have been developed in parallel and, in the past, they have had very different needs and goals.

Early PLs such as FORTRAN aimed at computing scientific formulas in much the same way that the normal human brain performs, only faster and more accurate. Thus, PLs support straightforward sequential execution of programs, with the exception of branching, which appear to be consistent with the von Neumann paradigm. In contrast, early language designers of HDLs were interested in modeling hardware at the level of digital gates. Their needs included (1) instantiating each gate in the circuit, (2) describing the behavior of the gates through Boolean equations, (3) specifying their interconnections, and (4) specifying the timing behavior of the gates through their propagation delays. While the early PLs could accommodate the requirement (2) superbly, they lack the notion of objects, object interconnection, and timing delays since there is no natural need for them. Baudet *et al.* [17] criticize PLs for completely ignoring timing. Even for real time program development that share the same characteristics as hardware development, PLs continue to utilize the traditional programming language constructs that lack any notion of timing requirements or characteristics. Furthermore, and more importantly, while the programs in PLs execute sequentially, except for the branching statements, the order of execution of the gates is determined by the timing of the external signals at the primary input ports of gates, timing of the signals generated by a few gates and asserted at the input ports of other gates, and the relevant gate delays, not necessarily the order of their arrangement in the circuit. The input stimulus to each gate, i.e. its logical value and timing, is dynamic, i.e. unknown a priori. Thus, the order of execution of the gates may not be known a priori and there is no known way to pre-arrange the program statements corresponding to the different individual gates in the circuit, as warranted by the PLs. HDLs impose a unique requirement—the development of a scheduler—which drives the wedge between PLs and HDLs even deeper. As expected, the structure of the scheduler and the HDL exert a profound influence on each other.

PLs and HDLs have differed in the past and will continue to differ in the foreseeable future because of their fundamentally different philosophies, emanating from their different goals. The goal of PLs is to hide the lower-level details such as hardware, delays, etc., from the user and provide a view into the behavior of the program at a higher, more conceptual level. There are good reasons underlying this goal. PLs do not advocate ignoring or hiding the lower-level details in a derogatory sense. They merely concentrate on the high-level view, given the limited comprehension ability of the human brain. Second, many of the lower-level details that may appear to pose a problem today may experience change or undergo transformation, led by technological improvements, etc., and disappear with the progress of time. For instance, conceivably, a hundred years from now or sooner, digital hardware may be superseded by optical hardware. In contrast, HDLs aim to help design computers and are naturally concerned with the hardware, timing of the hardware modules, the temporal interactions between the modules, and other low-level details. By necessity, HDLs must model hardware faithfully and accurately. The resolution of precision, if such a term may be coined, is much finer for HDLs than those of the PLs.

The common objective between PLs and HDLs is that they both need to describe behavior, unambiguously, precisely, and accurately. Thus, later HDLs [17], such as ADLIB-SABLE (Algol Derivative Language for Indicating Behavior), and PLs such as Pascal and C, both appear to embrace the notion of data structure, user defined types, strong data typing, structured programming, procedures and functions, exception handling constructs, and a host of loop control statements such as "while," "for," "do," "repeat," and "loop" coupled with "exit." Some of the modern PLs, such as SIMULA, Modula, Ada, and C++, have even embraced the concept of objects and object instantiation to keep pace with the increasingly popular paradigm of object oriented programming. The emergence of the object oriented paradigm may be attributed to three reasons: (1) the desire and need to tackle very large and complex software programs, (2) our improved comprehension of abstract software objects and their manipulation, and (3) the emergence of powerful processors that quickly execute the hundreds to thousands of lines of code, inherent in an object, providing a meaningful and realistic feel for object manipulation.

As hardware became more complex and extensive, encompassing thousands of gates, the hardware design paradigm embraced hierarchical design, i.e. one starts with a high-level description and then successively refines it to the lower levels until, at the lowest level, a VLSI chip is realized. Consequently, HDLs were forced to evolve and incorporate hierarchical representation, for which there appears to be no natural need among PLs at this time.

A key observation is that the sheer number of programmers in the world who use PLs, relative to the HDL users, has caused tremendous resources to be focused on PLs. As a result, PLs are more sophisticated in that they embody the advances in compiler optimization and even the instruction sets of many commercially available processors are designed to execute them fast. They are viewed as readily available and more stable. Since HDLs lack such widespread use, it is logical for the HDL community to take advantage of PLs, wherever possible.

3

Fundamental Requirements of Behavior-level HDLs

The fundamental goal of a behavior-level digital HDL is to model in a host computer any hardware system, synchronous or asynchronous, at any given level of abstraction, and the model or description must be as faithful, accurate, natural, and easily comprehensible to the designer as possible. Following development, the description serves as an accurate replica of the original hardware and the results of executing the model on a host computer must match those of the actual hardware system. To understand the nature of behavior-level HDLs, it is critical to first understand the fundamental characteristics of hardware and the current view of hardware. These characteristics also constitute the fundamental requirements of behavior-level HDLs. The characteristics of hardware are defined through the concepts of (1) entity, (2) connectivity, (3) concurrency, and (4) timing, while the designer's current view of hardware embraces the notion of (5) hierarchy.

3.1 ENTITY

The concept of entity is fundamental in this universe and also to our understanding of the universe. Although the universe and all knowledge about it may be one continuous thread to the creator, our understanding is that the universe, at any level of abstraction, consists of entities. The American Heritage Dictionary [1] defines the word entity: "The fact of existence, being. Something that exists independently, not relative to other things." Thus, an entity exists and its existence is guaranteed independent of all other entities. Because it exists, an entity must be self-contained, i.e. its behavior, under every possible scenario, is completely defined within itself. Because the entity exists independent of all other things, its behavior is known only to itself. Unless the entity shares its behavior with a different entity, no one has knowledge of its unique behavior. Conceivably, an entity may interact with other entities. Under such conditions, its behavior must include the scope and nature of the interactions between itself and other entities. The notion of entity is deeply rooted in our philosophy and culture. The Yoga Vasistha [30] writes, "In the infinite consciousness, there are countless universes which do not know of one another's existence." The idea finds support in the present day findings of modern cosmology. The principle of individualism, upon which the US Constitution is founded and one that believes in each individual human being as an important person in society with an independent will, is synonymous with the concept of entity. Individuality [31] reflects a separate and distinct existence—a sense of I—a continuous current that is atomic at a given level of representation.

In the discipline of digital systems, hardware is organized through entities, at any level of abstraction. Examples include an AND gate at the gate level, an ALU at the register-transfer level, and the instruction decode unit at the architecture level. For each entity, its behavior must include all that is relevant at that level of abstraction, such as the logic, timing, control, exceptions, etc. Since every entity is independent of others and no one entity has explicit knowledge of another entity's behavior, conceivably each entity may possess its own notion of timing including clocks, timing constraints, delays, etc. Thus, for an HDL to successfully describe a number of such entities constituting a system, it must necessarily be capable of supporting asynchronous behavior. Philosophically, asynchrony appears to be a manifestation of the concept of entity. However, one may argue that asynchrony is the more basic of the two and the point is therefore debatable.

In the event that the control and timing discipline of all entities are identical, as in the case of a synchronous design, an HDL capable of modeling only synchronous behavior may be adequate. Examples include CDL, DDL, AHPL, and ISP. Under this condition, the individual hardware units are not treated as entities since a single scheduler, operating under the synchronous finite state machine paradigm, controls all of them. Similarly, in the gate-level simulators, although every gate instance is distinctly enumerated, gates are not viewed as true entities since the underlying scheduler has complete access to the behavior of every gate. In contrast, the effort to model hardware units in the behavior-level HDLs such as ADLIB-SABLE [32] and VHDL [33] (Very high speed integrated circuit Hardware Description Language) reflects a superior, not necessarily complete, understanding of the concept of entities, as we will examine subsequently.

Although we may desire the behavior of every entity to be self-consistent, it does not necessarily follow from the concept of entity. However, one may argue that the issue of existence, independent of time, provides the motivation for self-consistent behavior of the entities. We will assume that in hardware systems, every entity is characterized by a self-consistent behavior.

3.2 CONNECTIVITY

Although they are independent, entities may choose to interact with one another. In general, a single entity all by itself is probably not all that interesting. A single gate is clearly not interesting. A single, stand alone, un-networked computer, though exciting a decade ago, is hardly considered useful today. Clusters of concurrent computers that we find fascinating today may come to be viewed as mundane in the future. Perhaps, as argued by many of the ancient philosophies, that is why the creator is said to have created diverse beings in the universe whose interactions with one another is synonymous with the gradual unfolding of the immense possibilities and creativity in nature. In the event that entities interact with one another, the scope and nature of entity A's interactions with other entities, B, C, . . . , must be completely specified, i.e. for all possible scenarios, within the behavior of A. The interactions between the specific entities may be captured through connectivity information. In general, an entity A may establish a connection with entity B for a limited duration, then establish another connection with entity C for a specific duration. However, in the current hardware design paradigm, the connectivity information is, in general, permanent and, therefore, static. If hardware modules evolve to acquire mobility in the future, appropriate HDLs may need to be designed. A key rule of connectivity between entities is as follows. Where A is connected to B via a link, neither A nor B can forcibly access the other's internal behavior. B can read only what A asserts on the link, and vice versa. There is no limitation on the nature of the information that may be placed on the link, extending from a simple integer as in gate-level HDL to a complex data structure as in ADLIB-SABLE and VHDL. However, it is logical to expect both A and B to understand each other's asserted data, without ambiguity. For digital hardware, a link between two entities represents the wire that connects the corresponding hardware modules. At a minimum, the link must contain, explicitly or implicitly, the electrical voltage values and the times at which the voltages are asserted, starting from time 0 up to the current simulation time. There are no fundamental limits placed on the direction of information flow in the links. A link may be permanently unidirectional either from left to right or right to left, or bidirectional. Given the static nature of the connectivity in hardware, a link must be declared unidirectional, along with the direction, or bidirectional, statically.

3.3 CONCURRENCY

Most research articles in the literature on HDL accurately observe that hardware is concurrent. This is indeed true. The American Heritage Dictionary [1] definition of concurrency is "simultaneous occurrence". From first principles, since every entity exists independent of others, entities must necessarily be concurrent. Therefore, in the digital design discipline, if a hardware system consists of *N* entities at a given level of abstraction, the potential concurrency is *N*. Where a host computer has available to it a total of *N* distinct processors, in theory, all *N* entities are potentially simultaneously executable, thereby achieving fast simulation of the entire system. However, the nature of concurrency in the HDLs is complex, requiring a careful analysis of the fundamentals.

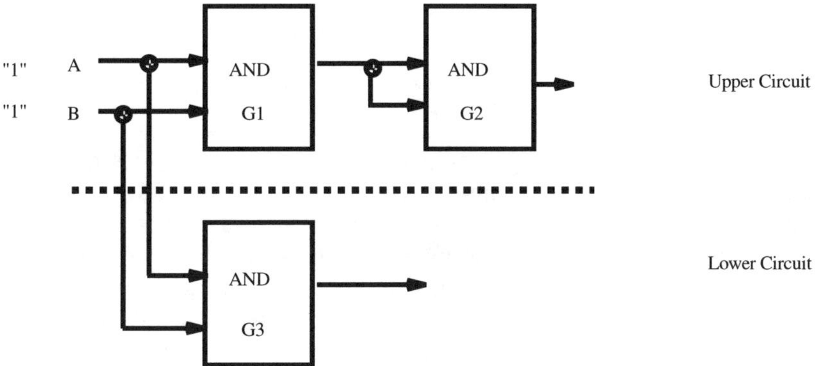

Figure 3.1 Understanding concurrency in hardware.

Consider a simple circuit consisting of three interconnected AND gates G1 through G3, organized through upper and lower circuits, as shown in Figure 3.1. Consider first the upper circuit and that a logical 1 value is asserted at both inputs A and B of G1. At first sight, to one familiar with HDLs, it may appear that gate G1 will execute first for the given external input stimulus, possibly generate an output, following which G2 will execute. In reality, however, as soon as power is turned ON, both gates G1 and G2 start to function simultaneously and this is a key basis for the claim that hardware is concurrent. Electrically, the gates operate continuously. It is not true that the gates are inactive when no signal is asserted at their inputs and that they "wake up" when a new signal is asserted at their input port(s). In Figure 3.1, G2 is not even aware that it is connected to G1 and G1 does not know that its input ports are primary, i.e. external signals may be asserted on them. For our convenience, we may view hardware through its activity but that view is not reality. The placement of the gate instances in a gate-level HDL in any order and the use of the guarded statements in the register-transfer level HDLs reflect their recognition of the concurrency of hardware.

The recognition that hardware is concurrent by itself is not of much value. The issue of concurrency gains importance when the hardware description is required to be executed on a host computer. For a given hardware system, if all of its constituent entities are potentially simultaneously executable, then, assuming the availability of adequate computing resources, they may all be actually executed simultaneously on multiple processors, thereby simulating the entire hardware system quickly. Fast simulation is important since it can help us detect and fix errors and design flaws, quickly and cheaply, before developing an expensive, full blown prototype. Of course, where the host computer is a single, sequentially executing computer, as in the case of all of the gate-level, register-transfer level, and architectural-level HDLs, any discussion of concurrency in the HDLs is severely limited.

If the starting and final goal of HDLs is simply to describe hardware, i.e. serve as documentation only, then the issue of concurrency has been fully addressed by LAMP, CDL, DDL, AHPL, and ISP. However, a key and useful goal of HDLs is to execute the model or description and, to this extent, an understanding of concurrency is important.

To understand concurrency, we will first focus on a different view of the execution of hardware descriptions under simulation, termed event driven simulation. Event driven simulation technique is the current choice of simulation due to its efficiency. We will present a simple understanding here and the issue will be addressed in detail in Chapter 9. In addition, this section will present two sources of concurrency in HDL simulation and a third source of concurrency will be discussed in Chapter 9.

$$\vec{O}_{t2} = f(\vec{I}_{t1}, \vec{S}_{t1}) \tag{3.1}$$

Consider equation (3.1), where the output vector, $\vec{O}$, at time $t2$, of a sequential hardware entity is expressed as a function of the input vector, $\vec{I}$, and state vector, $\vec{S}$, at time $t1$. Where $t1 = t2$, the output is considered to be generated instantaneously, as is the manner in which the gravitational law—action at a distance—is expressed. However, according to relativistic physics, any action or output is limited by the speed of light and, in the digital hardware discipline, the output of an entity requires a finite time, termed the propagation delay, to be realized at the output port. Thus, for digital designs, $t1 \neq t2$. In the digital design discipline, instantaneous output has been modeled through zero delays and it has been the source of much difficulty. Also, $t2$ cannot be less than $t1$. Therefore, $t2$ must be greater than $t1$ which is consistent with reality.

In equation (3.1), consider that a new input signal, $\vec{I}_1$, is asserted at the input port of the entity. Assume that an output $\vec{O}_1$ is generated. Where the output is identical to its previous value, the event driven simulation considers the case to be insignificant since there is no change at the output of the entity. In the case that $\vec{O}_1$ is different from its previous value, the event driven simulation considers the case significant and the new value is propagated to other entities that are connected to the entity in question. Thus, in event driven simulation, only changes in the logical value of input and output ports of the entities are important.

In Figure 3.1, assume that the circuit is powered, the circuit is stable, and that the starting point of the circuit operation is considered as time 0. Clearly, new signals are asserted at the input ports of gate G1 and it must be executed. In contrast, gate G2 does not have a new signal asserted at its input at time 0, so it need not be executed. Thus, at time 0, only gate G1 needs to be executed and the potential concurrency is 1. Consider that gate G1 generates a new signal value at time 10 which is then asserted at the input of gate G2. Also assume that a new signal value is asserted at the primary input ports of gate G1 at time 10. There is no activity between time 0 and time 10, so none of the gates need to be executed and concurrency is 0. At time 10, however, both G1 and G2 must be executed, implying a potential concurrency of 2. This describes the first source of concurrency. That is, if the host computer had two separate processors available, they could simultaneously execute G1 and G2 and accelerate the overall task of simulation. It is pointed out that, despite G2's dependence on G1 for signal values, both G1 and G2 may be executed simultaneously at time 10. An automobile assembly line manufacturing scenario may serve as an excellent analogy. Consider that worker W1 attaches a car body onto a chassis and, upon completion, the conveyor belt forwards it to the next station where worker W2 inserts the headlamps onto the body and chassis. While W2 is working on the headlamps, W1 is working to attach another body onto the subsequent chassis. Clearly, both workers W1 and W2 are working simultaneously, despite W2's dependence on W1 for a body attached to a chassis, and as a result, the overall rate of auto production is high.

In Figure 3.1, now consider both the upper and lower circuits. At time 0, gates G1 and G3 both receive new signals at their input ports. They do not depend on one another for signal values. Therefore, they may be executed simultaneously, yielding a potential concurrency of 2 at time 0. Thus, if a host computer makes two distinct processors available, G1 and G3 may be executed simultaneously, thereby speeding up the simulation. This describes the second source of concurrency. This scenario is analogous to two automobiles traveling simultaneously from point A to point B along two parallel lanes of a highway.

Although the potential concurrency for a hardware system with N entities is N, in general the available concurrency is limited. For high concurrency, every entity must be ready for execution on the processors of a host computer during every available execution cycle. Since a new signal at the input ports of an entity triggers its execution, and as input signals are received either from the external world or other entities, the available concurrency of the entire hardware system is limited by the connectivity of the entities, the propagation delays, and the external stimulus asserted at the primary inputs of the hardware system.

3.4 TIMING

The dictionary [1] definition of time is the non-spatial continuum in which events occur in apparently irreversible succession from the past through the present to the future. In the digital design discipline, the correct functioning of systems is critically dependent on accurately maintaining the relative occurrence of events, thereby underscoring the importance of timing. Barbacci [34] correctly observes that the behavior of computer and digital systems is marked by sequences of actions or activities while Baudet *et al.* [17] wisely view the role of time in HDLs as an ordering concept for the concurrent computations. Piloty and Borrione [35] propose the BCL time model for HDLs where real time is organized intc discrete instants separated by a single time unit and the beginning of each time unit contains an indefinite number of computation "steps" identified with integers greater than zero. Steps provide only a before/after relationship. The concept of "delta delay" in VHDL [33] is derived [36] from Piloty and Borrione's notion of steps and we will discuss it in detail in Chapter 7.

Since timing is conceptually an element of the hardware behavior, it may be modeled, as correctly observed by Baudet *et al.* [17], either implicitly or explicitly in an HDL. When timing is modeled implicitly, as will be explained in Chapter 6, the HDL design is simplified. However, the user's task of developing the hardware description becomes difficult. In contrast, where timing is modeled explicitly in the HDL, through its syntax and semantics, the result is greater clarity and understanding, although the implementation is possibly complex.

To correctly describe the timing of an asynchronous system with entities as its constituent elements, the timing constructs of an HDL must permit the accurate description of the timing of each entity and the relative timing constraints of the signals exchanged between the interacting entities. The HDL being executable, the language constructs must be unambiguous and precise. That is, the timing constructs must be capable of expressing the timing behavior of every entity and every possible timing constraint between the interacting signals of the entities. According to our current knowledge, while the combination of inertial and transport delays is capable of expressing the timing behavior of digital devices at the gate level and beyond, the combination of the signal attributes in VHDL [33] and the richness of a representative high-level language is capable of expressing explicitly all timing constraints encountered at the gate level and beyond. The notion of inertial delay is defined in Section 3.4.4 and its use is elaborated in Section 3.4.4.4.

At a given level of abstraction, although each entity may have its own notion of time, for any meaningful interaction between entities A and B, both A and B must understand at the level of a common denominator of time. We will term this as universal time, assuming that the system under consideration is the universe. Otherwise, A and B will fail to interact with each other. Entertain the following scenario. GG1 and GG2 are two great gods of time. While the length of the interval of GG1's opening and closing of the eye is 1 million years of our time, that for GG2 is 1 nanosecond. Thus, the resolutions of the messages emanating from GG1 and GG2 are 1 million years and 1 ns, respectively. Clearly, during the million years that GG1 has the eye closed, all messages from GG2 to GG1, sent at intervals as low as one nanosecond, will be ignored by GG1. In contrast, assuming that GG2 has a finite life span, when GG1 opens the eye and sends a message to GG2, GG2 has long been dead. Even if GG2 is not dead, there is no known way for GG1 to be certain that this is the same GG2 that GG1 had sent an earlier message to, prior to opening the eye. There is also the strong possibility that GG2 will view the message from GG1, that is unchanging for billions of its timesteps, as inert and lacking of any interesting information. In the Star Trek episode titled, "Wink of an Eye," the beings from the planet below the Enterprise are somehow transformed into an "accelerated" state. While the crew of the Enterprise appear to be in slow motion to these beings, the beings themselves are invisible to the crew of the Enterprise. As expected, there is no meaningful interaction between the two life forms until the captain and the science officer of the Enterprise are also transformed into the "accelerated" state.

Thus, at any given level of abstraction in hardware, the entities must understand events in terms of the universal time and this time unit sets the resolution of time in the host computer. The host computer, in turn, executes the hardware description, expressed in an HDL, of a digital system. Consider a hardware module A with a unique clock that generates pulses every second connected to another hardware module B whose unique clock rate is a millisecond. Figure 3.2 shows a timing diagram corresponding to the interval of length 1 s between 1 s and 2 s. Figure 3.2 superimposes the 1000 intervals each of length 1 ms corresponding to the clock of B. Clearly, A and B are asynchronous. Module A is slow and can read any signal placed on the link every second. If B asserts a signal value v1 at 100 ms and then another value v2 at 105 ms, both within the interval of duration 1 second, A can read either v1 or v2, but not both. The resolution of A, namely 1 second, does not permit it to view v1 and v2 distinctly. Thus, the interaction between A and B is inconsistent. If A and B were designed to be synchronous, i.e. they share the same basic clock, A would be capable of reading every millisecond and there would be no difficulty. In reality, microprocessors that require substantial time to generate an output of a software program are often found interfaced asynchronously with hardware modules that generate results quicker. In such situations, the modules and the microprocessor understand the universal time, i.e. they are driven by clocks with identical resolutions although the phases of the clocks may differ, thereby causing asynchrony.

Thus, the host computer which is responsible for executing the hardware descriptions corresponding to the entities must use the common denominator of time for its resolution of time. When the host computer is realized by a uniprocessor, the underlying scheduler implements this unit of time. When the host computer is realized by multiple independent processors, as will be discussed in Chapter 9, each local scheduler, associated with every one of the processors, will understand and implement this unit of time.

To recapitulate, each entity may possess unique clocks and generate events at their own unique rates. However, they all are characterized by the same resolution of time which is reflected by their understanding of the universal time. One module, say a floating point ALU, may be characterized by a delay of 1 microsecond, implying that it may generate events at microsecond intervals. A second module, say a CMOS EXOR gate, may generate events at 10 ns intervals. A third module, say an ECL NOT gate, generates events at 200 ps intervals. The lowest resolution of time of all of the three modules is picoseconds which therefore constitutes the universal time of the host computer. Note, however, that the time at which the events are generated by a module may differ from its resolution or universal time. Thus, a gate that is continuously functional, i.e. computing every picosecond, may be characterized by a gate delay of 2 ns, which implies that new values or events may be generated at its output at intervals of 2 ns.

Although the arguments leading to the notion of universal time are very important, for hardware systems, the universal time may be understood intuitively. Recall that we had stated earlier that hardware executes continuously. In a digital host computer, "continuous" must be manifested through discrete steps. These discrete steps reflect the universal time.

The notion of events in the event driven simulation of HDLs is an artifact of HDLs, including the need to execute hardware descriptions of digital systems on a host computer. Its development is motivated by two reasons. First, it is observed that at the gate level and higher,

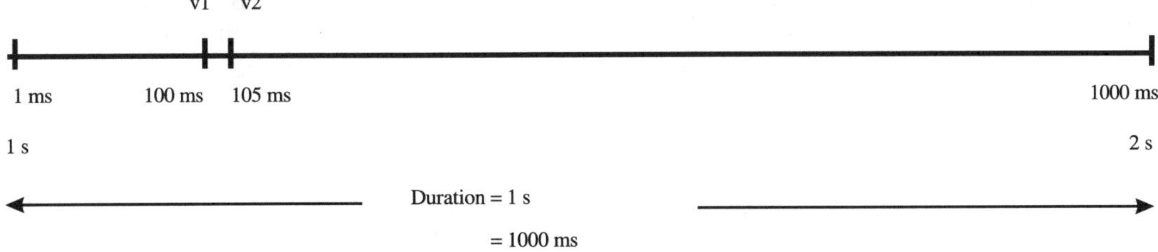

Figure 3.2 The concept of universal time.

the activity is low. That is, for a given external stimulus, asserted at the primary input ports, the changes of the logical values at the input and output ports of gates at any time instant are few. For efficient use of the available computing resources of the host computer and fast simulation, event processing appears a logical choice. Second, the opportunity to view the functioning of hardware through a limited number of events, as opposed to examining the operation of hardware devices at every one of the large number of universal time steps, is a definite advantage.

3.4.1 Is Timing Fundamental in HDLs?

The use of the timing occurrence of two or more activities to uncover a physical relationship may not yield correct results under all scenarios. Two or more actions may occur together or in a sequence and cause an unsuspecting mind to conjure up an erroneous relationship between them. For instance, consider that an individual, ignorant of the gravitational law and the fact that lighter things tend to go up, observes that coconuts fall to the ground during the day and the flame of a fire points up at night. If the individual concludes that coconuts fall down during the day and fire points up at night, the conclusion is only half truth and the reasoning is clearly wrong. Neither of these relationships have anything to do with the time of day.

In relativistic physics, the simultaneity of two or more actions at the same instant of time may be subjective. While two actions may appear as simultaneous to one observer, they may appear to take place in a sequence to a different observer. Thus, time cannot be the fundamental law in the universe.

While a hardware design on the earth is unlikely to be subject to the enormous speeds that warrant relativistic considerations, there is an important implication in that the use of time as fundamental may lead to inconsistencies. Indeed, time is but a manifestation of the principle of causality which is the fundamental law in the physical universe and the fundamental doctrine of philosophy. The principle states: "For every cause there is an effect and for every effect there must have been a cause". Since the speed with which an action may occur is limited by the finite speed of light, there is a finite duration, however infinitesimal its interval, between a cause and its effect along the positive time axis. The principle of causality provides the basis for the distinction between the past and the future. Cause is the unmanifested condition of the effect while effect is the manifested state of the cause.

Given a closed system, S, the activity of the constituent entities may be understood purely from cause and effect. The activity of an entity, A, is triggered by a stimulus from another entity, B, which serves as the cause. The stimulus from B is its effect and is the result of the activity of B, possibly caused by the effect of a different entity, and so on. To understand how S is initiated, two possibilities must be considered. Either S is active forever or S is inactive initially and is triggered by an externally induced event(s) at time = 0, defined as the origin of time. An example of the former is an oscillator while examples of the latter are more common in the digital design discipline.

Although timing is a key consideration in HDL design, as reflected throughout the literature including the most recent language VHDL [33], should any inconsistency arise, as will be discussed in Section 3.4.4, one must resort to the principle of causality as the fundamental law.

3.4.2 The Different Views of Time in HDLs

At any level of abstraction, the universal time is determined, as explained earlier, by the lowest common denominator of the resolutions of time of all the constituent entities. At the gate level, given that the fastest commercial ECL gate operates at approximately 100 ps to 200 ps, the universal time may be expressed in terms of 100 ps. At a higher level where a set of processors execute software and generate results at 5–6 ms intervals and interact with one another at 5–10 ms intervals, the universal time may be expressed in terms of ms. The motivation to select as coarse a timing granularity as possible arises from the desire to minimize the number of activities that must be simulated and analyzed.

When a hardware system is simulated on a host computer, the rate at which the entities are executed will depend on the details of the behavior of the entities and the speed of the host computer. In general, the time duration required by a host computer to execute an entity will differ from the execution time of the entity under actual operation, i.e. in terms of the universal time. The time required by the host computer is referred to as the wall clock time since the time may be measured with respect to the clock on the wall. If the host computer is extremely fast relative to the process that it models, then the wall clock time is shorter than the universal time. Where the process being modeled operates much faster than the host computer, as is frequently the case for hardware, the wall clock time is longer. Thus, while a 200 MHz processor may undergo one operation cycle in 5 ns, a simulation of it may require a wall clock time of 1 ms on a host computer. The reason for simulation, of course, is to assist us in the detection of errors and design flaws.

During simulation of a hardware system, events may occur at intervals defined in terms of integral multiples of the universal time step. Assume that gates G1, G2, G3, and G4 in Figure 3.3 need to be executed corresponding to input stimulus at 1 ns, 5 ns, 5 ns, and 10 ns, respectively. Under event driven simulation, first gate G1 is executed corresponding to a simulation time of 1 ns. Then, gates G2 and G3 are executed simultaneously for time 5 ns. Since there is no activity between the simulation times 1 ns and 5 ns, the simulator advances the simulation time to 5 ns. Similarly, the simulator next advances the simulation time to 10 ns and schedules gate G4 for execution. While the advancement of the simulation time is relatively straightforward for a uniprocessor host computer, it is more complex where the host computer consists of a number of independent and asynchronously executing processors and is the subject of discussion in Chapter 9.

Assume that each of the gates require 2 ms of wall clock time for execution on the host computer. Although G1 will execute on the host computer for 2 ms which is much longer than the 5 ns at which gates G2 and G3 are scheduled for execution, the progress of the simulation is guided by the principle of cause and effect within the world of simulation and the progress of simulation time is indifferent to the wall clock time.

Figure 3.3 presents the above scenario graphically. The level of abstraction is defined by the gates G1 through G4. The universal time progresses in terms of 1 ns and is shown here to range from 1 ns to 10 ns. The simulation time is shown advancing non-linearly from 1 ns to 5 ns to 10 ns and the wall clock time progresses from 1 ms to 9 ms, given that the host processor requires 2 ms of execution time for each gate.

3.4.3 Timing Delays

An analysis of the manufacture's timing specifications of gates and higher-level hardware modules reveals that the propagation delay plays a key role. Propagation delay is

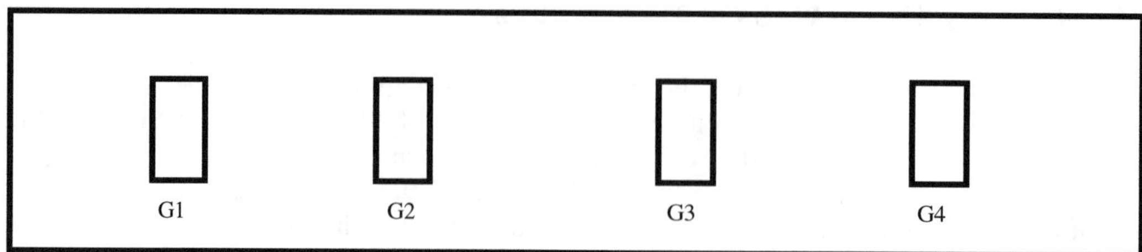

Gate-level Abstraction

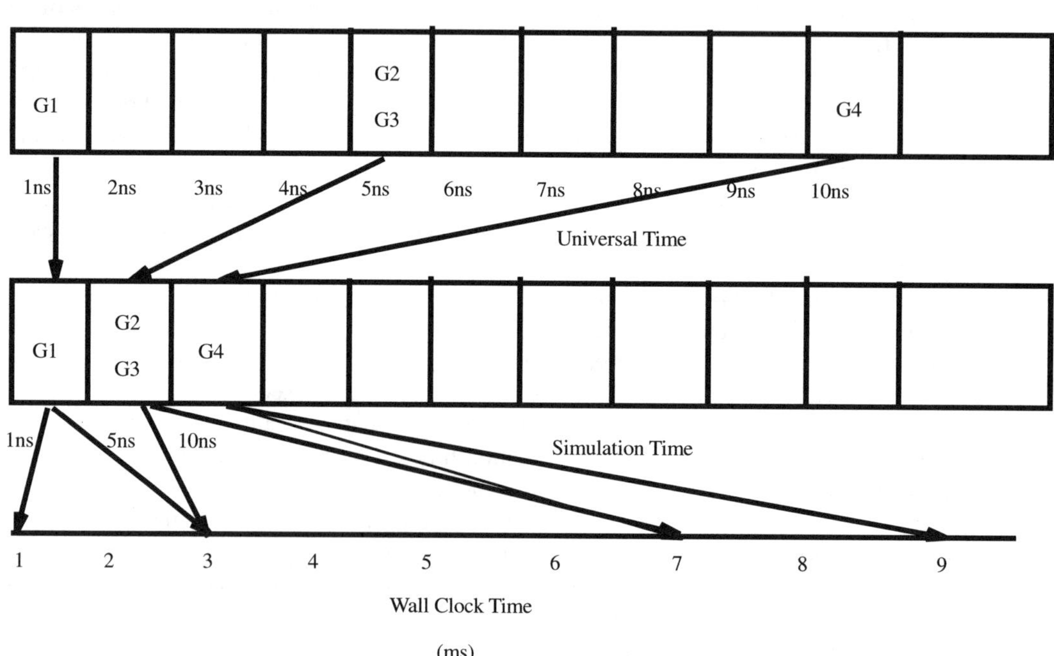

Figure 3.3 Understanding the different views of time.

defined [37] as the time between the specified reference points on the input and output voltage waveforms with the output changing from one defined level (high or low) to the other defined level. For the output changing from logical low to logical high, the propagation delay is referred to as rise time and denoted by t_{plh}, while for the output changing from logical high to low, it is referred to as fall time and denoted by t_{phl}. The values of the rise and fall delay may differ from one another by as much as a factor of 3. Thus, hardware is unique in that, unlike in queueing theory, the delay is a function of the state of the input stimulus. The terms rise and fall are used here strictly as a convenience and, for scientific precision, they must refer to t_{plh} and t_{phl} respectively.

Early gate-level HDLs permitted the specification of a single delay for a gate. Later gate-level HDLs such as LAMP [29] allowed designers to specify maximum and minimum rise and fall delays and detected and ignored input pulses with widths shorter than the minimum transition delay. The register-transfer level HDL, DDL, allows the designer to specify the time required by an operation in terms of a basic time unit. However, it lacks a mechanism to specify the duration of the clock duration and it is unclear how setup and hold constraints may be checked. CDL, AHPL, and ISP lack explicit delay specification of the individual gates and operations. A plausible reason is that, since they focus on synchronous designs, the specification of the individual gate delays is viewed as unimportant. The width of the clock, by design, must ensure that the cumulative delay of the gates does not cause the violation of setup and hold constraints.

The behavior-level HDLs such as ADLIB-SABLE [38] and VHDL [33] incorporate propagation delays. ADLIB-SABLE includes anticipatory timing delays which may result in an inconsistency leading to incorrect results. In Section 3.4.4, we will examine this limitation and study a proposed solution—preemptive scheduling mechanism—that detects and preempts inconsistent events. VHDL utilizes the notions of inertial and transport delays. While the concept of inertial delays is based on the preemptive scheduling principle, the concept of transport delays is observed to suffer from inconsistency and is addressed in detail in Chapter 10.

It is pointed out that the manufacturer-listed delay for a digital module reflects either the maximum, minimum, or a typical value for the module. While the delay value of each instance of a given generic hardware module is unique, in general, for a sample of manufactured devices, the delay values are observed to be randomly distributed. Magnhagen [39] reports that the analysis of TTL devices manufactured by RIFA and ASEA-HAFO in Sweden and TI in USA reveals that the stochastic distribution is approximately normal. Figure 3.4 presents a representative histogram of the variation of the tp_{LH} propagation delay of a sample of 200 74LS151 multiplexer devices manufactured by Texas Instruments Corp. in 1975. The range of delays for digital devices gives rise to considerable complication in the timing specifications of circuits in HDLs, and will be discussed in Section 3.4.4.

3.4.4 Preemptive Scheduling: Principle, Philosophy, and Implementation

3.4.4.1 Introduction. In the HDL simulation systems such as ADLIB-SABLE [38], DABL [40] (DAISY Behavioral Language), and LAMP [41], the timing behavior of digital devices is expressed in the respective hardware description languages ADLIB [32], DABL, and LSI-LOCAL using assignment statements coupled with a timing delay. The delay may be either explicitly associated with the assignment statement such as in ADLIB, or specified in an input–output delay table such as in DABL. An assignment statement coupled with a timing delay will be referred to as the timed assignment statement. Consider the following timed assignment statement in ADLIB, "assign X to Y delay t;" where "X" represents a constant, variable, or expression, "Y" the signal path identifier, and "t" the physical time delay associated with the assignment. The signal path, also referred to as the net in this chapter, corresponds to a connecting point of a hardware module through which it interacts with the external world. It may be of any arbitrary legal data type and its value is synonymous with the contribution of the hardware module to the external world or vice versa. Also the datatype of X must legally match with that of the signal path Y and the physical time delay may either be a constant, variable, or expression in the simulation time units. The semantics of the timed assignment statement implies that the the value of X will be asserted at the signal path Y following *t* simulation time units starting at the execution of the statement or the current

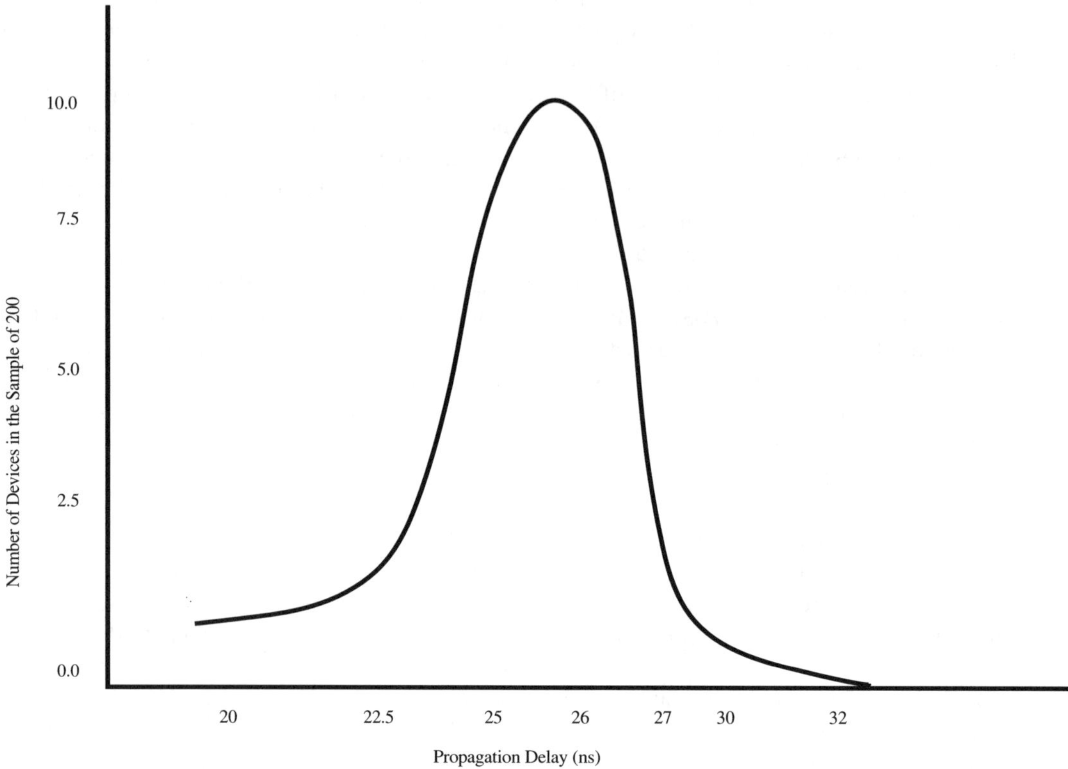

Figure 3.4 Histogram of propagation delays of a sample of 200 74LS151 devices.

simulation time. When X represents an expression, it is evaluated at the current simulation time. This semantics is termed anticipatory because execution of a timed assignment statement anticipates the value at the signal path Y in t time units in the future that may not be overridden prior to the current simulation time being equal to t.

In order to facilitate understanding of the different approaches introduced here, the event driven simulation mechanism that frequently implements the timing semantics in the HDL simulators mentioned earlier is first briefly reviewed.

An event driven simulation of a hardware description consists of iteration of three major phases. Phase I consists of the execution of the behavior descriptions corresponding to the hardware modules that are scheduled for simulation. When phase I is executed for the first time, all behavior descriptions are simulated. In the subsequent execution of phase I only those behavior descriptions are executed that are scheduled for simulation. Execution of each component description generates one or more projected events that correspond to the executed timed assignment statements in the behavior description. The projected events represent the predictions of the components' contribution to the external world. In phase II, the projected events generated by all behavior descriptions are placed in an event list. Placement of the projected events in the event list may be achieved in a number of ways and this chapter assumes an unique event list where the events are arranged in accordance with the simulation time at which they are intended to affect the environment. Phase III consists of selecting the event or set of events with the lowest value of simulation time from the event list and scheduling those component descriptions for execution that qualify. A description qualifies for execution when it is not suspended and one or more of its inputs are affected by the selected projected events in phase III.

An important characteristic of timing constructs in hardware description languages is that a description corresponding to a component must generate an event during execution after it has processed adequate and necessary information at the input ports. The anticipatory timing constructs in the conventional hardware description languages, however, may predict and generate an output event corresponding to an input transition at $t = t_1$ without taking into consideration input signal transitions at $t > t_1$. Consequently, such hardware descriptions may produce unreliable simulation results as explained in Section 3.4.4.2. Section 3.4.4.3 presents the switching behavior of TTL, NMOS, and CMOS components in the presence of input signals with different pulse widths and proposes empirical measures for T_{mip} relative to the t_{phl} and t_{plh} values of a component. T_{mip} is the inertial delay of a component and may be defined as the minimum duration for which an input change must persist for the component to change states. The use of T_{mip} is essential for accurate simulation. A high to low (t_{phl}) propagation delay is the elapse time between the specified reference points on the input and output voltage waveforms with the output changing from logical 1 to logical 0. A low to high (t_{plh}) propagation delay is the elapse time between the specified reference points on the input and output voltage waveforms with the output changing from logical 0 to logical 1. Section 3.4.4.4 presents the preemptive scheduling mechanism that detects and discards inconsistent events and generates correct behavior simulation results, and its underlying philosophy. Section 3.4.4.5 presents an alternate implementation of preemptive scheduling.

3.4.4.2 Inadequacy of Anticipatory Timing Semantics for Reliable Behavior-level Simulation. First, the problem associated with anticipatory timing constructs in the conventional behavioral simulators is presented through an example.

Figure 3.5(a) shows an inverter with high-to-low and low-to-high propagation delays of 14 ns and 10 ns respectively. The difference between the individual delay values, 4 ns for the inverter, is realistic. For MOS components [42], the difference between the high-to-low and low-to-high propagation delays may be as high as twice that of the smallest of the two values.

For the input high pulse of duration $t = 0$ ns to $t = 3$ ns as shown in Figure 3.5(b), the correct output should be a 1 throughout as shown in Figure 3.5(c), qualified by a spike at $t = 14$ ns. Figure 3.6 describes the most obvious use of the anticipatory timing semantics of ADLIB to describe the behavior of the inverter.

The body of the inverter description is executed whenever there is a transition at the input "inport". For the input signal transition from low to high at $t = 0$ ns, as shown in Figure 3.5, the statement at label L1 is executed and, consequently, a low value is scheduled at the output "outport" at $t = 14$ ns. Corresponding to the subsequent input transition from high to low at $t = 3$ ns, the statement at label L2 is executed and, as a result, a high is scheduled at the output for $t =$ current-simulation-time $+10 = 3 + 10 = 13$ ns. At this point, the scheduler has two events E1 at $t = 14$ ns and E2 at $t = 13$ ns. E2 is executed first at $t = 13$ ns followed by E1 at $t = 14$ ns with the consequence that the output signal value is high up to $t = 14$ ns and then goes low. The output is inconsistent with the correct output as shown in Figure 3.5(c) and is therefore incorrect.

The LAMP [41] simulator proposed a spike rejection algorithm that was designed to detect and discard input signals where the pulse width is less than the smaller of the high-to-

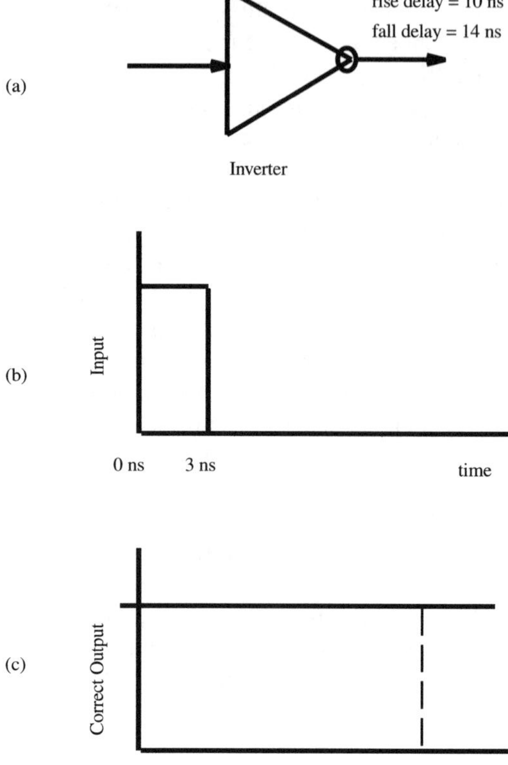

Figure 3.5 Simulation of an inverter with unique rise and fall propagation delays.

low and low-to-high propagation delays. However, this technique fails to resolve the problem of inconsistent events as explained through Figure 3.7.

Figure 3.7 describes an inverter with high-to-low and low-to-high propagation delay values of 6 ns and 18 ns respectively. Given a low pulse between $t = 0$ ns and $t = 7$ ns at the input of the inverter, while the correct output should consist of a low signal throughout, LAMP fails to identify the input as a spike and reject it because its width of 7 ns is greater than the high-to-low propagation delay and generates an incorrect output signal—high at $t = 18$ ns and thereafter.

```
nettype two_value_net = (high, low);
comptype inverter;
  inward inport: two_value_net;
  outward outport: two_value_net;
 begin
  if inport = high then
L1: assign low to outport delay 14
  else
L2: assign high to outport delay 10;
  end;
```

Figure 3.6 Behavior-level model of inverter in ADLIB.

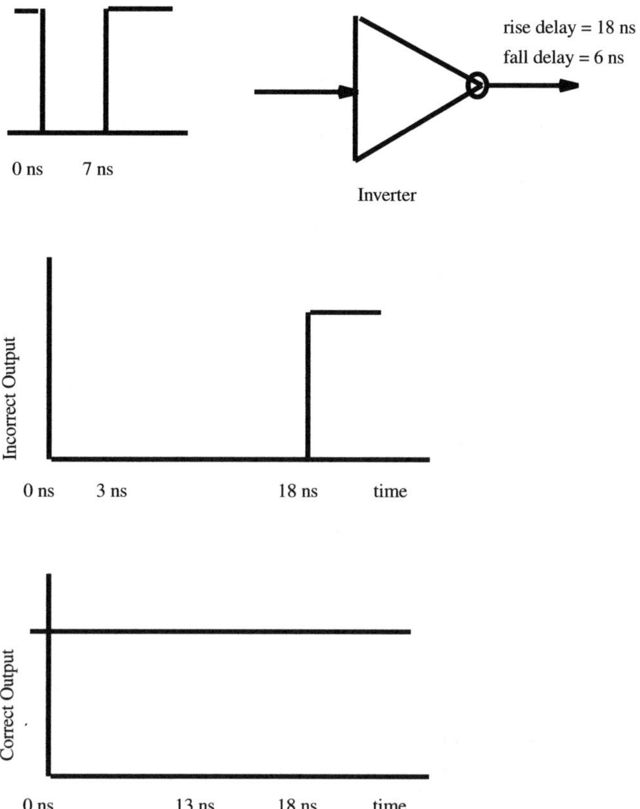

Figure 3.7 Inability of the spike rejection algorithm to detect inconsistent events.

3.4.4.3 Switching Behavior of Components and Input Pulse Widths. The concept of inertial delay is essential for accurate simulation and its use is further elaborated in Section 3.4.4.4. The values of T_{mip} for digital components, however, are not easily available, as observed by their absence in the manufacturer's specification handbooks [37]. This section presents the results of SPICE simulation that represent the switching behavior of TTL, NMOS, and CMOS inverters in the presence of input signals with pulse widths that bear specific relationship to the appropriate propagation delay value—t_{plh} and t_{phl}—as detailed subsequently. Figure 3.8(a) through 3.8(c) describes the output behavior of a TTL inverter in the presence of three sets of input signals with logical value 1 and pulse width duration equal to one-tenth, one, and five times the the value of t_{phl}. The upper and lower threshold voltages are 2 V and 0.8 V, respectively, and the values of t_{phl} and t_{plh} are 0.15 ns and 1.32 ns, respectively.

Observations of the switching behavior indicates that signals with logical value 1 and pulse width equal to or greater than the t_{phl} value may contain sufficient energy to switch the inverter while a signal and pulse width smaller than the t_{phl} value may not successfully switch the inverter. Consequently, the value of T_{mip} of a TTL inverter for high input signal may be assumed equal to t_{phl}. It may be similarly shown that the value of T_{mip} for a TTL inverter for low input signal is t_{plh}, and the value of T_{mip} for a non-inverting TTL buffer is t_{plh} for a high input signal and t_{phl} for a low input signal. For complex TTL components, the value of T_{mip} is

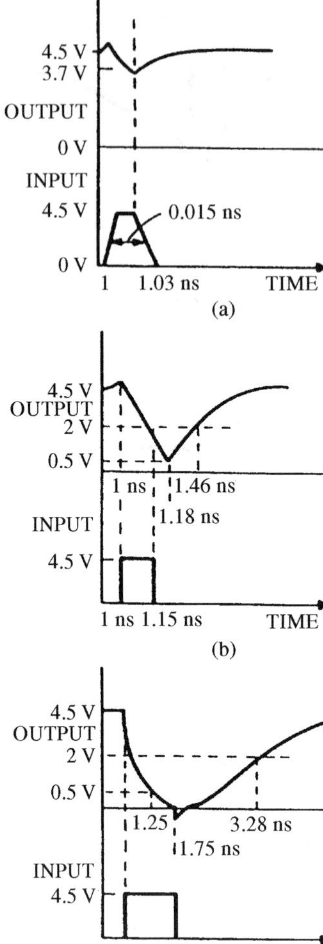

Figure 3.8 Switching behavior of TTL inverter, (a) pulse width equals one-tenth of t_{phl}, (b) pulse width equals t_{phl}, (c) pulse width equals five times t_{phl}.

either t_{phl} or t_{plh} for a high or low input pulse where the output signal of the component is inverting with respect to the input signal and t_{plh} or t_{phl} for a high or low input pulse where the output signal of the component is non-inverting with respect to the input signal.

Figures 3.9(a) through 3.9(c) describe the output behavior of an NMOS inverter in the presence of three sets of input signals with logical value 1 and pulse width duration equal to one-tenth, one-half, and five times the value of t_{phl}. The threshold voltage is 3 V and the values of t_{phl} and t_{plh} are 0.23 ns and 0.5 ns, respectively.

Figures 3.10(a) through 3.10(c) describe the output behavior of a CMOS inverter in the presence of three sets of input signals with logical value 1 and pulse width duration equal to one-tenth, one, and five times the value of t_{phl}. The threshold voltage is 2.5 V and the values of t_{phl} and t_{plh} are 3.5 ns and 3 ns, respectively.

Observations of the switching behavior in Figures 3.9(a) through 3.9(c) and 3.10(a) through 3.10(c) indicate that the pulse width durations of input signals of logical value 1 must

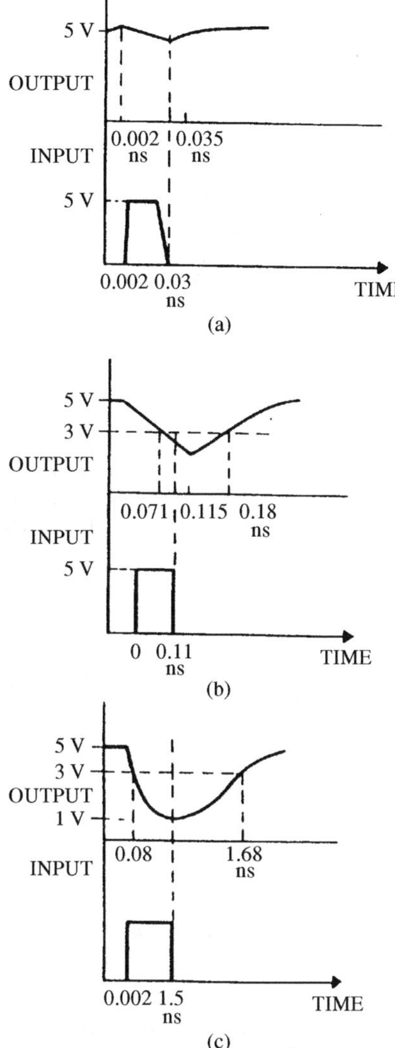

Figure 3.9 Switching behavior of an NMOS inverter, (a) pulse width equals one-tenth of t_{phl}, (b) pulse width equals one-half t_{phl}, (c) pulse width equals five times t_{phl}.

exceed one-half and one times the t_{phl} value to switch NMOS and CMOS inverters, respectively. Consequently, the value of T_{mip} of NMOS and CMOS inverters for high input pulses is $0.5t_{phlnmos}$ and $t_{phlcmos}$, respectively. It may be shown similarly that the value of T_{mip} of NMOS and CMOS inverters for low input signals is $0.5t_{plhnmos}$ and $t_{plhcmos}$, respectively, and those of non-inverting NMOS and CMOS buffers for high and low input signals is $0.5t_{plhnmos}$ and $t_{plhcmos}$ and $0.5t_{phlnmos}$ and $t_{phlcmos}$, respectively.

For complex NMOS components, the value of T_{mip} is either $0.5t_{phl}$ or $0.5t_{plh}$ for a high or low input pulse where the output signal of the component is inverting with respect to the input signal, and $0.5t_{plh}$ or $0.5t_{phl}$ for a high or low input pulse where the output signal is non-inverting with respect to the input signal. The value of T_{mip} for complex CMOS components may be obtained similar to those for TTL components.

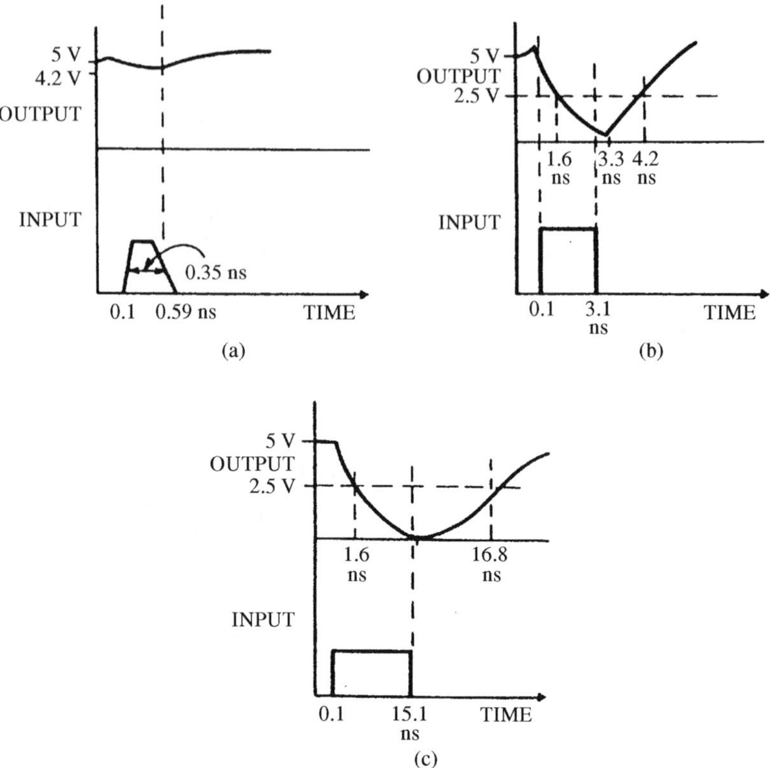

Figure 3.10 Switching behavior of a CMOS inverter, (a) pulse width equals one-tenth of t_{phl}, (b) pulse width equals t_{phl}, (c) pulse width equals five times t_{phl}.

3.4.4.4 The Fundamental Principle of Causality as a Basis for Preemptive Scheduling. The principle of preemptive scheduling was first introduced by the author in 1984 [43]. A subset of it was included in a critique of the VHSIC hardware description language [44], and in the IEEE ICCD conference [45], and finally a detailed presentation appeared in [46]. While the VHDL language reference manual, version 5.0, dated August 1984, utilized the anticipatory semantics as in ADLIB-SABLE, version 7.2, dated August 1985, finally included the preemptive semantics.

The essence of the concept of preemptive scheduling mechanism in digital simulation may be expressed as follows. When a decision (d_1) made later in time influences the environment at an earlier time in the future as compared to a decision (d_2) that is made earlier in time but influences the environment at a later time in the future, d_1 must preempt d_2 if they conflict with each other.

Assume a scenario where the effects e_1 and e_2 due to the causes c_1 at $t = t_1$ and c_2 at $t = t_2$ ($t_2 > t_1$), respectively, influence the environment at $t = t_3$ and $t = t_4$ ($t_4 < t_3$). From the principle of causality, c_1 and c_2 must produce effects. The effects, however, may influence the environment only after propagating to the destination in finite time. Because e_2 influences the environment first even though it is caused by a later cause and is inconsistent with e_1, e_1 must be preempted. The delay between the cause c_2 and effect e_2 being lower, if the most recent cause persists the effect e_2 will persistently influence the environment and cancel the effect e_1. The strength of the effect is a function of how recent is the cause, the nature and strength of the cause, the resolution of time, the current state of the system, etc. Assuming all other factors being constant, the more recent effect of a more recent cause implies greater influence than the later effect of an earlier cause. As an analogy, assume that a person A is abruptly pushed by person B from behind. While A is gradually falling down but before he has completely fallen to the ground, a third person C very quickly grabs A by the collar and pulls him up. As a result, A's fall to the ground, which was triggered by B and would have otherwise been true, is preempted by C.

In digital simulation, when an event E is generated at port X as the result of executing a timed assignment statement, if there exists other events at X caused by previously executed assignments such that these events are projected to affect the environment at a future time later than E, then E must preempt these events. In addition, input signals to a component with pulse width less than T_{mip} are detected and discarded. The software structures and mechanisms to realize the implementation are presented subsequently.

As one implementation of the preemptive scheduling principle, the anticipatory timing semantics in the conventional hardware description languages may be corrected to preemptive by modifying the scheduler without appreciably degrading the simulation performance. Consequently, descriptions of component behavior in this approach are simple and similar to those derived for the conventional hardware description languages and simulators. For instance, even the component description of the inverter derived in Figure 3.6 will generate accurate results in this approach. For the input signals shown in Figure 3.5, execution of the component description may be described as follows.

For the input transition from a low to a high at $t = 0$ ns, the statement at label L1 is executed with the consequence that the event E_1 that asserts a low at the output at $t = 14$ ns is generated. Corresponding to the subsequent input transition from high to low at $t = 3$ ns, the statement at label L2 is executed and an event E_2 that asserts a high at $t = 13$ ns is generated. Since the projected time for E_2 is $t = 13$ ns that is less than the corresponding time $t = 14$ ns for E_1, the intelligent scheduler will recognize E_1 as inconsistent and permit E_2 to preempt it. Consequently, the output is high throughout and is correct.

In the modified scheduler, all events generated by the component descriptions are queued into an event list that is organized such that the earliest and latest events are placed at the head and tail of the queue, respectively. When a component description C_1 generates an event—assignment at port O_1—a field in a two dimensional array of records F(C_1, O_1) flag is set to true.

In Figure 3.11, F is a two dimensional array of records, where the first and the second dimensions span over the range of components and maximum number of output ports of a component, respectively. The first field in the record is a Boolean "flag" that in essence remembers whether an event has been generated by a component description C1 at an output port O1 and has not been processed yet. The second field is a pointer that points to the event

```
F: array[1..TotalNoOfComponents, 1..MaxNoOfOutputPorts] of
       record
         flag      : boolean;
         pointer : ^EventListElement;
       end;
```

Figure 3.11 Software structures for inconsistent event preemption.

in the event queue generated by C1 at port O1 when flag is true. When flag is false, the pointer points to nil.

When event E_1 is extracted from the queue and its effect processed and there are no outstanding events in the event queue that are generated by C_1 at O_1, the corresponding flag field is reset to false. Otherwise the pointer is directed to point to the outstanding event in the queue. When a subsequent assignment is made from the same component description to the same output port, thereby generating an event E_2 while E_1 is still unprocessed in the event queue, a test is performed to verify whether the time associated with E_2 is greater than that associated with E_1. For a positive result, E_2 and E_1 are not inconsistent and E_2 is inserted appropriately in the event queue. Otherwise E_1 is inconsistent with E_2 and the preemptive semantics require E_1 to be deleted from and E_2 inserted into the event queue. In addition, the pointer is redirected to point to E_2. At any time instance during simulation, where events E_1 through E_n are outstanding and E_m arrives, given that E_1 through E_n and E_m are generated by C_1 at port O_1, the following test is performed. The pointer field of the structure F[C_1, O_1] points to E_1 such that the time associated with E_1 is less than those associated with E_2 through E_n. If the time associated with E_m is less than that corresponding to E_1, then events E_1 through E_n are inconsistent and deleted from the event queue. Event E_m is inserted into the event queue and the pointer now points to E_m. As a result, this approach results in a scheduler that may be characterized by an additional ability of detecting and descheduling inconsistent events during simulation.

The pulse width duration of a signal asserted at an input port of a component description is compared against the value of T_{mip} for the component as described subsequently. In addition to other structures, every component description contains a data structure with three fields to store a logical value, an assertion time, and an input port identifier.

When a component description is activated due to a signal transition at $t = t_1$ at input port I_1, the output value v_1 is determined by executing the description under the assumption that $t_{plh} = t_{phl} = 0$ ns. The logical value v_1, causal time t_1, and the input port identifier I_1 where the cause was asserted are stored in the data structure assuming that the description is executed for the first time. When the component description is executed again at $t = t_2$ due to an input cause at port I_2, the output value v_2 is determined similar to the previous instance. Where the logical values v_1 and v_2 differ and the difference $(t_2 - t_1)$ is smaller than T_{mip} for the component, the resulting effect of the combination of input transitions at $t = t_1$ and $t = t_2$ lacks sufficient energy to switch the component and must be discarded. Where the logical values of v_1 and v_2 differ but the difference $(t_2 - t_1)$ exceeds T_{mip} for the component, the resulting effect of the combination of input transitions at $t = t_1$ and $t = t_2$ may switch the component and v_2, t_2, and I_2 are written into the data structure. Where v_1 and v_2 are identical, the data structure continues to store the previous information. For the case $I_1 = I_2$, the input transitions at $t = t_1$ and $t = t_2$ simply imply a pulse at the input port I_1 of the component. The mechanism is illustrated through the following example.

Figure 3.12 describes a TTL AND gate with $t_{phl} = 0.23$ ns and $t_{plh} = 0.5$ ns and the input signals at ports A and B. Although both of the input signals at ports A and B are logical 1 and 2 ns in width, i.e. much larger than t_{plh} of 0.5 ns, the overlapping region where both inputs are high constitutes a pulse width of only $(2 - 1.95)$ ns $= 0.05$ ns, i.e. much smaller than $t_{plh} = 0.5$ ns and, consequently, may not switch the AND gate. The proposed mechanism correctly identifies this situation and discards the effects of the input transitions at $t = 1.95$ ns and $t = 2$ ns as shown subsequently.

Corresponding to the transition at port A at $t = 0$ ns, an output value 0 is determined by evaluating the AND gate description under the assumption that $t_{plh} = t_{phl} = 0$ ns and, as this

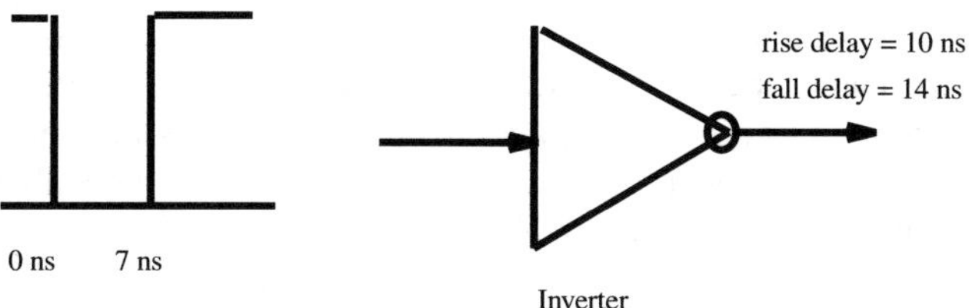

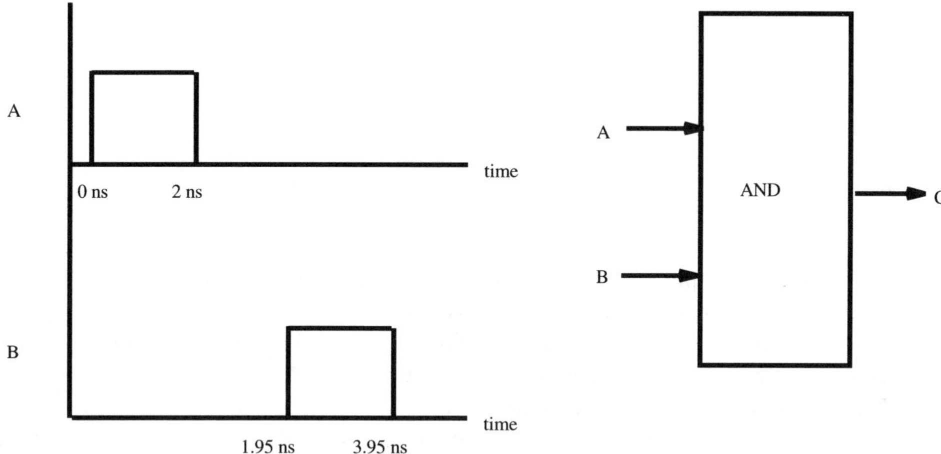

Figure 3.12 Preemptive scheduling.

does not indicate a change from the previous value of the AND gate, the data structure continues to store the previous information. Corresponding to the second transition at port B at $t = 1.95$ ns, an output value 1 is determined similar to the previous instance and stored in the data structure along with $t = 1.95$ ns and the input port B. Corresponding to the third transition at A at $t = 2$ ns, an output value 0 is determined. Since a pulse of value 1 is implied at the output caused by the input transitions at $t = 1.95$ ns and $t = 2$ ns and the difference $(2 - 1.95)$ ns $= 0.05$ ns is smaller than $t_{plh} = 0.5$ ns, the resulting effect of the combination of input transitions at $t = 1.95$ ns and $t = 2$ ns is discarded. Consequently, the mechanism correctly predicts no change in the value at the output port.

3.4.4.5 An Alternate Implementation of Preemptive Scheduling in HDLs. Alternatively, the principle of preemptive semantics may be realized in digital simulation through deferred scheduling of output assignments, which was also introduced by the author in [43].

When the model representing a component C is executed as a result of an input vector v_1 asserted at an input port at $t = t_1$ and an output assignment S_1 is generated for $t = t_2$ where $t_2 > t_1$, S_1 is stored internally unlike in the conventional simulator where it is immediately asserted at the output port. The input vector may be an externally applied signal or the result of execution of other models. When the model corresponding to C is subsequently executed as a result of another input vector v_2 at an input port at $t = t_3$ and the two conditions (1) $t_3 \geq t_2$ and (2) S_1 is consistent with all other output assignments that may have been generated as a result of execution of C but not yet asserted at the output port, are satisfied, S_1 may be asserted at the output port. Under these circumstances, S_1 can no longer be preempted and its assertion is guaranteed to be correct. Where S_1 is inconsistent with a previously generated signal, S_0 say, S_0 must be discarded according to the preemption principle elaborated earlier.

Consider the simulation of a two-input AND gate as shown in Figure 3.13. The input stimulus consist of high at $t = 0$; ns, low at $t = 100$ ns, high at $t = 120$ ns, and low at

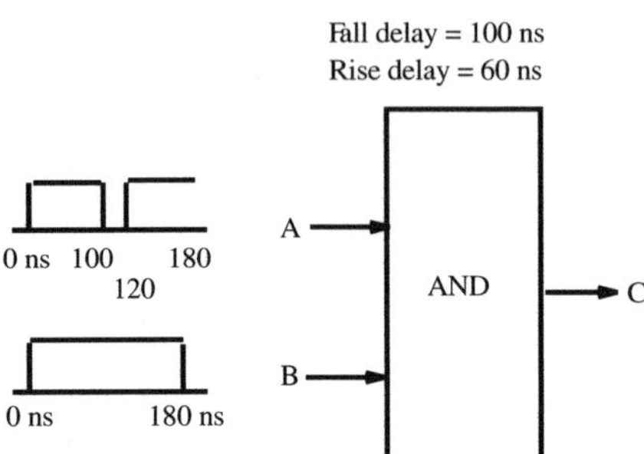

Figure 3.13 Deferred output assignment mechanism.

$t = 180$ ns at the input port A and high at $t = 0$ ns and low at $t = 180$ ns at input port B. Consequently, the model representing the AND gate may be executed first at $t = 0$ ns, then at $t = 100$ ns, thirdly at $t = 120$ ns, and finally at $t = 180$ ns. Corresponding to the first execution, a high value is generated for $t = 60$ ns and stored within the model description. When the model description is executed again at $t = 100$ ns, a low signal value for $t = 200$ ns is generated and also stored within the model description. An input signal at current time may only cause the generation of an effect in the future and because neither the previously generated signal, high at $t = 60$ ns, has been preempted nor any input signal at $t \geq 100$ ns may ever preempt it, the signal may be asserted at the output port with certainty. Corresponding to the third execution of the model at $t = 120$ ns, a signal value of high at $t = 180$ ns is generated and is stored within the model description. The previously generated signal of low at $t = 200$ ns is inconsistent with the recent output signal and is consequently preempted by the latter. During subsequent execution of the model at $t = 180$ ns, a signal of low for $t = 280$ ns is determined at the output and the previously generated signal of high at $t = 180$ ns is asserted at the output. Since $t = 180$ ns defines the end point of simulation such that no new vectors will be asserted, the single outstanding output signal of low at $t = 280$ ns will certainly influence the output port and is asserted appropriately. Consequently, the final output consists of a low at $t = 280$; ns and simulation of the AND gate in this approach has generated accurate results.

3.4.4.6 Comparative Analysis of the Two Implementations of the Preemptive Scheduling Principle. The implementation in Section 3.4.4.4 requires modification to the scheduler and is ideally suited for the scenario where the hardware descriptions corresponding to the different components of the system are all executed sequentially on a uniprocessor. For the same reasons, this implementation strategy is incompatible for asynchronous, distributed execution of the hardware descriptions on multiple processors which is characterized by the lack of a centralized scheduler. The approach in Section 3.4.4.5 localizes the detection and preemption of inconsistent events within every individual hardware description and is, therefore, ideally suited for an asynchronous, distributed execution of the hardware descriptions. In the IEEE Standard VHDL language reference manual [33], the semantics of inertial delays reflects the more advanced approach in Section 3.4.4.5. However, given that the current VHDL scheduler is limited to execution on uniprocessors or synchronous multiprocessors, at best, the advantages of the advanced implementation may not be exploited.

3.4.5 Asynchronous Timing Behavior and Asynchronous Interactions

For an HDL to successfully model the asynchronous behavior of a hardware system, the fundamental requirements are as follows. First, the language must be able to model the intrinsic, concurrent nature of hardware. Second, the language must enable the hardware system description to synchronize, during execution, with respect to an external, obviously asynchronous, signal. To realize this, it may be necessary to suspend the sequential execution of the hardware system until a specific timing condition involving the external signal is satisfied. Third, the language must permit the hardware system to include asynchronous delays in the course of its execution. This will enable the hardware description to pause at any point during its execution and for any arbitrary length of time, in universal time units. Fourth, the language must enable the hardware system description to assign signal values to output and bidirectional ports asynchronously, with respect to other entities.

3.4.6 Timing Constraints

Along with the propagation delays, timing constraints between two or more signals constitute the complete manufacturer's timing specifications of gates and higher-level hardware modules. The Texas Instruments TTL Databook [37] lists setup, hold, minimum pulse width, and maximum clock frequency as the key timing constraints. The gate-level HDLs, register-transfer level HDLs, and the architectural HDL completely lack any explicit language constructs to model timing constraints. The ADLIB [32] language lacks explicit constructs to express constraints and leaves it to the user's ingenuity to model them through the high-level language constructs. DABL [40], the Daisy Behavioral Language, includes timing check blocks to allow the user to explicitly state constraints between signals, as shown in Figure 3.14. The ASSERT construct permits the specification of timing constraints in the Conlan [35] HDL but is cryptic and non-intuitive. In contrast, VHDL utilizes the "signal attributes" and other language constructs and permits a convenient specification of timing constraints. Chapter 7 addresses the issue of timing constraints in VHDL in detail.

To express timing constraints, fundamentally, the language must permit an entity to access the complete (or partial, as appropriate) history of every signal, up to the current simulation time, that serves as an input to it. At a minimum, the set of logical values and the corresponding assertion times of every input signal must be available to the entity. The entity

```
CHECK
 ABUS = STABLE for 30 BEFORE CLK -> 1; /* Setup check */
 ABUS = STABLE for 10 AFTER CLK -> 1;  /* Hold check */
 CLK -> 1 OCCURS 100 AFTER CLK -> 1;   /* Minimum Pulse Width check */
 CLK -> 0 OCCURS 60 AFTER CLK -> 1;    /* Clock High Width check */
```

Figure 3.14 Timing constraints in DABL.

generates the values for its output signals, so by definition it has complete access to the history of all of its output signals.

Timing constraints may involve one, two, or more signals. For a single signal, issues such as minimum high and low durations are of concern. Where constraints involve two or more signals, they may be reorganized into sets of checks, with each set involving two signals. Fundamentally, timing constraints may assume one of two forms. Consider two signals S1 and S2. First, relative to a specific instant of time T1 in S1, it may be necessary to check the past behavior of S2, i.e. before T1. The notion of setup check in a flip-flop constitutes an example wherein one focuses on the active clock edge and examines whether the D input has been stable for setup time units prior to the clock edge. Second, relative to T1 of S1, it may be required to check the future behavior of S2, i.e. beyond T1. Clearly, this is impossible at time instant T1 since any behavior is known, with certainty, only up to the present, i.e. T1. The future is unknown at the present. Therefore, the verification must be realized at an appropriate future time. The issue of hold time check, again in a flip-flop, constitutes an example. Relative to the active clock edge, the D input must remain stable hold time units into the future.

3.5 HIERARCHY

The hierarchical design methodology is an effective technique to address the increasing size and complexity of hardware systems. In this approach, the entire system may be expressed at different levels of abstraction. At the higher levels, the description is condensed and the behavior is comprehensible at the high level, while at the lower level, the description is less compact and the details are increasingly visible. The basic character of hierarchy is that, at each level of abstraction, the description of the hardware system must be complete, self-contained, consistent, and natural relative to the actual hardware at that level. That is, the behavior must be understandable to the designer without resorting to the lower-level details. In general, the resolution of time is coarser and the simulation time decreases as one advances to the higher levels of the design abstraction.

In the hardware design process, efficient debugging often requires one subsection, say A, of the hardware system, S, to be represented at a lower level while the remainder of the system, (S − A), is represented at the higher level. Thus, while the designer is able to examine the functioning of A in greater detail, the overall simulation is fast since the remainder of the system is executed at the higher level. Under these circumstances, two issues surface. First, although the individual timing resolutions of A and (S − A) may differ, the simulator must ensure that they both understand the common denominator of time to facilitate interaction between them during the simulation. Second, the elaboration of A at the lower level of hierarchy relative to (S − A) may be achieved either statistically or dynamically. Under static elaboration, the subsection A must be identified a priori and elaborated before initiating the simulation. Under dynamic elaboration, the user is permitted to elaborate, during simulation, selected subsections of the hardware system at a lower level. The motivation for dynamic elaboration may lie in the observations and intermediate results that warrant further examination of the subsection in greater detail. Both ADLIB-SABLE [38] and VHDL [33] strictly adhere to static elaboration. In Chapter 6, we will address the issue of dynamic elaboration. In discussing the future of simulation, Breuer and Parker [47] had predicted the timing timewarp phenomenon where the entire digital system would proceed under a coarse time axis and then dynamically zoom for more detail, slowing down time and simulation.

3.6 CONSTISTENCY OF HDLs WITH THE UNDERLYING HARDWARE

In addition to being consistent with the characteristics already outlined in the previous sections of this chapter, the language constructs of a behavior-level HDL must be natural. That is, a hardware description in terms of these constructs must correspond to reality at the corresponding level of abstraction and be intuitively comprehensible to the designer. Under these circumstances, the learning and teaching of HDLs will be greatly facilitated.

The naturalness of hardware language constructs may be understood by examining the difficulties posed by a few unnatural constructs. For instance, in ADLIB [32], while the use of the SYNC construct to describe a clock is acceptable within a comptype, defined subsequently in Chapter 4, the declaration of SYNC within the main program, i.e. across comptypes, immediately limits the description to a synchronous design and opposes ADLIBs goal of a behavior-level HDL. In VHDL [33], the introduction of shared variables, while a programming convenience, denies the entities their independence and poses a serious difficulty. The LOGLAN HDL [48], proposed by Pawlak and Wrona, permits the dynamic creation and removal of objects. Since objects, at any level of abstraction, must correspond to realistic hardware modules, unless a hardware system realizes the creation and destruction of hardware units during actual operation, the corresponding LOGLAN language constructs are unnatural and inconsistent.

4

The First Behavior-level HDL—ADLIB-SABLE

ADLIB [32], an acronym for a design language for indicating behavior, is the first behavior-level hardware description language, designed by Dwight Hill and Willem vanCleemput [49]. ADLIB is complemented by the structural design language SDL [26] that realizes the interconnection of the digital modules, and the event driven simulator, SABLE [38]. ADLIB has pioneered the concept of comptypes to model generic digital components and each instance of a comptype is viewed to exist independently in the simulator. The notion of an independent unit was solidified by the requirement that the inward, outward, and bidirectional ports of a comptype be declared explicitly. As a superset of the Pascal programming language with strong typing, ADLIB offered the designer enormous power to describe the high-level behavior, as opposed to the highly restrictive, register-transfer level HDLs. Unlike Pascal, ADLIB pioneered the use of delays in assignment statements. These assignments, coupled with the view of independent instances of comptypes, and the SABLE event driven simulator, empowered the designer to describe asynchronous hardware systems using the ADLIB-SABLE system. Thus, ADLIB-SABLE was the first effort at describing true asynchronous systems. The ADLIB-SABLE system also pioneered the concept of a "net", i.e. a high-level abstract interconnection between two or more modules. The net is organized as a typed data structure to enable the communication of abstract information between the modules.

In summary, ADLIB-SABLEs contributions are

- Emulates the concurrency, inherent in hardware, through the notion of comptypes.
- Permits the description of the unique timing of each comptype.
- Empowers the designer with the programming power of Pascal.

4.1 MODELING ASYNCHRONOUS SYSTEMS IN ADLIB-SABLE

As an example, consider a simple asynchronous system, shown in Figure 4.1 and borrowed from [32], that consists of two object instances of two different comptypes: Dealer and Player. The Dealer has a single output port while the Player has a single input port. While Joe is

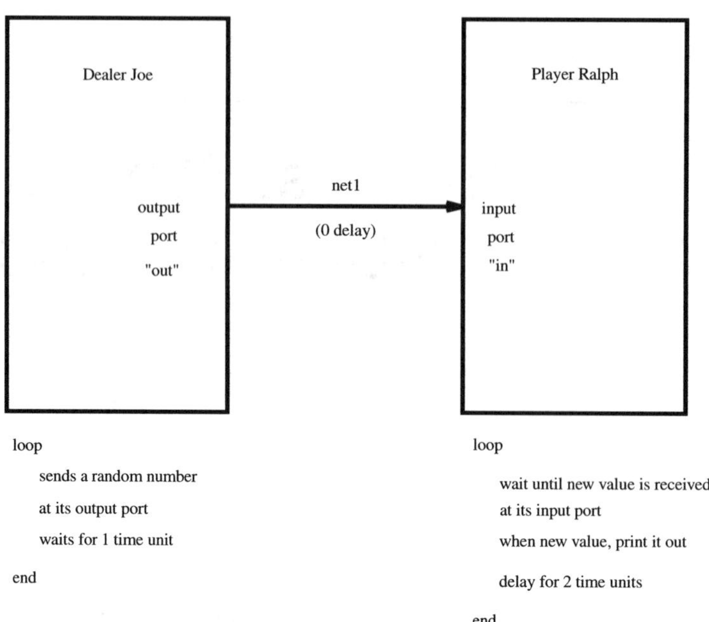

Figure 4.1 Modeling an asynchronous system in ADLIB-SABLE.

viewed as an instance of Dealer, Ralph is an instantiation of Player. Joe and Ralph are connected through a net that can carry a single integer value. The behavior of the Dealer is as follows. The Dealer generates and sends a random integer number, between 1 and 10, at its output port. It then waits for 1 unit of time and sends a second random value, and so on. The Player behaves as follows: It waits until it receives a new value at its input port. When it receives a new value, it prints it out, waits for 2 units of time, and then checks for new values at its input port. Assume that the net requires 0 units of delay to propagate a value from the sender to the receiver. Given that Joe and Ralph are driven by their unique delays and neither is aware of the other's timing, the system is asynchronous. Also, there is the implicit assumption that both Joe and Ralph are aware of the universal time in which the delays are expressed. In this section, we will provide the essential characteristics of ADLIB-SABLE. For a complete description of ADLIB-SABLE, the reader is referred to [50].

While Figure 4.2 presents the ADLIB descriptions of the modules of the asynchronous system in Figure 4.1, Figure 4.3 shows their connectivity through the SDL language. In Figure 4.2, the randomly generated integer, between 1 and 10, by the Dealer is manifested through the function "rndint" which accepts 1 and 10 as its arguments. Clearly, the times at which Joe asserts the randomly generated numbers is controlled by Joe and are unknown to Ralph. Similarly, Joe is completely oblivious of the times at which Ralph checks for new values at its input. In fact, neither Joe nor Ralph are aware of the existence of one another.

Joe and Ralph's obliviousness of one another leads to an interesting scenario. Since the ADLIB-SABLE system implements event driven simulation, if Joe generates integers v_i and v_j ($v_j = v_i$) in succession, the simulator will ignore v_j, i.e. not propagate it. If we assume that the successive values generated by Joe are distinct, then Joe will assert new values once every time unit. Assume that the values $v_0, v_1, v_2, \ldots, v_N$, are asserted at times $0, 1, 2, \ldots, N$, by Joe, The simulator stores these values along with the corresponding times. However, since

```
Program Example;
  nettype
    intnet = integer;
  Comptype Dealer;
    outward out: intnet;
  begin
    while true do begin
      assign rndinit(1,10) to out;
      waitfor true delay 1.0;
    end;
  end;
Comptype Player;
  inward in: intnet;
  begin
    while true do begin
      waitfor true check in;
      writeln(tty, in);
      waitfor true delay 2.0;
    end;
  end;
```

Figure 4.2 ADLIB descriptions for the modules.

```
Name: Test;
Types Dealer, Player;
Joe: Dealer;
Ralph: Player;
end;
netsegment;
net1 = Joe.out, Ralph.in;
endnets;
```

Figure 4.3 Interconnection of the modules in SDL.

Ralph waits for 2 time units after every read, clearly it will only succeed in reading the alternate values generated by Joe. That is, during its execution, Ralph will request the simulator whether there are new values at its input at times 0, 2, 4, ..., to which the simulator will respond affirmatively with the values v_0, v_2, v_4, etc. The behavior, described here, is consistent with the hardware.

In the above paragraph, we assumed that the simulator successfully stored the values asserted on the net by Joe along with the corresponding simulation times. Thus, if the delay value in the behavior of Ralph were changed to a large number such that it reads from its input at very long intervals, the simulator will need a large storage area to store the large number of values asserted by Joe between Ralph's successive reading. In reality, however, any storage facility is finite and this implies that the semantics of nets in ADLIB-SABLE incurs practical limitations.

Figure 4.4 shows a simple hardware circuit consisting of two two-input AND gates, G1 and G2, and an OR gate G3. The outputs of G1 and G2 are connected to the inputs of G3. Figure 4.5 presents the ADLIB descriptions while the connectivity is encapsulated in Figure 4.6. The generic gates are characterized by a parameter that reflects its propagation delay. The

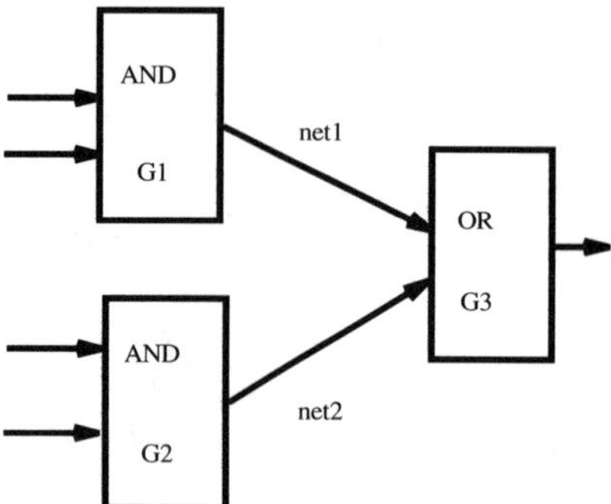

Figure 4.4 Modeling a simple circuit in ADLIB-SABLE.

```
Program Example;
  nettype
    Boolnet = Boolean;
  Comptype AND2 (PD: integer);
    inward in1, in2: Boolnet;
    outward out: Boolnet;
    begin
      waitfor true check in1, in2;
      assign in1 and in2 to out delay PD;
    end;

  Comptype OR2 (PD: integer);
    inward in1, in2: Boolnet;
    outward out: Boolnet;
    begin
      waitfor true check in1, in2;
      assign in1 or in2 to out delay PD;
    end;
```

Figure 4.5 ADLIB descriptions for the AND and OR gates.

```
Name: Test;
Types AND2, OR2;
AND2: G1 (PD="5"), G2 (PD="6");
OR2: G3 (PD="3");
end;
netsegment;
net1 = G1.out, G3.in1;
net2 = G2.out, G3,in2;
endnets;
```

Figure 4.6 Interconnection of the gates in SDL.

actual values of the parameters are supplied through the SDL descriptions and binding with the corresponding instances occur at instantiation time.

4.2 THE KEY LANGUAGE CONSTRUCTS OF ADLIB

A few of the important ADLIB constructs that underlie its ability to describe high-level behavior are borrowed directly from Pascal and include the (1) if then else construct, (2) case construct, (3) while construct, (4) repeat until construct, (5) for construct, and (6) goto construct. However, two unique and powerful constructs that qualify ADLIB as a behavior-level HDL are the (7) assign construct and (8) waitfor construct. The "transmit" and "upon" constructs of ADLIB are not primary and may be derived from the "assign" and "waitfor" constructs.

4.2.1 Assign Construct: assign ⟨expression⟩ to ⟨net name⟩ ⟨timing clause⟩

When the statement is executed, the expression is evaluated immediately and stored away in the simulator. The evaluated value is asserted to the appropriate net at a later point in time according to the timing clause. Thus, where the statement "assign 1 to out delay 15 ns" is executed at the current simulation time of X ns (say), the value of 1 is asserted to the net labeled "out" at time (X + 15) ns. To determine the value of time at which a statement in a given ADLIB description is executed, consider the following code segment shown in Figure 4.7.

```
Program Example;
  ....
  Comptype XYZ;
    ...
    begin
S1:   waitfor true delay 1.0 ns;
S2:   ...
S3:   assign 1 to out delay 2.0 ns;
S4:   ...
S5:   assign 0 to out delay 3.0 ns;
S6:   ...
    end;
```

Figure 4.7 An ADLIB code segment.

When SABLE has completed initialization, every instance of comptype XYZ will be ready for executing statement S1 and simulation time is set to 0 ns. The execution of S1 at simulation time 0 ns will imply that the subsequent statements may not be executed until time $0 + 1.0 = 1.0$ ns. Thus, S2 will execute at time 1 ns and, assuming that S2 does not contain any waitfor or upon constructs, statement S3 will also execute at a simulation time of 1 ns. Following the execution of S3, the simulator will contain an assignment of 1 at the net "out" scheduled for the current time plus 2 ns, i.e. 1 ns + 2 ns = 3 ns. Statement S4 will also execute at a simulation time of 1 ns and, assuming no waitfor nor upon constructs, so will

```
Program Example;
  nettype intnet = integer;
  clock clk1 (4,2);

  Comptype XYZ;
    inward in1: intnet;
    outward out: intnet;
    begin
      ...
      assign in1 to out SYNC clk phase 2;
      ...
    end;
```

Figure 4.8 An ADLIB code segment.

statement S5. The execution of S5 will also cause the simulator to hold an assignment of 0 at the net "out" scheduled for the current time plus 3 ns, i.e. 1 ns + 3 ns = 4 ns.

Figure 4.8 presents a different type of timing clause in the waitfor construct, one where the assignment is synchronized relative to a clock pulse, clk1, that is declared in the body of the main program. Clearly, clk1 is global in that it is visible to every instance of every comptype defined within "Program Example." That is, there is no need to explicitly define clk1 as an input to every instance of each comptype. In Figure 4.8, the value at the input port in1 is asserted at the output port out when the phase 2 of clk1 is active.

4.2.2 Waitfor Construct: waitfor ⟨boolean expression⟩ ⟨control clause⟩

The function of the waitfor construct is to halt the progress of execution of the behavior description, subject to the control clause and the Boolean expression. The arguments of the Boolean expression include any variables, constants, and values of the nets. The ADLIB-SABLE system supports three basic types of control clauses:

A: waitfor voltage > 0.001 delay 2.0 ns;
B: waitfor acknowledge = 1 SYNC clock1 phase 2;
C: waitfor old_clock = 0 check clock;

For the waitfor statement under label A, the statement is executed at the current time and the Boolean expression (voltage – 0.001) is evaluated utilizing the current value of the variable, voltage. If the expression evaluates to true, subsequent statements in the ADLIB code segment are executed following a delay of 2 ns. Otherwise, the simulator exits the ADLIB code segment and the statement is re-executed again at 2 ns following the current simulation time. The re-execution of the statement would involve evaluating the Boolean expression (voltage – 0.001), and so on.

For the statement under label B, when the phase 2 of clock1 is active, the value of the Boolean expression (acknowledge = 1) is computed using the current value of acknowledge. If the expression evaluates to true, subsequent statements in the ADLIB code segment are executed. Otherwise, the simulator exits the ADLIB code segment and the statement is re-executed at the subsequent active period of phase 2 of clock1.

For the statement under label C, the statement is executed at the current simulation time. If no new value has been assigned to clock, relative to its previous value, the simulator exits the ADLIB code segment. In contrast, if a new value has been assigned to clock, the Boolean expression (old_clock = 0) is evaluated. If the expression evaluates to true, subsequent statements in the ADLIB code segment are executed. Otherwise, the simulator exits the ADLIB code segment. The statement is re-executed at the subsequent simulation time.

4.2.3 Abstract Nets

The ability to model the interconnection nets through high-level abstract data structures constitutes an important factor in ADLIB's capacity to model complex behavior. The successful use of ADLIB-SABLE to support timing verification [51], fault simulation [52], and formal verification [53] reflects the value of the abstract nets. Figure 4.9 presents a representative use of complex nets to model a RS232 connection between two modules. The net, rs232, is a complex data structure with four fields: fg, td, rd, and rts. In the body of comptype XYZ, a neg_12v value is assigned selectively to the td field of the outgoing net, tty_line.

4.2.4 Internal Nets

In addition to inward, outward, and bidirectional nets, ADLIB permits the use of internal nets to represent an intermediate signal. Internal nets are strictly visible within the comptype, they permit both reading from and writing onto them, and are useful in simplifying code. For instance, where a complex Boolean function such as ((a and b) or c) requires computation, internal nets may simplify the ADLIB code, as shown in Figure 4.10.

```
Program Example;
  type widerange = (neg_12v, pos_12v);
       grounded = (zero)
  nettype
       rs232 = record
                  fg: grounded;    (* frame wground *)
                  td: widerange;   (* transmit data *)
                  rd: widerange;   (* receive data *)
                  rts: widerange   (* request to send *)
               end;

  Comptype XYZ;
    outward tty_line: rs232;
    begin
      ...
      assign neg_12v to tty_line.td delay 3.0 ns;
      ...
    end;
```

Figure 4.9 Modeling a RS232 connection through complex net types in ADLIB.

```
Program Example;
  nettype Boolnet = Boolean;

  Comptype XYZ;
    inward a, b, c: Boolnet;
    outward d: Boolnet;
    internal x: Boolnet;
    begin
      ...
      Transmit a and b to x delay 15 ns;
      Transmit x or c to d delay 14 ns;
      ...
    end;
```

Figure 4.10 Illustrating internal nets in ADLIB.

4.3 A KEY CONTRIBUTION OF ADLIB-SABLE

ADLIBs biggest contribution to the discipline of CAD was in its ability to realize a working behavior-level simulator. Although ADLIB was a concurrent language, it was designed to emulate multiprocessing, not to execute on an actual parallel machine which was lacking anyway. To achieve its goal, the ADLIB-SABLE system had proposed software techniques that were highly innovative. In ADLIB, the comptypes were concurrent objects that required, in theory, concurrent execution. Although the Simula-67 programming language and environment from Sweden allowed users to declare, instantiate, and manipulate objects, it lacked the notion of timing between the execution of objects. Thus, ADLIB-SABLE could not be trivially realized through SIMULA. The lack of a reliable Simula-67 environment forced Dwight Hill, the designer of ADLIB-SABLE, to seek Pascal as a platform on which to compile the ADLIB descriptions. However, Pascal is a sequential programming language and lacks any notion of objects, and this led Dwight to come up with an amazingly creative solution.

Consider, for simplicity, a set of N instances, $I_1, I_2, \ldots, I_N$, all of the same comptype C. In theory, all N instances are concurrent, i.e. they may potentially execute simultaneously. Since Pascal lacked any notion of objects, Dwight replaced the input, output, and bidirectional ports and the parameters of each instance with unique data structures, and for all instances, Dwight substituted a common procedure. Thus, to emulate the execution of an instance, the states of the corresponding ports and parameters were first restored, the procedure executed, and the results stored back in the data structure. To fake the concurrent execution of the instances on a single sequential CPU, Dwight utilized the notion of co-routines. The execution of an instance, I_i, continues until a waitfor or upon statement is encountered which causes the execution to be suspended. The thread of control is then transferred to a different instance, I_j, that is ready for execution. At a later point during simulation, when the conditions are right, the thread of control is returned to instance I_i and execution is resumed. Clearly, the instance, I_i, may contain multiple waitfor or upon statements and at the time of resuming execution, the thread of control must be transferred back to the correct statement where the instance was previously suspended. A further complication is that the exact number and location of the waitfor or upon statements in an ADLIB description are not known a priori to the designers of ADLIB-SABLE. Dwight organized the Pascal procedure corresponding to the body of a comptype through a huge case construct with the individual case statements defined by the waitfor and upon constructs of the ADLIB description. Dwight introduced delta as the selector variable for the case construct. A case statement allowed for the thread of control to exit the body of the procedure during suspension of the corresponding instance and to enter the body of the procedure upon resuming execution. The delta value for a given instance is stored along with the parameters and ports and serves as the vehicle to enter and exit the procedure body at precise locations.

Figure 4.11 presents an ADLIB description of two modules: sender and receiver. Only the receiver module contains two waitfor statements. Figure 4.12 presents the Pascal code compiled from ADLIB corresponding to sender and receiver. First, it is pointed out that procedures sender and receiver correspond to the ADLIB comptypes. Second, the Pascal var parameter, external, is a record of type "z9basecomptr" and contains the values of the ports, parameters, and delta. In ADLIB-SABLE, the delta values 0 and 2 correspond to initialization and debugging and will not be considered here. Note, however, that following the completion of initialization in procedure sender, the value of delta is set to 1. As a result, when sender is executed again, the thread of control passes directly to the code starting at label L1, and the body of the sender comptype is executed.

In the comptype receiver, the statements following "waitfor (inn = 1) check inn;" may be executed only when the net inn receives a new value and the Boolean expressions (inn = 1) evaluates to true. Otherwise, execution of the comptype receiver is suspended. In the procedure receiver in Figure 4.12, the statements starting at label L2 realize the following. If a new value is not received at inn, the thread of control exits the procedure while leaving the value of delta unchanged at 1. Thus, when the procedure is activated again at a later point in

```
Program Test;
  nettype
    intnet = integer;

  comptype sender;
    outward out: intnet;
    var m, n: integer;
    Begin
      m := 10;
      n = m * n;
      Assign 20 to out delay 2.0;

    End;
  comptype receiver;
    inward inn: intnet;
    var y: integer;
    Begin
      waitfor (inn = 1) check inn;
      y := inn;
      writeln('The value received is', inn);
      waitfor true check inn;
      writeln('Done');
    End;
```

Figure 4.11 An ADLIB description.

```
  procedure sender (var external: z9basecomptr);
  ...
  case external^.delta of
    0: initialization
       external^.delta = 1;
L1: 1: m = 10;
       n = m * n;
       assign(20, time+ 2);
    2: debugging
  ...

  procedure receiver (var external: z9basecomptr);
  ...
  case external^.delta of
    0: initialization
       external^.delta = 1;
L2: 1: examine whether new value asserted at inn;
       if negative, exit;
       if affirmative then
         begin
           is (inn = 1) equal to TRUE
           if negative, exit;
           if affirmative then
              begin
                y gets the value of inn;
                writeln('The value received is', inn);
                external^.delta = 3;
              end;
         end;
L3: 3: examine whether new value asserted at inn;
       if negative, exit;
       if affirmative then
         begin
           writeln('Done');
         end;
    2: debugging
  ...
```

Figure 4.12 Compilation of the ADLIB description into Pascal.

time, the thread of control passes directly to the code at label L2. In contrast, if a new value is indeed received at inn, then the comparison is made between the logical value of inn and 1. If the result is negative, the thread of control exits the procedure while leaving the value of delta unchanged at 1. Thus, when the procedure is activated again at a later point in time, the thread of control passes directly to the code at label L2. However, if the result of the comparison is affirmative, the statements following the waitfor statement are executed and the value of delta is set to 3.

In the comptype receiver, the last statement is "waitfor true check inn;" and it will cause the execution of the comptype to be suspended until a new value is received at the inn port. In the procedure receiver in Figure 4.12, the statements starting at label L3 realize the following. If a new value is not received at inn, the thread of control exits the procedure while leaving the value of delta unchanged at 3. Thus, when the procedure is activated again at a later point in time, the thread of control passes directly to the code at label L3. In contrast, if a new value is indeed received at inn, the statement following the waitfor statement is executed and, in this case, execution of the comptype receiver is complete.

4.4 LIMITATIONS OF ADLIB-SABLE

The development of ADLIB-SABLE marks a significant turning point in the evolution of HDLs and was a brilliant effort. It triggered the launching of the VHDL effort. However, with our deeper understanding of HDLs today and their needs, we conclude that ADLIB-SABLE had incurred a few significant limitations.

We will critically examine the characteristics of ADLIB-SABLE in the light of the fundamental requirements of a behavior-level HDL, as laid down in Chapter 3.

4.4.1 Comptype as an Incomplete Entity

ADLIB permits the declaration of variables and clocks in the body of the main program and outside of the comptypes. These variables and the clock are global, i.e. they are accessible to all instances of all comptypes defined in the program. While any comptype instance can read and write a global variable, the behavior of a comptype may synchronize with respect to the clock. Any global variable implies that all the component instances are, in theory, connected to each other through that variable. Such an implication contradicts the very notion of an entity which, by definition, must be self contained. The notion of global variables assumes that any update to any global variable is immediately, i.e. in zero time, visible to all instances of all comptypes. This clearly violates the basic principle of physics. There is also no known way to build a hardware to correspond to the notion of global variables. Last, the use of a global clock implies a synchronous digital design—a serious setback from ADLIB-SABLE's ambition to model asynchronous systems.

There is another serious limitation with ADLIB-SABLE. Conceivably, an entity may possess memory which, in turn, may be expressed through a state vector. By definition, the state vector must continue between successive executions of the entity. In ADLIB-SABLE, ideally, the state vector should be expressed through a variable, defined within a comptype. As explained earlier, in ADLIB-SABLE, a comptype is compiled into a Pascal procedure which is executed corresponding to the activation of all instances of the comptype. Therefore, following the execution of the procedure, corresponding to an instance, the content of the variables must be saved. Later, during the subsequent activation of the same instance, the contents of the variables must first be restored and then the procedure allowed to execute. The Stanford version of ADLIB-SABLE, as of 1985, that includes the ADLIB compiler, contained in `ADLIB.PAS`, and the SABLE environment contained in `SABLE.PAS`, does not view the variables defined within the body of a comptype as states. That is, between two successive executions of a component instance, there is the absence of continuity of the contents of the variables. Therefore, to express a state variable withina comptype, one needs to define it as a global variable within the main ADLIB program, create as many distinct variables as the number of instances of the corresponding comptype, and access the correct variable from the appropriate instance. Alternatively, one may intercept the Pascal code generated following compilation of the ADLIB code and manually include the state variables into the record structure of type "z9basecomptr" corresponding to the comp-type. As a result, the variables will be treated like the delta and the values of the nets. This scheme was utilized successfully by the author in the context of the fault simulation study [52].

The limitation, though conceptually severe, is relatively easy to fix by modifying the ADLIB compiler.

4.4.2 Connectivity in ADLIB-SABLE

The specification of the interconnection of instances of comptypes in ADLIB-SABLE is accomplished efficiently through utilizing the SDL language. The instantiation of the generic comptypes including the specification of the parameters and the enumeration of the nets are both natural and intuitive. The use of the abstract data types to represent the nets and the strong typing is one of ADLIB-SABLE's significant contribution to the HDL discipline. The consistency checking in SDL, including the detection of two output ports of two separate instances inadvertently connected through a net, reflects good design and is more natural and intuitive than in VHDL.

4.4.3 Concurrency in ADLIB-SABLE

ADLIB-SABLE pioneered the concurrency of comptype instances by emulating the multiprocessing of the instances. Although the instances appear to execute simultaneously to the user, in its current form, ADLIB-SABLE may not be executed on a real parallel processor. ADLIB-SABLE's scheduler is designed solely for uniprocessor execution.

4.4.4 Incompleteness of Timing in ADLIB-SABLE

As discussed earlier in Section 3.4.4, the delays in ADLIB-SABLE are realized through anticipatory scheduling. As a result, it fails to detect and delete inconsistent events and runs the risk of generating erroneous results.

ADLIB-SABLE does not provide any special language constructs to facilitate the verification of timing assertions relative to one or more signals such as setup, hold, minimum clock width, and maximum clock frequency. It relies on the designer's ingenuity in utilizing the ADLIB constructs to verify the timing assertions. Even if we were to assume that ADLIB's belief that all timing assertions may be verified through the existing ADLIB constructs is correct, it raises the concern that such coding would be non-intuitive, unnatural, and cumbersome. Consider the problem of verifying the setup constraint between the D signal and the clock, both input to a flip-flop. ADLIB offers two basic language constructs that deal with timings of signals: waitfor and assign. The "assign" clause simply asserts a value on a signal at a later time. It cannot be used directly to verify the relative timing between the D signal and the clock. In the "waitfor" clause, the timing sub-clause either checks for a signal experiencing a change in its logical value, pauses for a delay, or synchronizes with a clock phase. This sub-clause cannot directly verify any timing assertion. The Boolean expression sub-clause of "waitfor" can compare any data values and may presumably be used to compare the timing values. Unfortunately, ADLIB-SABLE does not provide any timing information of a signal at the user level. That is, to a designer synthesizing an ADLIB code for a hardware module, access to the scheduler and any timing information of signals is denied.

Thus, any attempt to verify any timing assertion appears doomed in ADLIB-SABLE. Recognizing this crucial limitation, Hill and Coehlo [50] proposed to provide the timing of signals to the user in Helix, a revised version of ADLIB-SABLE. However, the author, armed with a firm belief that human creativity has no limits, had posed this question as a problem in the take home midterm examination in the CSE517 course in semester I of 1996–7 at ASU. As if to prove the point, the class invented a simple yet creative approach to verify timing constraints in ADLIB-SABLE, as outlined subsequently. The resulting code, however, is cumbersome.

In this approach, the value of the setup delay is examined and it is noted that it must constitute an integral multiple of the basic simulation time. The setup delay value is known a priori to the designer and assume that it is 4 ns. In the ADLIB comptype for the flip-flop, D and clock are input signals representing the D input and the clock. In addition, internal signals D1, D2, D3 and D4 are defined. These internal signals are derived from the D signal, subject to delays by 1 ns, 2 ns, 3 ns, and 4 ns, respectively. At the appropriate active clock transition, i.e. low to high say, the ADLIB code examines whether the value of the signals D1 through D4 are all identical, i.e. either all are low or all are high. If affirmative, the setup condition is satisfied. Otherwise, the setup requirement is violated. We need to ensure that the D input is stable, i.e. either at 0 or at 1, for at least 4 ns prior to the active clock edge. This is achieved by examining its value at the current simulation time and each of the points 1 ns, 2 ns, 3 ns, and 4 ns, prior to the current simulation time.

Figure 4.13 presents the ADLIB code to verify the setup timing assertion in a flip-flop. To simplify, a part of the code is presented in pseudo code. Whenever the D input undergoes a change, all of D1 through D4 must be updated, as reflected in the code under label L1. Should the D input change at the same time that the clock experiences a transition to the active edge, both setup and hold conditions are violated, as reflected in the code starting at label L3. The code, starting at label L4, is initiated corresponding to the active clock edge. When the logical values at all of D through D4 are identical, the setup condition is honored. Otherwise, it is violated.

The verification of the hold constraint may be achieved in a similar manner, not exactly identical, and is left as an exercise to the reader. However, it is pointed out that only assignments to the future are permitted, i.e. the delay values can only be positive, never negative. Thus, given a signal defined up to time T, delayed replicas of it may be synthesized only to view its past behavior. The future behavior of the signal is unknown at the current time. At the active edge of the clock, verification of the hold time requires knowledge of the D input for hold time units beyond the current time. Therefore, the ADLIB code to verify the hold assertion is complex.

4.4.5 Inability to Represent Hierarchy in ADLIB-SABLE

Although SDL is stated to be a hierarchical language, ADLIB lacks the language constructs to express a digital design in a hierarchical form, while SABLE is not organized to accommodate hierarchical simulation. The key difficulty with ADLIB is its inability to incorporate comptypes within comptypes. ADLIB-SABLE does offer multi-level simulation wherein different sub-circuits of the hardware system are statically expressed at different levels of abstraction, and then simulated. Thus, while the sub-circuit that is simulated at a

```
Program Test;
  nettype
    intnet = integer;

  comptype flip-flop;
    inward D, Clock: intnet;
    outward Q: intnet;
    internal D1, D2, D3, D4: intnet;
    ...
    Begin
      assign D to D1 delay 1 ns;
      assign D to D2 delay 2 ns;
      assign D to D3 delay 3 ns;
      assign D to D4 delay 4 ns;
      waitfor (clock = 1) check D, clock;
Li:     if (D had changed) then begin
               assign D to D1 delay 1 ns;
               assign D to D2 delay 2 ns;
               assign D to D3 delay 3 ns;
               assign D to D4 delay 4 ns;
        end;
        if (clock changed but clock = 0, i.e. inactive edge) then
               do nothing;
L3:    if (both D and clock had changed and clock = 1, i.e. active edge)
               then setup and hold conditions are clearly violated;
L4:    if (clock changed and clock = 1, i.e. active edge) then begin
               if (D = D1 = D2 = D3 = D4) then
                 setup condition is honored;
               else
                 setup condition is violated;
  ...
  End;
```

Figure 4.13 Verifying timing constraints in ADLIB and pseudo code.

higher, more abstract level, executes quickly, the remainder subcircuit that is simulated at a lower level of abstraction executes slowly but provides detailed results. Thus, one may envision a scenario where a few comptypes model ALUs and register banks, while others model NAND, NOR, and NOT gates.

Clearly, ADLIB-SABLE fails to offer the user the opportunity to initiate simulation at a high level and then dynamically choose to elaborate one or more subsections of the hardware system at a lower level to observe the detailed simulation results.

4.4.6 Unnatural and Non-intuitive ADLIB Language Constructs

Although Hill [54] states that he included shared variables in ADLIB to facilitate debugging, the concept of global variables is counter intuitive for an HDL and does not correspond with reality.

The purpose of internal nets, according to Hill [32], is to help specify the behavior of a component with inertial delays or other internal timing characteristics, model an internal bus, or to store an intermediate value. Internal nets are part of the behavior specification only and do not appear in the structural description. As an example, Hill organizes the behavior of a complex combinational function block through two sub-functions where the output of the first function feeds into the second function as an argument via an internal net. Such an effort, namely to provide the underlying structure of a behavior description, poses a difficulty with the fact that comptypes cannot contain comptypes. ADLIB-SABLE lacks hierarchical representation. Thus, while potentially convenient to the user, as evident in Section 4.2.4, internal nets do not correspond to reality.

There is one other difficulty with a key ADLIB language construct. Consider the waitfor clause of type C: "waitfor ⟨Boolean expression⟩ check signal1, signal2, . . . ;". Upon examination, one would think that the designer of ADLIB intentionally separated out those elements that varied with time, i.e. signals, and placed them under the check sub-clause. Thus, when the values of these signals changed with time, that fact could be detected by the check sub-clause. In contrast, the aim of the ⟨Boolean expression⟩ sub-clause was to capture only the logical behavior of constants and variables and those of signals without the element of time. In essence, while the check sub-clause examined the timing dependence, the Boolean expression analyzed the combinational behavior. If this hypothesis is correct, then it leads to the following natural semantics of the check clause. Examine the list of signals under the check sub-clause and when any of them has undergone a change, at that simulation time instant, evaluate the Boolean expression. If the result is affirmative, continue on to execute the subsequent statements. Otherwise, the comptype should be suspended and the execution resumed when any of the signals do change at a future time instant. Thus, the evaluation of the Boolean expression is immediate, at the simulation time instant when the check sub-clause "passes".

However, Hill [32] clearly states that the semantics is one of waiting. That is, following the successful passing of the check sub-clause, the simulator evaluates the Boolean expression continuously until it evaluates to true. There are three serious difficulties with this semantics. First, continuous evaluation of the Boolean expression implies that the comptype instance must be evaluated every simulation time step, which immediately raises questions about SABLE's claim of implementing an event driven simulator. Second, if the Boolean expression is to be continuously evaluated, clearly one or more of its arguments must be time varying and, if so, it should appear under the check sub-clause. That is, once the simulator detects a change in this argument, the Boolean expression evaluation is triggered and the semantics of the waitfor clause would be highly efficient.

To understand the third difficulty, consider the waitfor clause of type A: "waitfor ⟨Boolean expression⟩ delay 2.0 ns;". According to Hill, the Boolean expression is evaluated every 2.0 ns. However, there has to be consistency between the waitfor clauses of types A and C. That is, the Boolean expression must be continuously evaluated until it evaluates to true. If so, the role of the "delay 2.0 ns" comes into question.

The above discussion underscores the need for the language designer to accept the burden to ensure that any user synthesized behavior description that utilizes the available constructs and is grammatically correct must be logically correct.

4.5 DRAFT REQUEST FOR PROPOSAL (DRFP) FOR VHDL

Following ADLIB-SABLE's successful introduction into the research community and observing a few of its limitations, the US DoD launched a major effort to design a new HDL to address the future very high speed integrated circuits (VHSIC) and named it VHDL. The requirements for VHDL were expressed in the draft request for proposal [55], DRFP F33615-83-R-1003, dated September 8, 1982. Although the DRFP is a 1-inch-thick document with over 120 pages, the fundamental requirements are described below and analyzed.

- 1. The VHDL language design must describe the electrical behavior and structure of digital systems in an unambiguous fashion. Behavior includes both logical and timing aspects.
  - 1a. The language must allow for assertions to be included, examples being the verification of setup and hold conditions and minimum or maximum pulse width.
  - 1b. Timing must be expressed through signals which are generally external. Within a design entity, internal signals may be defined simply as an artifact to simplify behavior description and with the understanding that it does not correspond to a real signal.
  - 1c. The language shall support both synchronous and asynchronous designs.
  - 1d. VHDL must allow the declaration of one or more global clocks.
  - 1e. The semantics of the timing constructs must be unambiguous and must not lead to conflicts.
- 2. VHDL shall be capable of describing components operating in parallel and cooperating with each other.
- 3. VHDL should support powerful and simple ways to express multiple instances of generic components and their connections.
- 4. The language design must be hierarchical, consistent, and user friendly.
  - 4a. Each level of hierarchical decomposition may be composed of I/O equivalent alternative sets of behaviors or further decompositions. For example, a microprocessor at the highest level may consist of a register, memory, alu, control, and program counter at the lower level. The alu, in turn, may consist of registers and an adder at the next lower level. The adder, in turn, may consist of a number of gates at the lower level, and so on, until one encounters a primitive gate which lacks further structural decomposition.
  - 4b. Where different components are at different levels of abstraction and are connected through signals, to execute a simulation, a transformation may be necessary to verify the interconnection of the signals. The transformation may either be generated automatically or by the designer.
- 5. The VHDL language design must incorporate Ada. However, when VHDL derives a structure from an Ada syntax, it should not be "similar" to Ada but adopted exactly as defined in Ada.
- 6. VHDL descriptions must be verifiable through simulation.
  - 6a. The simulator must be capable of accepting variable timing delays.
  - 6b. The simulation must allow for parallel processing to minimize simulation time and space.

6c. The semantics of the built-in scheduling algorithm must be first-in-first-out within priorities. A process may alter its own priorities.

- 7. The VHDL should be designed with a view toward automatic synthesis.

The requirements 1b, 1c, 1d, 2, 3, and 6 are inspired by ADLIB-SABLE. A few of the requirements, namely 1a, 1e, 4, 5, 6a, and 6b, address ADLIB-SABLE's weaknesses and reflect VHDLs desire to transcend them. As we will examine later, requirements 1e and 4 have not yet been fully addressed in the current version of VHDL. Although requirement 5 may have been motivated by other reasons, a few of the Ada constructs hold tremendous potential for any HDL, as we will examine in Chapter 6. Requirement 1c, as stated, is unnecessary since synchronous designs are a subset of asynchronous designs. Requirement 1d, however, paves the way for future problems in VHDL as we will examine in Section 7.2. Currently, requirement 7 appears to confront a number of important advances of VHDL as we will observe in Chapter 11.

5

Verilog HDL

The Verilog hardware description language was originally conceived in 1985 [56] by Gateway Inc., which was later merged with Cadence, Inc. It is currently used by a number of designers and it is a strong competitor of VHDL. A unique characteristic of Verilog HDL is its compact size which implies a simpler and easier to learn HDL, from the designers' point of view. In addition, it claims to offer better performance than the more elaborate VHDL. However, it lacks a number of important characteristics that may cause the generation of erroneous results and hinder the correct representation of hardware. This chapter critically analyzes the key concepts of Verilog HDL in the light of the fundamental requirements. The aim is not to solely criticize it but to enable a better and deeper understanding of the HDL discipline. This chapter focuses on the Verilog HDL grammar and semantics, as contained in [56] and [57] and corroborated by [58] and [59].

5.1 ENTITY

In Verilog HDL [56, 59], the basic unit of hardware description is termed a "module." Every module is identified by a unique identifier and the ports must be declared within the body of the module. Verilog HDL supports input, output, and inout ports. The types of the ports are predefined and limited to either wire(s) or register(s). The semantics of Verilog HDL correctly claim that, while a wire represents a connection between modules, a register represents a memory element, i.e. one with the capacity for storage. A system, composed of one or more sub-systems, may be described in Verilog HDL through a Verilog file consisting of the corresponding Verilog modules. Towards describing the behavior of a module, Verilog HDL claims to offer virtually all of the language constructs of the C programming language plus a few additional constructs that will be discussed subsequently. One special construct is "task" that is identical to a C function except that it is permitted to contain timing assignment and timing control clauses. Figure 5.1 presents a skeleton module corresponding to a 1-bit full adder that contains three inputs—a, b, and cin—and two outputs—s and cout. All of the I/O pins are defined as individual wires since they imply merely connections and Verilog HDL

```
module _1bitfulladder (a, b, cin, s, cout);
  input a, b, cin;
  output s, cout;
  wire a, b, cin;
  wire s, cout;
  ....
endmodule;
```

Figure 5.1 A skeleton Verilog module corresponding to a 1-bit full adder.

restricts their values to 1 (logic 1), 0 (logic 0), X (unknown or conflict), and Z (high impedance or tristate or floating).

The values of a wire may be read by a module and utilized in arithmetic expressions as an unsigned integer and, thus, a description of the module _1bitfulladder of Figure 5.1 may be developed as shown in Figure 5.2. In Figure 5.2, the module is executed only when the wire values of a, b or cin incur a change. In Verilog HDL, a block of code that is encapsulated by "always begin" and "end" is termed a behavioral instance. A module may contain multiple behavioral instances, each of which executes concurrently. Verilog HDL [56] does not precisely define the term concurrency. This chapter assumes a general definition for concurrency, as explained in Chapter 3, where two or more concurrent units are independent, asynchronous, execute simultaneously, and are characterized by the absence of a global scheduler.

```
module _1bitfulladder (a, b, cin, s, cout);
  input a, b, cin;
  output s, cout;
  wire a, b, cin;
  wire s, cout;
  always @(a or b or cin) begin
    s = (a and b and cin) or (a and b and not (cin)) or
        (a and cin and not (b)) or (b and cin and not (a));
    cout = (a and b) or (a and cin) or (b and cin);
  end;
endmodule;
```

Figure 5.2 A Verilog module corresponding to a 1-bit full adder.

In reality, however, the multiple behavioral instances may not be successfully executed concurrently on a multiple processor system. The Verilog HDL neither requires each behavioral instance to be self contained nor does it restrict the designer from operating on variables that are shared between two or more behavioral instances. As a result, in an actual distributed computing environment, two or more behavioral instances, executing on unique processors, may observe two different values of the same variable, leading to inconsistency. Thus, in contrast to its claim, behavioral instances in Verilog HDL cannot model concurrent units of hardware since there is no correspondence between shared variables and reality.

Verilog HDL restricts the input, output, and inout ports to wire(s) and register(s) and does not permit user defined types. This restriction is puzzling since ADLIB-SABLE [32] had clearly established as early as 1979 the advantages of typed ports. User defined port types facilitate the implementation of complex simulation systems such as deductive fault simulation and timing verification and will be amenable to more abstract descriptions of complex systems in the future. Verilog HDL [56] offers no reasons for the restriction. Perhaps, the Verilog HDL designers wanted to limit the choices and prevent designers from inadvertently developing erroneous descriptions. If so, their objective appears to have been unsuccessful as explained in Figure 5.13 in Section 5.5.

5.2 CONNECTIVITY

Although the top level unit in Verilog HDL, for the purpose of simulation, consists of a single module, Verilog HDL permits the inclusion of one or more modules within any module to develop a static, hierarchical representation. Verilog HDL refers to such descriptions as structural. Unlike ADLIB-SABLE [32], connectivity in Verilog is through implication, as shown in Figure 5.3.

```
module adder4 (a, b, s);
  input a, b;
  output s;
  wire [3:0] a;
  wire [3:0] b;
  wire [4:0] s;

  _1bitfulladder u1(a[0], b[0],  0, s[0], c0);
  _1bitfulladder u2(a[1], b[1], c0, s[1], c1);
  _1bitfulladder u3(a[2], b[2], c1, s[2], c2);
  _1bitfulladder u4(a[3], b[3], c2, s[3], s[4]);
endmodule;
```

Figure 5.3 A structural description in Verilog.

In Figure 5.3, a 4-bit adder with no carry input is expressed structurally in terms of four individual modules, each representing a 1-bit adder and constituting an instance the _1bitfulladder that is defined in Figure 5.2. Thus, the structural style of description utilizes Verilog HDL's instantiation capability. The first two inputs of each 1-bit adder are read from the corresponding wires of the input wire collections: a and b. The intermediate carry values are propagated from one module to the subsequent one through the use of the undefined, implicit wires—c0, c1, and c2—as permitted by Verilog HDL. While Figure 5.3 represents a two-level hierarchy, a three-level hierarchy may be easily visualized through the conceiving of a module "alu", of which one of the constituent modules is adder4.

While the connectivity in Verilog HDL is clearly adequate, the absence of the explicit interconnection nets between modules, unlike in ADLIB-SABLE [32], assumes instantaneous propagation of signals between modules, which opposes reality. The lack of explicit nets also precludes the specification of the timing characteristics of the nets which are increasingly becoming critical. In addition, while it is difficult to detect errors such as an inadvertent connection of two or more outputs, it is hard to identify whether a connection of an inout port with an output port is intentional or an inadvertent error. Verilog HDLs failure to require the explicit declaration of c0 through c2 further aggravates the problem. A user may erroneously specify a variable name in place of c0 through c2, causing unnecessary difficulty in debugging.

5.3 TIMING

5.3.1 Timing Assignments

Verilog HDL supports the usual assignments to variables and "continuous assignments" to output and inout ports. Only the continuous assignments are associated with delay values. Figure 5.4 presents four forms of assignment statements supported by Verilog HDL. The assignment statement at label L1 implies that the result of adding a and b is assigned to nout, 3 time units after the current simulation time. The assignment statement at label L2 is identical to that at L1 except that the word assign is dropped. The assignment statement at label L3 implies that the result of adding a and b is assigned to nout, 0 time units after the current simulation time. The assignment statement at L4 differs from that at label L3 in that "#0" is missing but is identical in every respect. The statement at label L4 implies an assignment with 0 time units delay.

```
    module xyz (a, b, m, nout);
      input a, b;
      output m, nout;
      wire a, b, m;
      wire [3:0] nout;
      integer temp;
      parameter tplh = 5,
                tphl = 9;
Li: assign #3 nout = a + b;
L2: #3 nout = a + b;
L3: #0 nout = a + b;
L4: nout = a + b;
    ...
L5: temp = a and b;
    if (temp == 1)
      #tplh m = temp;
    else if (temp == 0)
      #tphl m = temp;
    endmodule;
```

Figure 5.4 Assignment statements in Verilog HDL.

The Verilog HDL code segment starting at label L5 describes the behavior of a simple two-input AND gate subject to rise and fall propagation delays. A temporary variable stores the result of AND'ing a and b. The output assignment utilizes the t_{plh} delay value when a 1 is scheduled at the output and the t_{phl} delay value corresponding to a 0 scheduled at the output.

In Verilog HDL, continuous assignments are associated with delays. Since the delay values are discrete in nature, the execution of the assignments in Verilog HDL must occur at discrete points in the simulation time. Thus, the rationale underlying the use of the word "continuous" in continuous assignment is imprecise.

The most significant limitation with the assignment statement construct in Verilog HDL is the absence of the inertial delay semantics [56] and the inability to detect and deschedule inconsistent events [60]. As a result, Verilog HLD may generate incorrect results, as was explained earlier in Section 3.4.4.

According to the Backus-Naur Form (BNF) of the Verilog HDL grammar [56], the description in Figure 5.5 is legal. The "#3" and "#6" implies 3 and 6 time units respectively, and are presumably associated with the 1-bit wire and the 4-bit wire, respectively. While the semantics of this delay value is unspecified, i.e. whether inertial or transport, there are two additional problems. First, conceivably, a wire delay of 3 time units associated with an output port implies that the outgoing signal is delayed by 3 units. However, the meaning of a wire delay of 3 time units associated with an input port is unclear. Second, any propagation delay value must be calculated based on the total length of the wire between the two modules. This, in turn, would require each module to be intrinsically aware of all modules to which it is connected, a requirement that is fundamentally difficult to achieve. The Verilog HDL does not address this subject adequately. It is also uncertain how one may succeed in preventing the wire delay from being accounted for twice—once in each of the two modules that the wire interconnects. Thus, a re-examination of the Verilog HDL design is warranted.

```
module xyz (a, b, m, nout);
  input a, b;
  output m, nout;
  wire #3 a, b, m;
wire [3:0] #6 nout;
  ....
endmodule;
```

Figure 5.5 The wire delay syntax warrants a re-examination of Verilog HDL.

A key requirement for the simulation of the descriptions expressed in Verilog HDL is that all events corresponding to a specific simulation time instant must be executed and removed before the simulator progresses to the subsequent simulation timestep. Unfortunately, Verilog HDL's permissiveness to use zero delays may violate this requirement even for the most simple asynchronous design—a latch. Assume a cross-coupled NAND gate latch with two NAND gates with inputs {a, x} and {y, b}, respectively. The outputs are y and x respectively. In the Verilog HDL code, presented in Figure 5.6, the two NAND gates are represented through two concurrent behavioral instances, each containing a zero delay assignment statement. An analysis of the actual hardware is presented in Section 7.2.3 under "Difficulties with delta delays" and it reveals an ambiguous scenario, yielding erroneous results. Furthermore, as a result of the ambiguity, the simulator may incur an indefinite number of events, all defined at the same simulation time.

5.3.2 Modeling Asynchronous Timing Behavior in Verilog HDL

To permit every behavior instance to exercise control of its execution independently, Verilog HDL provides three forms of control—#expression, event-expression, and wait

```
module latch (a, b, x, y);
  input a, b;
  inout x, y;
  wire a, b, x, y;

  initial begin
    x = 1;
    y = 1;
  end;
  always @(a or x) begin
    y = not (a and x);
  end;
  always @(b or y) begin
    x = not (b and y);
  end;
endmodule;
```

Figure 5.6 Ambiguity in a Verilog module corresponding to a latch.

{expression}. According to Verilog [56], while #expression is intended for synchronous control, event-expression is meant for asynchronous control, and wait {expression} for level sensitive control. Consider the Verilog code shown in Figure 5.7 that is copied from [56].

In Figure 5.7, the behavioral instances b1 and b2, corresponding to initial, are intended to perform initialization sequences. In Verilog HDL [56], the execution of the description must yield the sequence of values as shown in Figure 5.8. The sequence, as predicted by Verilog HDL, is based on the assumption that the progress of the simulation time during execution is uniform in all of the concurrently and asynchronously executing instances.

```
module time_control;
  reg [1:0] r;

  initial begin : b1;
    #10 r=1;
    #20 r=1;
    #30 r=1;
  end

  initial begin : b2;
    #5  r=2;
    #20 r=2;
    #30 r=2;
  end
  always @r begin : b3
    $display ("r=%d at time %d",r $time);
  end;

endmodule;
```

Figure 5.7 Timing control in behavioral instances.

```
r = 2 at time  5
r = 1 at time 10
r = 2 at time 25
r = 1 at time 30
r = 2 at time 55
r = 1 at time 60
```

Figure 5.8 Output predicted by Verilog HDL.

However, as examined subsequently, this assumption is incorrect and conflicts with actual hardware.

In an actual hardware, when the behavioral instances b1 and b2, labeled initial, and b3, labeled always, are executed concurrently on separate processors, each processor will maintain its own scheduler and the progress of the simulation time in each scheduler may be different. There is no notion of a global scheduler and hence there is no concept of global simulation time. Each of b1, b2, and b3 is sequential within itself. Since there is no explicit connection between instances b1 and b2, the rate of progress of the simulation in each of the processors will clearly be a function of the speed of the underlying processor, i.e. their rates of progress may be different. Thus, it is conceivable that the instance b1 may execute faster and generate "r = 1 at time 10" before instance b2 may generate "r = 2 at time 5," and the instance b3 receives r = 1 from b1 before receiving r = 2 from b2. In the absence of the notion of global simulation time, instance b3 cannot determine whether it should have incurred first r = 1 from b1 or r = 2 from b2. The terms "#10" in b1 and "#5" in b2 appear as mere parameters to b3 and they possess no timing significance. Therefore, instance b3 must display the values in the order that it intercepts them and, thus, a partial picture of the sequence generated by actual hardware is shown below, which differs from the output predicted by Verilog in Figure 5.8.

r = 1 at time 10
r = 2 at time 5

To obtain the output predicted by Verilog, shown in Figure 5.8, the execution of the instances b1, b2, and b3 must proceed in lockstep, controlled by the centralized scheduler of a synchronous, distributed algorithm, that maintains the global simulation time. Thus, b1, b2, and b3 are subject to forced synchronization and are no longer independent, asynchronous, and truly concurrent. Furthermore, where b1, b2, and b3 execute simultaneously on concurrent processors, they must all maintain explicit connections to the centralized scheduler, which would significantly diminish the simulation performance.

An example of the use of @event-expression for the independent control of behavioral instances has already been shown in Figure 5.6.

The opportunities of developing incorrect hardware descriptions in Verilog HDL is amplified through the Verilog HDL code in Figure 5.9 that is copied from page 143 of [56]. The description constitutes a sample program that is intended to verify timing assertions.

According to Verilog HDL, the behavioral instance labeled L1 in Figure 5.9 detects changes in the signal rd and stores the current simulation time values in rd_low and rd_high corresponding to rd values of low and high, respectively. The behavioral instance L2 detects a positive edge of signal rd, reads the value from rd_low, and verifies the low pulse width of the signal. However, analysis reveals that the code is likely to generate erroneous results. Given

```
module xyz;
  time rd_high, rd_low;

  always @rd begin : L1
    if (rd == 1)
           rd_high = $time;
    else if (rd == 0)
           rd_low=$time;
  end;

  always @(posedge rd) begin : L2
    checktiming ("tRR", tRRmin, rd_low);
  end;
endmodule;
```

Figure 5.9 Erroneous Verilog code for checking timing violations.

that the instances L1 and L2 may execute concurrently and independently, conceivably a latency may exist between the computation of rd_low at L1 and its receipt at L2 following propagation from L1. Consider the following likely scenario. The instance L2 has just detected a positive edge of the signal rd but the underlying processor of L1 is slow and had not yet successfully computed the most recent value of rd_low. Thus, the instance L2, unaware that it has not yet received the most recent value of rd_low, utilizes its old copy of rd_low which corresponds to the previous clock edge, and generates erroneous results. The principal reason for this discrepancy is Verilog HDL's inability to accurately model real hardware.

5.3.3 Timing Constraints

Except for the Verilog XL simulator supported predicate, $time, that permits the user to read the current simulation time, Verilog HDL lacks language constructs to help verify timing assertions. Thus, the task of checking for setup, hold, minimum pulse width, and maximum clock frequency constraints is cumbersome. Consider the verification of the setup constraint of a D-flip-flop. Since Verilog HDL lacks the signal attributes provided in VHDL, it is impossible to focus on the positive edge of the clock signal when it coincides with the current simulation time and then examine the state of the D-input signal, setup time units prior to the current simulation time. Thus, the behavior description of the D-flip-flop must continuously watch for any changes in both the clock and D-input signals. In Figure 5.10, the statement at label L1 checks if both the clock and D-input signals incur changes at the same time. If true, both setup and hold violations occur. The statement at label L2 checks whether the clock incurs a transition. If true, the current simulation time is noted. If the D-input signal is a 0, the current time is compared against the stored time value when the D-input signal had incurred a high to low transition. If the difference exceeds setup time, the setup constraint is satisfied. Otherwise, it is violated. When the D-input signal is a 1, the current time is compared against the stored time value when the D-input signal had incurred a low to high transition. If the difference exceeds setup time, the setup constraint is satisfied. Otherwise, it is violated. The statement at L5 verifies if the D input incurs a change to 0. If true, the time value of the high to low transition of the D-input signal, D_low, is updated to the current simulation time.

```
module xyz;
  integer D_old, Clk_old;
  time D_high, D_low, Clk_high;

  initial begin
     D_old  = 0;
     Clk_old = 0;
  end;

  always @ (D or Clk) begin
L1:  if (Clk_old==) and Clk==1) and ((D_old==0 and D==1) or
         (D_old==1 and D==0))
       print violation;
L2:  else if (Clk_old == 0 and Clk == 1) then begin
       Clk_high = $time;
L3:  if (D==0)
         if (Clk_high - D_low) >= setup)
           print setup ok;
         else print setup failed;
L4:  if (D==1)
         if (Clk_high - D_high) >= setup)
           print setup ok;
         else print setup failed;
     end;
L5:  else if (Old_D == 1 and D == 0) then begin
         D_low = $time;
         Old_D = D;
     end;
L6:  else if (Old_D == 0 and D == 1) then begin
         D_high = $time;
         Old_D = D;
     end;
  end;
endmodule;
```

Figure 5.10 Checking for setup constraint violation.

Similarly, the statement at L6 verifies whether the D input incurs a change to 1 and, if true, the time value of the low to high transition, D_low, is updated to the current simulation time.

While the Verilog HDL code in Figure 5.10 will work accurately, it is extremely inefficient when the D input incurs a large number of signal transitions well before the positive edge transition of the clock signal. In contrast, the VHDL signal attributes are highly efficient. To verify the hold time constraint, a Verilog HDL designer may adopt one of two schemes, both highly inefficient. First, following the detection of the positive clock edge, the behavior description focuses on every subsequent simulation clock timestep for hold time units and examines the state of the D-input signal. If the D-input value remains unchanged, the hold constraint is satisfied. Otherwise, it is violated. In the second scheme, following the detection of the positive clock edge, the behavior description focuses on detecting the next transition of the D-input signal. It then computes the difference between the transition time of the D-input signal and the stored value of the clock positive edge, and compares against the hold time. if the difference is positive, the hold constraint is satisfied. Otherwise, it is violated. The inefficiency with this scheme is high if the next transition of the D-input signal occurs much later relative to the positive clock edge.

5.4 HIERACHY

The ability to instantiate and include modules within a module provides the mechanism for static hierarchical representation in Verilog HDL. Figure 5.11 presents a skeleton Verilog structural description of an ALU. Prior to initiating simulation, the Verilog simulator must first elaborate the module alu into the instances k1, k2, k3, that correspond to specific modules—adder4, multiplier, and barrel_shifter. Second, the simulator must elaborate each of the instances k1 through k3, if warranted, into their constituent modules. For example, the module adder4 must be elaborated into its four constituent modules—_1bitfulladder—utilizing the description in Figure 5.3. The elaboration process terminates when no successive expansion into the lower-level modules is possible and is labeled "static."

```
module alu;
  input ...;
  output ..;
  wire [3:0] ...;
  wire [3:0] ...;
  wire [4:0] ...;

  adder4 k1(...);
  multiplier k2(...);
  barrel_shifter k3(...);
endmodule;
```

Figure 5.11 A skeleton description of an ALU in Verilog.

Verilog HDL supports mixed-mode simulation where select sub-systems are elaborated into their lower-level descriptions for detailed and time-consuming simulation while the remainder of the system is simulated quickly at the higher level. However, it is unable to provide the zooming capability that is explained in Section 6.7.

5.5 NATURALNESS AND CONSISTENCY WITH UNDERLYING HARDWARE

Of all the difficulties of Verilog HDL discussed in this chapter, perhaps the most crucial one is its permissiveness with zero delay usage and the temptations that the designers feel in using it to check for logical correctness, regardless of timing. While logic devoid of timing cannot yield correctness, zero-delay hardware is not reality. The use of zero delay is the principal cause for the confusion in the Verilog HDL simulator which, in turn, causes designers to invent ad hoc solutions to problems without a solid understanding of operational hardware.

As recently as 1996, Becker [61] encouraged the use of zero delays to improve the simulation performance. Becker argues that, since the values on the wires and registers will change at the clock transition instant, latching a value at this instant may lead to non-determinism, and requires latching to commence at a time instant just prior to the clock edge, termed "pre-clock" point. Becker's argument and requirement are both due to the zero-delay usage and constitutes a direct violation of the fundamentals of hardware operation and hold time.

Consider the Verilog HDL description in Figure 5.12, adopted from page 201 of [56]. The intent is to execute the module comb for a number of input test patterns and print out the output result. A top-level module, test, is synthesized for this purpose. It contains an instantiation u1 of module comb, a behavioral instance b1 of type initial that generates the test patterns, and a behavioral instance b2 of type always that prints every new output as it is generated. The authors of [56] observe that the description as written in Figure 5.12 does not execute as intended. Only the output corresponding to the last test pattern is printed out. The authors of [56] reason that the instance b1 executes too fast, as a single event, and monopolizes the underlying processor. As a result, neither u1 nor b2 gets a chance to execute until b1 is complete. This assessment appears puzzling because the 5000 zero-delay assignments that are generated as a result of repeated execution of the statement at label L2 of instance b1 must be queued in the scheduler and available for the instance u1 to utilize during its execution. The observation noted in [56] leads one to suspect that the scheduler design may be erroneous. A solution, proposed in [56], suggests replacing the "inreg = i" statement in b1 with "#0 in reg = i" which will prevent the instance b1 to execute in its entirety as a single event.

```
module comb (in1, in2, in3, in4, in5, out1, out2);
    input in 1, in2, in3, in4, in5;
    output out1, out2;
L1: assign (out1, out2) = in1 + in2 + in3 + in4 + in5;
endmodule;

module test;
  reg [4:0] inreg;
  wire [1:0] outwire;
  integer i;
  parameter testsize = 5000;

  comb u1 (inreg[4], inreg[3], inreg[2], inreg[1],
               inreg[0], outwire[1], outwire[0]);

  initial : b1;
    for (i = 0; i < testsize; i = i + 1)
L2:     inreg = i;

  always @outwire : b2;
    $display("outwire = %b", outwire);
```

Figure 5.12 Incorrect Verilog HDL code.

```
module adder4 (a, b, s);
  input [3:0] a;
  input [3:0] b;
  output [4:0] s;
  reg [4:0] s;

  always @(a or b) begin
    s = a + b;
  end;
endmodule;
```

Figure 5.13 Behavioral description of a Verilog module for a 4-bit adder.

The reasoning by the authors of [56], unfortunately, reveals a fallacy and the solution does not reflect a true understanding of the operational aspects of hardware. Despite introducing a delay of 0, all individual executions of the loop in the instance b1 still execute at simulation time 0 and therefore correspond to a single event. The value zero added to zero, any number of times, still yields zero. The principal problem here is that the designer of module test appears to have acquired an erroneous understanding of the real hardware that is being modeled here. Clearly, the module comb requires a non-zero time to execute in reality. To test its operation for different sets of input test vectors, "comb" must be allowed adequate time to generate the output for an input vector and propagate the result to instance b2 before accepting the subsequent input test vector. Therefore, assuming that the actual propagation delay of comb to be 5 time units (say), the statement at label L1 must be modified to #5 assign (out1, out2) = in1 + in2 + in3 + in4 + in5." The statement at label L2 must be modified to "#10 inreg = i," which allows the comb 10 time units to generate and propagate the output corresponding to an input. With these modifications, even if the instances are executed on concurrent processors with the processor executing u1 operating much slower than the processor underlying instance b1, the output assignments of b1 will be queued and ready for u1 to utilize them successively, i.e. it will process a subsequent input test vector following the successful processing of the preceding test vector.

The description in Figure 5.13 is adopted from page 25 of [56]. The authors of [56] develop a behavioral description of a 4-bit adder and argue that the output port, s, is declared as a register to permit behavioral statements to assign values to it. According to Verilog HDL, a register reflects a memory element. Thus, to describe an adder, a purely combinational device, at the behavior level in Verilog HDL, one is required to use registers which are but sequential devices. Clearly, Verilog HDL may benefit from a redesign effort.

6

Design of a Concurrent HDL in Ada from First Principles

6.1 INTRODUCTION

Simulation of a digital design prior to prototyping is a useful tool because it can assist in early detection of design errors and consequently reduce production costs. In order to simulate a design, it must be modeled in a hardware description language either at the behavior, function, or gate level. While the conventional hardware description languages such as DDL [24] and ISP [25] are capable of modeling circuits at the architecture and instruction levels, ADLIB-SABLE [38] is capable of simulating at the gate instruction, and behavior levels. The principal limitation with these languages is that they differ from their simulation environments, with the consequence that the approach is non-uniform. Furthermore, in virtually all of the current simulators, including those associated with DDL, ISP, and ADLIB, communication among the hardware modules and scheduling of the hardware descriptions are centralized and, consequently, inappropriate for execution on multiprocessor systems.

Simulation at a lower level such as the gate level is usually time consuming while functional simulation is faster at the cost of less detailed results. Conventional multi-level simulators have addressed this problem by permitting simulations at more than one level in the same environment. In ADLIB-SABLE, for example, gate- and functional-level simulations may be carried out simultaneously. Diana [62] and SPLICE [63] permit mixed-level simulations at the gate and circuit level. These multi-level simulators are static, however, and can often be ineffective and uneconomical. The level at which a subsystem is simulated is decided prior to initiating simulation and can therefore be based on the designer's intuition and experience or on the results of a previous simulation run.

In the approach introduced in this chapter, Ada is used to describe hardware at the behavior, function, and gate levels, and as a distributed simulation environment, which results in a uniform approach to the simulation of digital systems. Ada's synchronization techniques coupled with new techniques for distributed modeling and scheduling result in the possibility of distributed execution of hardware descriptions on parallel processors. The approach, termed RDV, has been verified through utilizing the Stanford University prototype ADAM compiler and environment that executes on a DEC-20 computer. There are, however, difficulties with the execution model of Ada, and VHDL's approach to address timing directly may be more natural than the indirect approach to timing that is adopted in this chapter.

6.2 CHARACTERISTICS OF AN HDL

The following characteristics are basic to any good hardware description language

- Generic modeling and instantiation are permitted.
- The model's logical behavior, timing, and synchronization may be expressed accurately.
- Communication among model instances is permitted.
- Distributed scheduling of the model instances is possible.
- Digital designs may be expressed hierarchically.

6.3 GENERIC MODELING

A digital system design contains a finite number of interconnected devices and may consist of more than one identical component. Because duplication in software is inexpensive and fast, a single parameterized model of each device type is sufficient and one such model may be written for a family of components. The parameters provide a means for specifying the identity of each device and other distinguishing features.

For any digital design, elements in RDV may require basic interconnection information during simulation. This information is manually stored in a database with predetermined data structures. In the future, an interconnection language would need to be developed so that the database may be automatically generated from a schematic description.

Every component type of a digital design is represented in RDV as an Ada task type. Task instances correspond to instances of a component type and are uniquely identified. Instances X and Y of the same type may be further characterized by a unique set of parameters. In a conventional simulator such as ADLIB-SABLE, instantiation takes place during the compilation phase and the exact number of components must be known prior to initiating simulation. Instantiation of tasks in Ada may take place either during the compilation or the execution phase. Further details of the execution-time instantiation are presented subsequently.

Figure 6.1 shows an RDV representation of a circuit consisting of two Motorola 68000 microprocessors and ten two-input AND gates that requires compile-time instantiation.

Each component and path in the digital design is identified by either a designer-specified or compiler-allocated identifier and the input and output pins (or ports) are assigned identifiers within that device. The database contains a mapping from every port to the path to which each port is connected and a mapping from every path to all ports connected by each

```
Package Circuit-A is
  Task type M68000 is ...;
  M : array (1..2) of M68000;
  Task type AND2 is ...;
  ANDX : array (1..10) of AND2;
End Circuit-A;
```

Figure 6.1 Instantiation of models.

path. Because identifiers are assigned arbitrarily to components, there is a mapping from every identifier to the component type. Every task has an identifier, and the database contains a mapping from every task identifier to the respective device identifier. With this organization, database access times are small and contribute to the overall speed of the RDV. For the digital design in Figure 6.2, the interconnection database is given in Figure 6.3. In Figure 6.2, "p" denotes the port number.

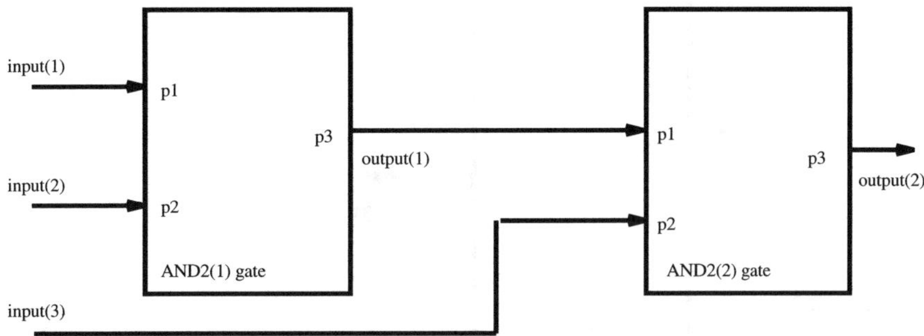

Figure 6.2 An example design for simulation.

```
ct(1):=AND2; ct(2):=AND2;
c(1,1):=1; c(1,2):=2; c(1,3):=4; --for comp # 1
c(2,1):=4; c(2,2):=3; c(2,3):=5; --for comp # 2

nt(1,1).portcomp:=1; nt(1,1).portno:=1; n(1):=1; --for inp1
nt(2,1).portcomp:=1; nt(2,1).portno:=2; n(1):=1; --for inp2
....
cti(1):=2; -- two tasks of AND2 gate type
```

Figure 6.3 Interconnection database for the design in Figure 6.2.

In the representation in Figure 6.3, the data structure "ct" stores the mapping from the two components to their types, "c" contains the mapping from the two components to all the nets they are connected, "nt" stores the mapping from a net to all components it connects, and "cti" contains the total number of tasks of each type.

6.4 LOGICAL BEHAVIOR, TIMING, AND SYNCHRONIZATION

The logical behavior, timing, and any required synchronization of a component type is expressed in Ada. Input and output ports of a component are typed such that they are capable of receiving and transmitting complex objects and detecting illegal transfers. For instance, the input and output ports of a two-input AND gate, shown in Figure 6.5, are represented in Figure 6.4.

The two input and output ports of the AND gate are represented by inward(1), inward(2), and outward(3) respectively. The datatype of the input and output ports of the AND gate is Boolean. The actual message objects that are transmitted through the two input and the output port are encoded using the port identifiers and are pointed at by F(inward(1)),

```
Task body AND2 is
  inward : array (1..2) of Boolean;
  outward : array(3..3) of Boolean;
  ...
End AND2;
```

Figure 6.4 Modeling inputs and outputs.

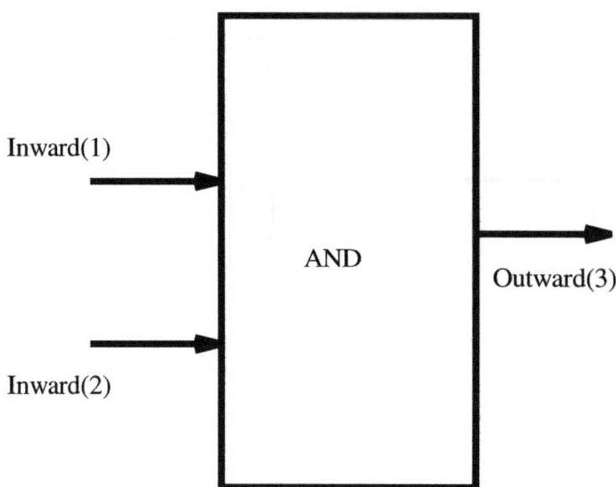

Figure 6.5 An AND gate.

F(inward(2)), and F(inward(3)) respectively, where F is a data structure defined by the nature of simulation. For behavior-level simulation, F assumes the form of a record with two fields. The first field, "value", stores the logical value and the second field, "assertion-time", contains the time at which the value is asserted.

The **logical** behavior of the two-input AND gate may be expressed as shown in Figure 6.6.

In the behavior description of the AND gate, the output logical value is the result of AND'ing the two input logical values. For complex models such as the Motorola 68000 and the AMD2903, the constructs of ADA such as "case", "if then else", and "loop" may be used and are not shown here. States of such complex descriptions are represented through Ada variables.

The **timing** characteristics of hardware such as propagation delays, setup, and hold times are modeled in RDV. For the two-input AND gate, if the signals at the two inputs are asserted at $t = t_1$ and $t = t_2$ ($t_2 > t_1$), the output signal is asserted at $t = t_2 +$ p-delay, as shown in Figure 6.7. The function "max" returns the maximum of its arguments and "p-delay" is the propagation delay in Figure 6.7.

For a D-type flip-flop shown in Figure 6.8, the assertion time of the data input must precede that of the clock by a minimum of "t-setup" as shown in the description in Figure

```
Task body AND2 is
....
  begin
  T(outward(3)).value :=T(inward(1)).value and
                            T(inward(2)).value;
  end;
End AND2;
```

Figure 6.6 Modeling logical behavior.

```
Task body AND2 is
  .....
  var
    p-delay : integer;
  begin
    T(outward(3)).assertion-time :=
         max{T(inward(1)).assertion-time,
          T(inward(1)).assertion-time} + p-delay;
  end;
End AND2;
```

Figure 6.7 Modeling timing characteristics.

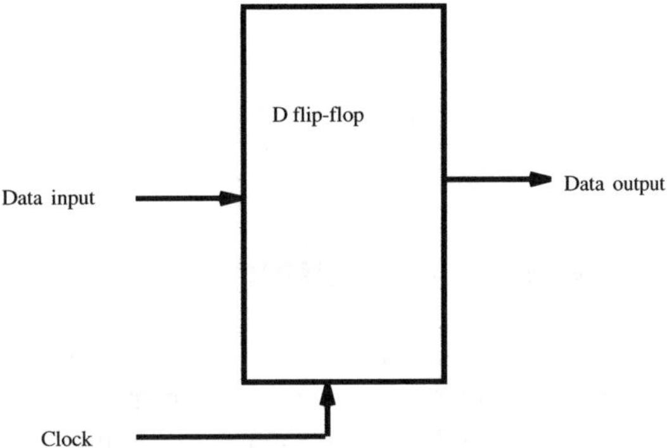

Figure 6.8 A D-flip-flop.

6.9. While inward(1) and inward(2) represent the data and clock inputs respectively, outward(3) reflects the output of the flip-flop.

When two or more signals must be synchronized, they should both necessarily be associated with a component where the synchronization is realized. Because synchronization may also be related to communication between task instances, a discussion will also occur later in this chapter. As an example of synchronization, in the model of the D-flip-flop (Figure

```
Task body D-flip-flop is
  inward : array(1..2) of Boolean;
  outward : array(3..3) of Boolean;
  var
    t-setup : intege
  begin
    ...
    if F(inward(1)).assertion-time -
            F(inward(2)).assertion-time
             < t-setup then put("Error in setup");
  end;
End D-flip-flop;
```

Figure 6.9 Timing characteristics of a flip-flop.

6.8), the output may only change when a transition takes place at the clock input. Consequently, the model description is not executed until a new value is received at the clock input as shown in Figure 6.10.

```
Task body D-flip-flop is
  inward : array(1..2) of Boolean;
  outward : array(3..3) of Boolean;
  begin
    ....
    accept-an-input-at-inward(2) do
    if new value at inward(2) differs
              from the old value then
                 assert-output-signal;
    ...
  end;
End D-flip-flop;
```

Figure 6.10 Modeling synchronization of signals.

6.5 COMMUNICATION AMONG MODEL INSTANCES

The flow of signal between the physical devices in an actual circuit is modeled in RDV as message objects between the corresponding task instances. The physical wire through which signal flows from one device to another is modeled as a path between the corresponding instances. Messages are sent and received over such paths. In Ada, entry procedures and calls form the basis of message passing. Messages are typed with well defined data structures. The data structure of the message may differ depending on the purpose and nature of the verification system. The advantages of typed-messages are that any attempt to pass illegal message data structures generated in a device is trapped, implementation of complex simulation system such as deductive fault simulation and timing verification is facilitated, and good documentation of the system specification is ensured.

Figure 6.11(a) shows a part of an actual circuit and consists of an inverter connected to the output of a two-input AND gate. The two inputs to the AND and the output of the inverter are connected to other devices that are not shown here. Each signal path is assumed to be identified by integers as shown in the figure. Each port of the AND and the inverter are also identified by integers. During operation, other devices assert stimulus on the signal paths 2 and 3. The AND gate receives them through its ports 1 and 2 that are connected to the signal paths 2 and 3. The signals are ANDed and the result is asserted through port 3 which is connected to signal path 1. The inverter receives the input signal, determines the output by inverting it, and sends it out on path 4 through its port 2 at a time instant that is determined by its propagation delay.

As elaborated earlier, corresponding to every device type, RDV contains a task type. For this example, task types AND2 and INVT model gates of type two-input AND and inverter, respectively. An instance of a task type models a physical device of the corresponding type. Here, AND(1) and INV(1) model the AND and the inverter gates of Figure 6.11(a), respectively. Figure 6.11(b) present the Ada model for the design in Figure 6.11(a) while Figure 6.11(c) describes the message record. Every message communicated between task

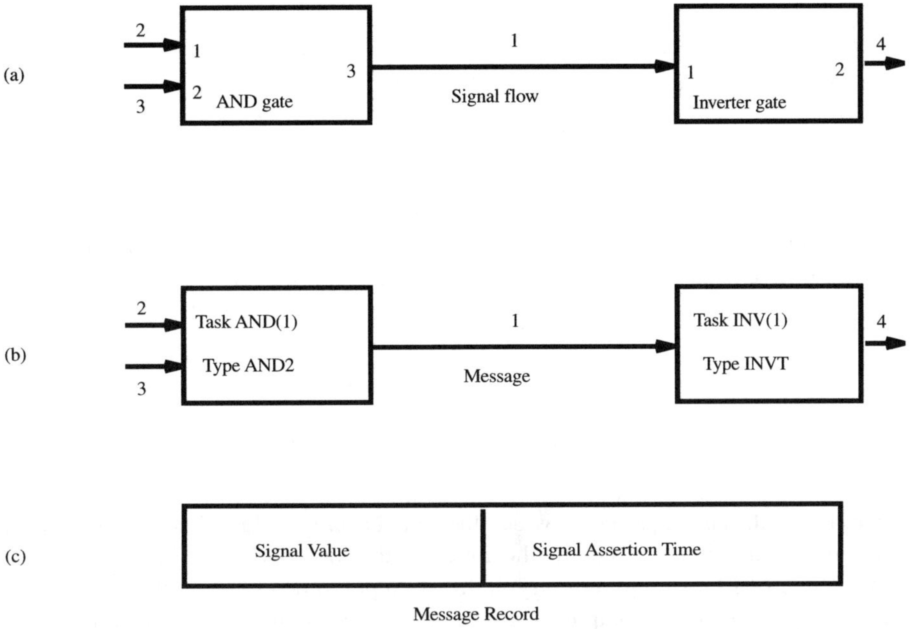

Figure 6.11 Communication between the physical devices: (a) an example design, (b) Ada model of the design, (c) message data structure.

instances utilizes this structure and the record consists of two fields, one containing the signal value and the other containing the time at which the signal is asserted on the corresponding signal path.

Because "entry" procedures constitute the basis of messages, when a device A is likely to receive a signal value from device B, the task type corresponding to A must contain an entry procedure and the task type for B must possess adequate knowledge of the procedure such that it can issue entry calls to it. In addition, the task body corresponding to A must contain an "accept" clause to intercept an entry call from the task body corresponding to B. The signal value, propagated from B to A, is encoded as a message which, in turn, is pointed at by the parameter of the entry procedure that is passed from the task body corresponding to B to the task body corresponding to A.

The table in Figure 6.12 presents the correspondence between each aspect of the actual circuit and its counterpart in the Ada model.

As described earlier, the connectivity of a design is stored in the form of an interconnection database in the Ada model. When a device asserts a signal value on its output path, the signal is propagated to all other devices connected to the output path. In the Ada model, the task instance corresponding to the device, first, retrieves the identifiers of all other task instances connected to its output, from the interconnection database. It then issues entry calls to each one of them, where the parameter being passed serves as a pointer to the message that encodes the signal.

Figure 6.13 presents the communication model through the task type INVT that models inverters. The entry procedure of task type INVT is c1 whose parameter DS is of data type integer and constitutes an actual pointer to the message. The value of the parameter is the

Actual Circuit	Ada model
component type	task type
component instance	task instance
signal path	message
connectivity	interconnection database
N input terminals of a component	input(1..N)
M output terminals of the component	output(N + 1..M)
propagate signal on output path	retrieve task instance identifiers from inter-connection database and execute entry calls for each such instance.

Figure 6.12 Correspondence between actual design and model.

identifier of the signal path over which the signal is propagated. Because the AND gate device may propagate signal values to the inverter, the task body AND2 must contain adequate knowledge of c1 to issue an entry call. The input and output terminals of the inverter are modeled as input(1) and output(2) which are ports through which communication takes place. Because ports 1 and 2 of the inverter are connected to signal paths 1 and 4, input(1) and output(2) access values 1 and 4, respectively. In the body of INVT, first, the task instance receives its identification. Thus, every task instance is uniquely identified and corresponds to the exact device that it models. Signals propagating into the inverter and out of it are modeled through messages. Specifically, the "accept" clause in the body of INVT receives the message sent to the task INV(1) by AND(1). The actual message, however, is pointed at by 1, which is the identifier of the corresponding signal path in Figure 6.11(a). The value of the signal is

```
  Task INVT is
    entry c1(DS : integer);
  end;

  Task body INVT is
  input : array(1..1) of Node;
  output : array(2..2) of Node;
  begin
    receive-task-identification (Y);
    accept c1(DS : integer);
    decode-message-and-obtain-input;
    determine output;
    encode-output-using-output(2);
    for every instance in the fanout
    do begin
        retrieve-task-instance (X);
        X.c1(output(2));
  end do;
end;
```

Figure 6.13 Communication model.

extracted from the first field of the message record. The time at which the signal is asserted is obtained from the second field. The signal value is inverted to obtain the output while the time at which the output will be asserted is determined using the input-assertion time and the propagation delay of the inverter gate. Because the body of task type INVT is generic, i.e. designed to reflect all instances, the specific connectivity information required by INV(1) to send the output message is not stored in INVT. The output signal value and the time of assertion are stored in the first and second fields of the message record which is pointed at by output(2)—the identifier of the corresponding signal path. Finally, the body of INVT accesses the database to find all other task instances on the fanout list of the inverter, INV(1), and entry calls are issued to each of these task instances with "output(2)" as the actual parameter.

The nature of the predetermined message structure is unique to its purpose. While the structure shown in Figure 6.11(c) reflects behavior-level simulation, for fault simulation or timing verification, other structures may need to be designed. One advantage of this approach is that the task instances communicate directly with each other. As a result, communication is distributed and potentially faster than the conventional technique, where a single, centralized scheduler is responsible for all communication. The approach also constitutes an accurate reflection of reality where a hardware component itself propagates signal values to one or more of the components on its fanout list. Since messages are encoded in data structures that may be predefined for each type of simulation, this approach is general and may be applied to fault simulation and timing verification.

6.6 DISTRIBUTED SCHEDULING

Scheduling is an important component of simulation. It determines the order in which the models of devices are activated in the course of simulation. In conventional simulators, a single entity is responsible for scheduling, making the approach strictly sequential. This is inefficient since more than one device may be ready for activation at any time during simulation. In RDV, models are implemented as concurrent tasks and the scheduler is distributed among the task instances. That is, each task determines when it is ready for scheduling. The advantage is concurrency while the penalty may assume the form of expensive task monitoring and debugging.

When simulation is initiated, all task instances are suspended until messages are received at all their input ports. As part of initialization, external signals are asserted at all of the primary inputs of the design. When a task instance has determined the output and is ready to send a message at one of its output ports, it uses the knowledge of the path identifier to which the port in question is connected and, using this identifier as a pointer, the set of devices and corresponding port numbers connected to the path are accessed from the interconnection database. The task then sends the outgoing message to the entire set of devices sequentially. The order in which the message is sent is arbitrary. Other tasks that subsequently receive messages at all of their input ports are executed and the process continues until no more task instances are scheduled for activation. In addition to the responsibility of scheduling itself, every task must ensure the detection and deletion of inconsistent events.

An integral component of distributed scheduling is the message structure used for signal propagation. The signal structure defined earlier is refined as follows. Except for those that constitute feedback paths, every signal contains the complete waveform, i.e. the logical transitions and their assertion times between the initiation and completion of the simulation.

Clearly, at initiation, only those signals of the components that are primary will be defined. As the simulation progresses, other signals will be gradually defined.

6.6.1 Combinational Designs

Figure 6.14 presents the distributed scheduling of a simple AND gate. The AND gate is scheduled for activation only when the signals at all of its input ports are defined, i.e. the complete waveform between the initiation and completion of simulation is available for every input port. The AND gate model first evaluates the output corresponding to the entire input waveforms. In the process, it identifies the inconsistent events, if any, utilizing the preemptive scheduling principle presented in Section 3.4.4. Next, it deletes the inconsistent events, constructs the complete output waveform, and propagates it to the components that are connected to its output port. In Figure 6.14, the intermediate outputs generated by the AND gate for the input waveform are: logic 1 at time 0 ns, logic 0 at time 200 ns, and logic 1 at time 180 ns. The model identifies the transition, logic 0 at 200 ns, as inconsistent, and deletes it. Thus, the correct final output is logic 1 between times 0 ns and 180 ns and logic 0 thereafter.

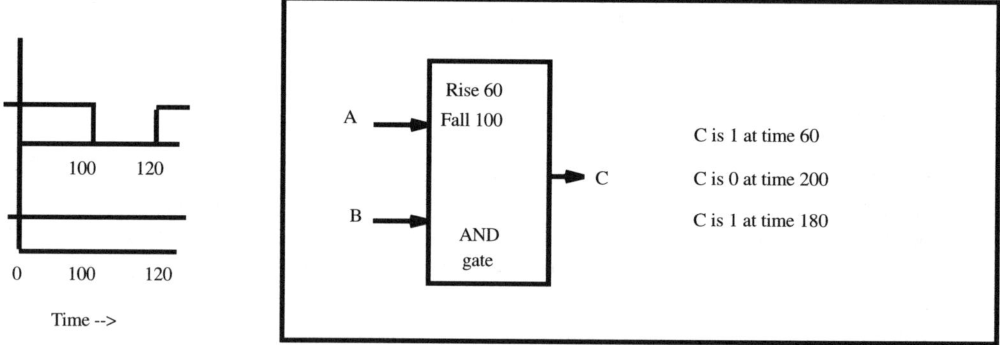

Figure 6.14 Distributed scheduling of combinational designs.

6.6.2 Sequential Designs

The exact details of distributed scheduling of sequential designs differ slightly from that of combinational designs. Figure 6.15 presents a sequential circuit consisting of an AND gate and an inverter. One input to the AND gate, P_3, is primary and, therefore, the entire signal waveform is available. The signal is defined from $t = 0$ ns, the origin, to $t = 30$ ns, the completion of simulation, as shown in Figure 6.15. The output of the AND gate is connected to the input of the inverter and the path is identified by P_1. The output of the inverter is connected to the second input of the AND gate and the path is identified through P_2. The signal paths are assumed to be initialized to value X (unknown) at $t = 0$ ns. The propagation delays for the AND and the inverter are arbitrarily chosen to be 5 ns and 10 ns, respectively.

Although both gates may be activated at the same time, we will first consider the execution of the AND gate. The AND gate is partially simulated, given that the signal at one of its ports is defined only partially. The signal output at P_1 is obtained as X between $t = 0$ ns

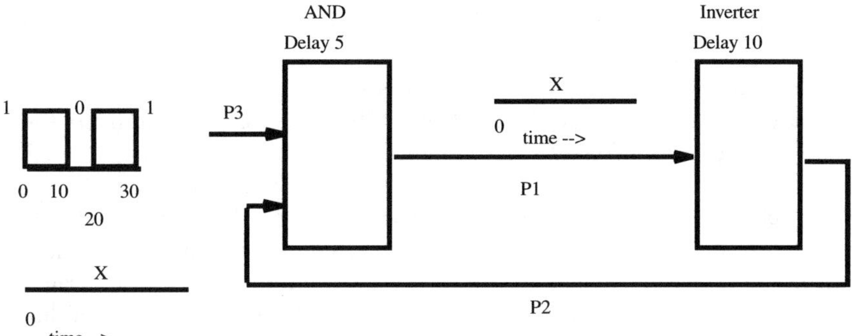

Figure 6.15 Distributed scheduling of sequential designs.

and $t = 5$ ns. The AND model stores within itself the knowledge that the input signals at $t = 0$ ns have been used to determine the output. The signal is examined within the model and, given the absence of any inconsistency, it is asserted at the output port. Next, the inverter is scheduled for activation because a new input has been asserted by the AND gate. The signal output at P_2 is determined to be X between $t = 0$ ns and $t = 15$ ns. The inverter model also stores the information that the input signal between $t = 0$ ns and $t = 5$ ns has already been exhausted. The output is examined and, as no inconsistency is implied, the signal is propagated to the output port. Next, the AND gate is activated with the new input at its second input port. Since the AND model has already been simulated for input signals defined at $t = 0$ ns, it may be subsequently simulated for input signals corresponding to $t > 0$ ns and up to $t = 15$ ns. The output at P_1 is obtained as X between $t = 5$ ns and $t = 15$ ns, and 0 between $t = 15$ ns and $t = 20$ ns. This process continues until the input signal at the primary input port, P_3, has been completely exhausted to determine the final, complete signals at P_1 and P_2.

6.7 HIERARCHICAL REPRESENTATION AND DYNAMIC MULTI-LEVEL SIMULATION (ZOOMING)

6.7.1 The Principle of Zooming

Dynamic multi-level simulation or zooming consists of the expansion of a behavior-level model of a higher-level component into the models corresponding to lower-level constituent components at run time. Also, it involves the initiation of simulation at the lower level to obtain results with greater detail. Zooming facilitates localizing the source of an error or discrepancy in a digital design at the lowest possible level of abstraction. Zooming assumes a complete or partially complete hierarchical representation of the design.

In this approach, the behavior-level model of a digital system is expressed as a single entity M^1 at the highest abstraction level. M^1 in turn consists of the models $M_1^2, \ldots, M_i^2, \ldots, M_n^2$ at the second level of abstraction. Each of M_i^2, $\forall i \in 1, \ldots, n$, may consist of the models $M_{i1}^3, \ldots, M_{il}^3$ corresponding to the third level of abstraction. Although this principle, in theory, extends to any arbitrary lower level of abstraction, RDV is limited to behavior-, function-, and gate-level representations.

Simulation is initiated at the highest level where the model M^1 is verified. When the results are unsatisfactory, the designer expands M^1 into $M_1^2, \ldots, M_i^2, \ldots, M_n^2$ and initiates simulation at level 2. For more details, a M_i^2 $\forall i \in 1, \ldots, n$ further expands into $M_{i1}^3, \ldots, M_{il}^3$ and simulation initiates at level 3.

The following are a few of the many example cases where a detailed investigation is necessary. When a 4-bit adder is simulated at the behavior level and the output is observed to differ from the expected results, zooming traces the cause of the error at the lower gate level. Consider a 4-bit asynchronous counter synthesized from four previously designed toggle flip-flops, each with a unique set of timing characteristics. Also, consider that the counter is an element of a larger digital design and is required to conform to the design's timing specifications. During simulation of the counter, if observations indicate that the output state transition is slower than permitted, zooming helps trace the cause of the timing problem to one of the flip-flops. Thus unlike in the conventional multi-level simulators, unnecessary simulation runs may be eliminated by using hierarchy dynamically and only when necessary. Since simulation at a lower level is usually more CPU intensive and only some components may require expansion at the lower level, dynamic multi-level simulation may be more efficient.

6.7.2 Zooming in Combinational Designs

In this section, zooming is explained through an example circuit—a 2-bit adder, subject to timing verification [64]. Figure 6.16(a) contains a functional model of a 2-bit adder A_1 (level 1). Figure 6.16(b) describes a lower-level representation of A_1—two 1-bit adders, A_{11} and A_{12} (level 2). When A_{11} is expanded at the gate level, the expanded circuit is shown in Figure 6.16(c), where the constituent gates are A_{111} through A_{116}. The inputs to A_1 are a0, b0, a1, b1, and cin and the outputs are s0, s1, and cout. The propagation delays from the inputs to the s0, s1, and cout outputs are 5 ns, 20 ns, and 20 ns respectively. The inputs to A_{11} are cin, a0, and b0 and the outputs are s0 and c1, where c1 is the intermediate carry. The propagation delay from the inputs to the s0 and c1 outputs are 5 ns and 15 ns, respectively. Similarly, the propagation delays from the inputs of A_{12} to its outputs s1 and cout are 5 ns and 5 ns, respectively. The propagation delays for A_{111} through A_{116} are 1 ns, 4 ns, 1 ns, 1 ns, 11 ns, and 4 ns, respectively. Before timing verification initiates at the behavior level, RDV requires that complete signal waveforms—signal transitions and assertion times between the origin and completion of simulation—are asserted at all primary input ports. During verification at level 1, a state value (S) is asserted throughout, starting at $t = 0$ ns at the input ports cin, a0, a1, b0, and b1. The output signal waveform is determined for the output ports s0, s1, and cout as shown in Figure 6.17. The signal at cout stabilizes at $t = 20$ ns but assume that such a slow signal is not acceptable because it may cause violation of timing constraints such as setup and hold. The designer may then zoom, wherein A_1 is expanded into A_{11} and A_{12}.

Verification initiates at this level (2) and first A_{11} and then A_{12} is verified. Using the same input values for cin, a0, and b0 as in the previous case, the signal values at s0 and c1 are determined and shown in Figure 6.17. The signal value at c1 stabilizes at $t = 15$ s which the designer considers unacceptable, and therefore decides to zoom further into A_{11}. When A_{11} expands into A_{111} through A_{116} and verification initiates at level 3, the values at the internal signal paths m1, m2, and m3 are determined, using the input signals at a0, b0, and cin from the previous case, and is shown in Figure 6.17. The signals at m2 and m4 stabilize at $t = 1$ ns and $t = 2$ ns, respectively, and are therefore not responsible for causing the output at c1 to be late. However, the signal value at m3 stabilizes at $t = 11$ ns which causes the signal at c1 to be late. The conclusion is that the AND gate A_{115} is much too slow and may need to be

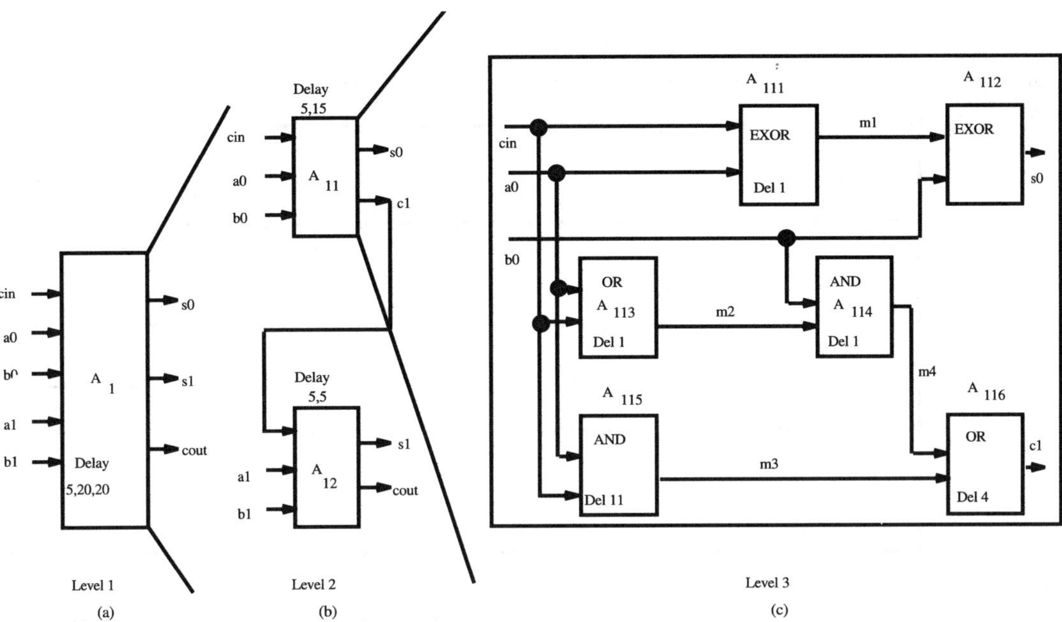

Figure 6.16 Zoom simulation of a 2-bit adder, delays are in nanoseconds. (a) A behavior-level 2-bit adder, (b) two 1-bit adders, (c) a gate-level circuit for the 1-bit adder, A_{11}.

replaced by a gate with a lower value of propagation delay. Thus, the source of the problem is localized to a faulty gate and the verification terminates.

6.7.3 Zooming in Sequential Designs

In zooming applied to sequential designs, the signal values at the inputs of the upper-level device are forced on the primary inputs of the lower-level circuit that is generated from expanding the upper-level device. The states of the lower-level components are initialized using the values of the states of the upper-level device. Sequential circuits may be modeled as Moore or Mealy machines. Here, a Mealy model is assumed without any loss in generality and the results are equally valid for Moore models. Mathematically, a sequential design (Mealy model) may be characterized by N inputs, M outputs, K internal states, and L initializing inputs, such that the output values at $t = t_2$ depend on the input and state-variable values at $t = t_1$ ($t_1 < t_2$). Physically, a sequential circuit consists of combinational logic and some form of memory such as latches and flip-flops. As previously stated, a device is not verified until either complete or partially complete timing lists are assigned to all input ports. The results of verification (output timing lists and/or error messages) are valid only up to the time instant where all input timing lists are defined.

6.7.3.1 Simulation of Sequential Designs in the Absence of Zooming. Figure 6.18(a) presents a generalized sequential design with N inputs, M outputs, K states and L initialization inputs that are responsible for initializing all memory within the circuit to a known state before verification is initiated. Figure 6.18(b) presents an equivalent circuit where

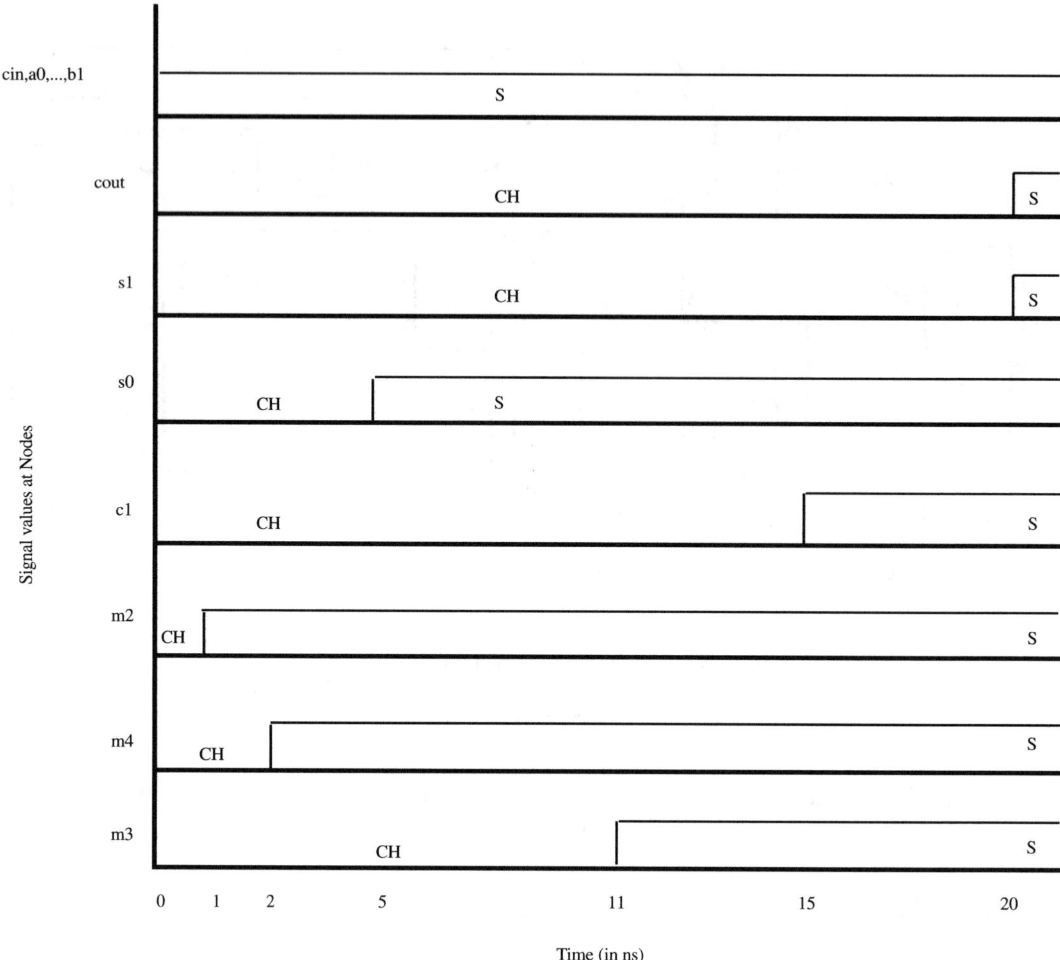

Figure 6.17 Signal values for the design in Figure 6.16.

the combinatorial and memory sub-components of the design are separate from each other. The combinational part has N inputs, M outputs, K state inputs and K state outputs. The state inputs and outputs are, respectively, the outputs and inputs of the memory part which is initialized to a known state by the initializing inputs.

Verification is initiated after timing lists are assigned to the primary inputs and the flip-flops initialized which, in effect, assigns partial timing lists to the state inputs. These partial lists are for a shorter duration than those corresponding to the primary inputs because they are only initializations. For example, the timing list of a primary input may be defined between $t = 0$ ns and $t = 500$ ns while the partial timing list of a state input may be defined only at $t = 0$ ns. For clarity, the timing model of the design in Figure 6.18(a) is assumed to consist of two parts, one for the combinational unit (CP) and the other for the memory unit (MP). Execution of the model corresponding to the combinational unit uses the input timing lists to determine the output timing lists between $t = 0$ ns and $t = D$ ns, where D ns is the

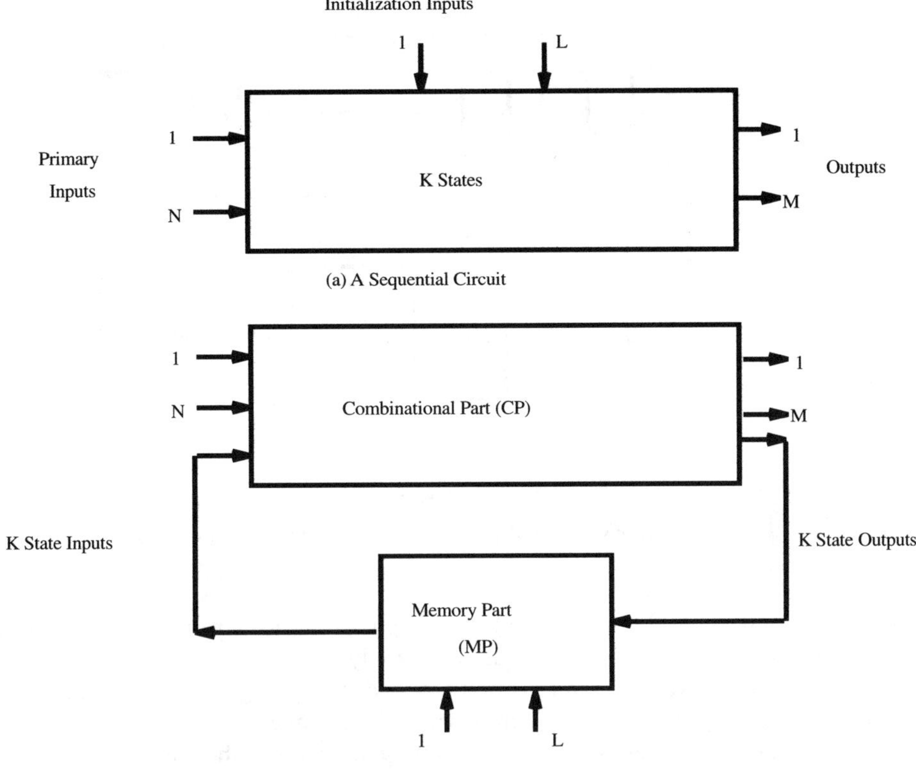

Figure 6.18 Simulation of sequential designs without zooming: (a) a sequential design, (b) Mealy representation of the design.

propagation delay of CP ($D \ll 500$ ns). The model then processes the timing lists at the state outputs corresponding to the memory unit to determine new state-input timing lists which will now span between $t = 0$ and $t = D + D1$, where $D1$ is the propagation delay for MP. Once again, the model corresponding to CP executes with the new timing lists and, thereafter, the model corresponding to MP executes. This process continues until verification is completed only up to the time beyond which the primary inputs are not defined.

6.7.3.2 Simulation of Sequential Designs in the Presence of Zooming. The combinational and memory units in Figure 6.18(b) each contain several lower-level devices at the lower level of the hierarchy. While the combinational unit may be comprised of adders, shifters, and multiplexers, the memory unit may consist of flip-flops and gates. Figure 6.19 presents such a representation of this system design where the combinational part (CP) contains the lower-level combinational circuits C1, ..., Ci and the memory part (MP) contains the sequential circuits F1, ..., Fj.

For each of the units C1, ..., Ci, the output vector depends only on the input vector and for each of the F1, ..., Fj, the output is determined by their inputs and internal states. The total number of the state variables must be greater than or equal to the number of states in the

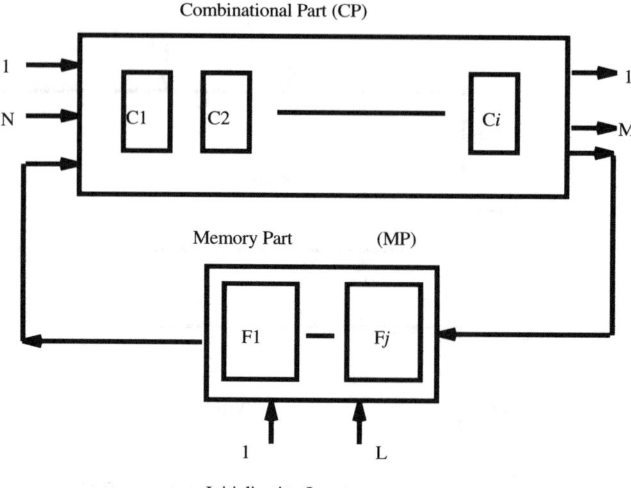

Figure 6.19 A different representation of the sequential design.

sequential circuit in Figure 6.18(a). The initializing vector(s), used for the functional representation in Figure 6.18(a), must also apply to the representation in Figure 6.19. If the number of state variables in F1, ..., Fj exceeds that in Figure 6.18(a), the excess variables must be initialized differently and are discussed subsequently.

Timing lists are assigned to the primary inputs and the initialization causes partial timing lists to be assigned to the state inputs. The CP is verified at the behavior level, and the partial-output timing lists are generated only between $t = 0$ ns and D ns. The user now has a choice of zooming whereby CP is expanded into C1, ..., Ci which are then verified. However, zooming at this point yields results that are only obtained between $t = 0$ and D. The MP uses the partial-output timing lists at the state outputs and is verified at the behavior level. During verification, the state variables (initialized at the start of verification) and the partial-input timing lists are processed to generate new partial-output lists between $t = 0$ and $D + D1$ ($D1 > D$) and the new values for the state variables in MP. $D1$ is the propagation delay for the memory unit, MP.

At this point, the user again has a choice of zooming whereby MP is expanded into F1, ..., Fj which are then verified. Because each F1, ..., Fj is a sequential circuit, their state variables must be initialized before verification can be initiated. The initialization must correspond to the state of the higher-level device, MP, just prior to zooming. If the number of state variables in F1, ..., Fj is equal to those in the memory part, there is a 1 : 1 correspondence between them and the initial values of the variables in F1, ..., Fj are directly obtained from those in the memory part, MP. When the number of state variables in F1, ..., Fj is larger, those in excess must be initialized independent of the upper level either by the models of F1, ..., Fj or by the interface between the two levels of verification. Verification at the level of F1, ..., Fj will yield results obtained between $t = 0$ ns and $(D + D1)$ ns.

To continue verification, control returns to the higher, behavior-level representation of CP and MP and performs another cycle of partial verification. The process continues until the input timing lists at the primary inputs have been completely exhausted. During every cycle, however, the user can zoom at the lower level and observe verification in detail for the duration of the cycle. Unlike in a completely combinational circuit where the user can zoom onto a lower level and observe detailed verification for the entire simulation duration, multiple zooms are required for sequential designs, each time to observe detailed verification over a small interval of the entire simulation duration.

6.7.3.3 Requirements. Requirements for the dynamic expansion of a component C at level A to its lower-level components represented by D at level B may be expressed as follows. It is necessary that the implementation of D be defined and a one-to-one correspondence exist between the input and output ports of C and D. In addition, where C is combinatorial design, the signals at the input ports of C must be mapped into signals at the corresponding inputs of D during zooming. The mapping function reduces to an identity function when the value systems at the levels A and B are identical. When component C is sequential, the values associated with the state variables of C must also be mapped into the values associated with the corresponding state variables of D. Both C and D are equivalent and, as a result, the function that maps the state variables must either be isomorphic or homomorphic [65]. In the event of isomorphism, a value associated with a state variable at level A is translated into the value system at the lower level and asserted to the corresponding state variable at level B. Given a homomorphism, the value associated with a state variable at level A may be asserted to more than one state variable at level B or vice versa.

6.7.4 Implementation of Hierarchy and Zooming

This subsection presents the software techniques that exploit the Ada constructs "packages" and "exceptions" to implement hierarchy and permit zooming in RDV [43]. Although the SDL language [26] permits hierarchical descriptions of digital systems, they are limited to static use as opposed to the dynamic use of hierarchy.

An Ada [66] "package" may be used to describe groups of related entities such as types, objects, and subprograms, whose inner workings are concealed and protected from their users. Elaboration of a package body consists in the elaboration of its declarative part followed by the execution of the sequence of statements.

An Ada "exception" is an event that causes suspension of normal program execution. When an exception is raised, control may be passed to a user-provided handler at the end of a block, subprogram, package, or task in which the exception is raised or may be propagated outside.

RDV requires that component behavior be described in Ada and hierarchical interconnection of components of a digital design be specified by the designer through the use of Ada packages. In this approach, the digital system, M^1 is expressed as a single task instance and a package PC^1 corresponding to M^1 consists of the lower-level component models $M_1^2, \ldots, M_i^2, \ldots, M_n^2$ expressed as task instances. The connectivity of the components at the second level is contained within PC^1. The task instances execute when the M^1 expands into level 2. For each of M_i^2 $\forall i \in \{1, \ldots, n\}$, a package PC_i^2 contains the lower-level component models $M_{i1}^3, \ldots, M_{il}^3$ implemented as task instances, as well as the connectivity. The task instances execute following the elaboration of M_i^2. To illustrate this approach, a simple example follows that consists of a circuit expressed as model M^1 at level 1 that, in turn, consists of a single model M_1^2 at level 2. M_1^2 contains a single component model M_1^3 at level 3. Consequently, package PM^1 and PM_1^2 consist of M_1^2 and M_1^3, respectively. The procedure "Main" in Figure 6.20 corresponds to M^1. Packages PC^1 and PC_1^2 are presented in Figures 6.21 and 6.22, respectively.

```
   Procedure Main is
   c : character;
   begin
 l1: behavior-level-simulation-of-M -at-level 1;
     put("User, you are at level 1");
 l2: put("Type a y to zoom to level 2");
 l3: get(c);
 l4: if c = 'y' then PC .bexp; end if;
   end Main;
```

Figure 6.20 Digital design at level 1.

Appropriate exceptions, which we will describe, are defined within every package. The highest-level component is represented as a procedure, not a package, and consequently contains no exceptions. When the designer desires control to return to a higher level following termination of execution at the lower level, an exception is deliberately raised. This exception returns control to the immediate outer scope, i.e. the procedure corresponding to the higher-level model. Exceptions are a logical choice in dynamic multi-level simulation. Careful use of them is imperative since, otherwise, exceptions cause unnecessary complications during execution.

When simulation is initiated, the procedure "Main" in Figure 6.20 activates. The statement at label l1 that corresponds to the behavior-level simulation of M^1 at level 1 executes. At this point, the system queries the user whether expansion into the lower level is desired. For a negative response, the simulation terminates and for a positive response, the package in Figure 6.21 is elaborated and the procedure "bexp" executes. Execution of statement l1 of Figure 6.21 corresponds to the simulation of M_1^2 at level 2. At this point, the

```
Package PC is
  procedure bexp;
  goto1 : exception;
End PC ;

Package body PC is
    procedure bexp is
    c : character;
    begin
      put("User, you are at level 2");
  l1:simulate-model-M ;
      put("Type a y to expand M to level 3");
  l2:get(c);
  l3:if c = 'y' then PC .aexp;
  l4:            else raise goto1; end if;
    exception
        when goto1 => put;("Returning control from level 2 to 1");
  end bexp;
End PC ;
```

Figure 6.21 Digital design at level 2.

```
Package Pc is
  procedure aexp;
  goto2 ; exception;
End PC ;

Package body PC is
  procedure aexp is
    c : character;
    begin
      put("User, you are at level 3");
  l1:simulate-model-M ;
     put("Type a y to return control from level 3 to 2");
     get(c);
  l2:if c = 'y' then raise goto2; endif;
    exception
        when goto2 $=>$ put("Returning control from level 3 to 2");
  end aexp;
End PC ;
```

Figure 6.22 Digital design at level 3.

user may either choose to return control to level 1 when statement l4 is executed or expand into level 3, in which case statement l3 executes. When statement l4 executes, an exception "goto1" arises and control returns to the procedure "Main." Corresponding to the execution of statement l3, the package PC_1^2 in Figure 6.22 is elaborated and the procedure "aexp" executes. First, the statement at label l1 executes, which corresponds to the simulation of component M_1^3 at level 3. At this point, the user may only choose to return control to level 2 as level 3 is assumed to be the lowest in the hierarchy. As a result statement l2 executes and an exception "goto2" arises with the consequence that control returns from level 3 to 2. The user continues further simulation at level 2 or returns control to level 1 and terminates the session.

6.7.5 Results of Conventional Simulation versus Simulation Under Zooming

In this section, an example design is simulated in RDV and the performance is compared against the execution time that would be required in a conventional simulator. In an experiment two identical systems are considered for simulation where the first design consists of a 2-bit adder represented at the behavior level and another 2-bit adder described at the gate level. This corresponds to a scenario in the conventional mixed-mode approach where the designer intends to verify the second adder at the gate level. The designer has made an intelligent guess that the results of behavior-level simulation of the second adder may be inadequate to localize a logical error. The second design consists of two 2-bit adders at the behavior level and, assuming that the results of behavior-level simulation are satisfactory, neither of them are expanded at the lower level. Consequently, this scenario corresponds to dynamic multi-level simulation where the decision against expansion is taken at run time and based on the actual results of behavior-level simulation. The CPU times required for the simulation in RDV of the first and second designs are 6.9 s and 2.5 s, respectively. Thus, as expected, zooming implies superior performance.

7

VHDL

In this chapter, we will examine VHDL in the light of the fundamental characteristics of behavior-level HDLs. First, we will concentrate on its key features and then on its limitations. We will not focus on enumerating every detail of the language since it is extensive and has been covered in [33] as well as in a number of books [67, 68].

7.1 CHARACTERISTICS OF VHDL RELATIVE TO THE FUNDAMENTAL REQUIREMENTS OF HDLs

7.1.1 Entity and Concurrency

VHDL utilizes the word entity to define the primary hardware abstraction. Although its mention occurs earlier, chronologically speaking, in the VHDL DRFP [55]. the use of entity is a significant step in the right direction. VHDL defines an entity: it represents a portion of hardware design that has well-defined inputs and outputs and performs a well-defined function.

An entity is organized into two main parts: entity declaration and entity body. An entity declaration specifies a unique name for the entity and contains the common interface which is shared by one or more entity bodies corresponding to the entity declaration. The entity declaration has three sub-parts: an entity header which enumerates the input, output, and bidirectional ports, and any generic parameters, an entity declaration that contains all items that are common to all corresponding entity bodies, and an entity statement part that contains assertions, procedure calls, and process statements [33] that are also common to all entity bodies. The second and third sub-parts merely reflect the common items between all entity bodies corresponding to the entity declaration. Unless a set of statements are organized explicitly through a process, all statements in VHDL are concurrent. By this token, the assertion, procedure, and process statements in the entity statement part are all concurrent. As with ADLIB-SABLE and Ada as an HDL, the ports—in, out, and inout—may assume any complex data structure and are strongly typed.

```
entity latch is
  generic (N: integer);
  port (Rin, Sin, Clock: in BIT; Qout: out BIT);
  constant min_duration := 12ns;
  begin
    if Clock'Event then assert (Clock'LAST_EVENT >= min_duration)
          REPORT "minimum duration violation"
          SEVERITY ERROR;
end entity latch;
```

Figure 7.1 VHDL entity declaration.

Figure 7.1 presents the entity declaration part corresponding to a latch. It specifies a generic variable *N*, a constant value for the minimum duration, the input and output ports, and an assertion to check the minimum high and low pulse widths of the clock. The type BIT is a predefined standard in VHDL and consists of two enumerated values "0" and "1". Regardless of the number of different architecture bodies that may be associated with the entity latch, the contents of Figure 7.1 will serve as a common interface to all of them.

The body of an entity is specified through the keyword architecture and each entity body, corresponding to a unique entity name, carries a unique identifier. Each body may be expressed either through a sequential or concurrent set of statements. According to [33], a sequential representation is referred to as behavior. The concurrent set of statements of an entity body may assume either the form of a dataflow description or a structural organization, i.e. in terms of lower-level entities that are referred to as components. Thus, while the dataflow and behavior style descriptions may be utilized either for an entity at the highest or lowest level of abstraction, the structural style of entity body is suited for describing a static hierarchical digital design. Consider the simple circuit in Figure 4.4 and assume that it constitutes an entity named Simple_circuit with input ports labeled in1, in2, in3, and in4, and output port out1. Figures 7.2 through 7.4 present three different entity bodies corresponding to the Simple_circuit.

Figure 7.2 presents an entity body written in the behavior style. The use of the process clause forces all of the three statements to be executed sequentially. The process is triggered for execution when any of the input ports in1 through in4 receive a new value. First, the variable temp1 receives an assignment followed by the variable temp2 and finally an assignment is made to the output port out1 after a delay of 10 ns. Clearly, a complete hardware system may be represented entirely as a behavior description.

```
architecture Behavior of Simple_circuit is
begin
  process (in1,in2,in3,in4)
    variable temp1, temp2: BIT;
  begin
    temp1 := in1 and in2;
    temp2 := in3 and in4;
    out1 <= temp1 or temp2 after 10 ns;
  end process;
end Behavior;
```

Figure 7.2 A sequential behavior entity body.

```
architecture Dataflow of Simple_circuit is
  signal net1, net2: BIT;
begin
L1: net1 <= in1 and in2 after 5 ns;
L2: net2 <= in3 and in4 after 5 ns;
L3: out1 <= net1 or net2 after 5 ns;
end architecture Data flow;
```

Figure 7.3 A concurrent dataflow entity body.

```
architecture Structure of Simple_circuit is
  signal net1, net2: BIT;
  component and2
       port (a,b: in BIT; c: out BIT);
  end component;
  component or2
       port (a,b: in BIT; c: out BIT);
  end component;
begin
  G1: and2 port map(in1, in2, net1);
  G2: and2 port map(in3, in4, net2);
  G3: or2 port map(net1, net2, out);
end architecture Structure;
```

Figure 7.4 A concurrent structural entity body.

Although the Simple_circuit consists of three concurrent gates, the behavior description in Figure 7.2 reflects a black-box representation with four inputs in1 through in4, a single output out1, and an input–output functional relationship. The black-box is triggered for execution when any of the inputs change since, in theory, a new value may be generated at the output. Thus, when the entity body "Behavior" is initiated for execution, it corresponds to a specific value of simulation time, say T_1. In the absence of any "wait" statement that suspends execution, all of the statements in the entity body are assumed to execute instantaneously at the simulation time. Any signal assignment statement is also executed instantaneously but the actual assignment of the value to the signal may be deferred. If the entity body contains a "wait," then all consecutive statements starting with the first one and up to "wait" are executed instantaneously, at the simulation time. Execution is suspended at the "wait" and when the entity body resumes execution later at a different simulation time, say T_2, the statements following the "wait" and up to the subsequent "wait" are again executed instantaneously, corresponding to T_2.

Consider the following scenario. Assume that new values are asserted at input ports in1 and in4 at times 0 and 10 ns, respectively. Assume further that new values are generated at the output port as a result of these input signals. The sequential process in Figure 7.2 is initiated for execution at simulation time 0 ns since in1 is updated. A new value is assigned to temp1 first and then a new value is generated for the signal out1. The assignment to out1 is scheduled for $0 + 10 = 10$ ns. Given that there are no activities between the current simulation time and 10 ns, the scheduler jumps the current simulation time to 10 ns. There are two activities—assigning the new value at out1, and re-initiating the entity body Behavior for execution corresponding to the new signal value at in4. The scheduler may choose to perform the two activities in any order. Consider that it completes the assignment to out1 first, assuming that it has not been preempted. When the process is executed for the current simulation time of 10 ns, the new value at input port in4 is consumed, a new value is asserted to temp2, and then the new value is scheduled for assertion at out1 at time $10 + 10 = 20$ ns. When the scheduler advances the current simulation time to 20 ns, it will perform the assertion at out1, assuming that it has not been preempted. Figure 7.5(a) presents a graphical view of the activities as a function of simulation time.

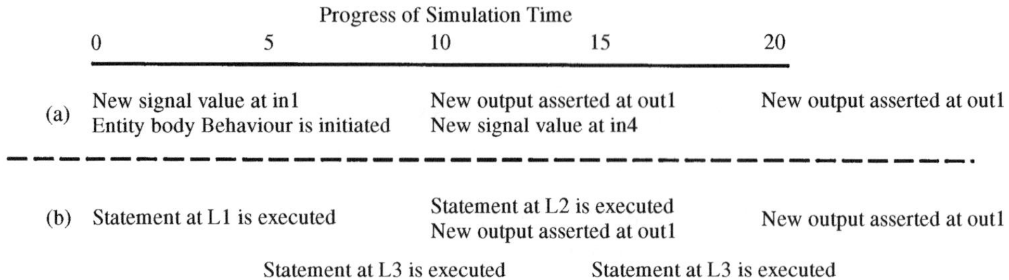

Figure 7.5 Timing of the execution of entity bodies in VHDL.

Figure 7.3 presents a dataflow description where all of the statements are concurrent, i.e. they may potentially execute simultaneously. Thus, the order of occurrence of the statements is not important. Two internal signals, net1 and net2, are defined and the first two statements correspond to assignments to the internal signals. An assignment may be made to net1 when either in1 or in2 or both experience a change. Similarly, when either in3 or in4 or both are subject to change, an assignment to net2 is realized. An assignment is made to output port out1 when either of net1 or net2 experiences a change in value. As with the behavior style description, a complete hardware system may be represented entirely as a dataflow description.

Although every statement in the entity body is concurrent in theory, in practice their execution is controlled by a single, centralized scheduler. Thus, the implication of concurrency is severely diminished. The reason for the use of a centralized scheduler is to maintain consistency and guarantee correctness. In Figure 7.3, the semantics of the statement at label L1 is that, following any change in in1 or in2, at a simulation time T_i, an assignment is scheduled to be asserted to net1 after a delay of 5 ns from T_i. Similar are the semantics of the statements at labels L2 and L3. Assume that both L1 and L2 are executed at time 0. Thus, net1 and net2 are scheduled to receive new values at $0 + 5 = 5$ ns and $0 + 5 = 5$ ns, respectively. In the absence of external control, L3 may initiate its execution utilizing only the newly asserted value at net1 although the correctness argument requires that updated values at both net1 and net2 at $t = 5$ ns must be considered. As a result, execution of L3 may yield erroneous results. The role of the scheduler is to ensure that all updates to all signals corresponding to a specific simulation time are completed before activities for a future simulation timestep are permitted. In Figure 7.3, if L1, L2, and L3 are scheduled for activation at simulation times T_1, T_2, and T_3, respectively, first the minimum of these time values, T_{min}, is computed and then only the corresponding statement(s) is allowed to execute. Following its completion, the minimum, T_{min}, is computed again and only the corresponding statement(s) is permitted to execute, and this continues until there is no new activity.

Consider that the entity body, Dataflow, is subject to the same input signals as with the sequential entity body, Behavior. At simulation time 0 ns, the entity body is initiated for execution. Only the statement at label L1 executes since in1 receives an update. As a result, a new value is scheduled for assertion on net1 at time $0 + 5 = 5$ ns. The scheduler advances the current simulation time to 5 ns and completes the assignment to net1, assuming that it has not already been preempted. At this point the entity body is initiated again and the statement at label L3 is executed since net1 has received an update. Upon execution of L3, a new value is scheduled to be asserted to out1 at time $5 + 5 = 10$ ns. Next, the scheduler advances the current simulation time to 10 ns. The entity body is re-initiated and the statement at label L2 is executed. As a result, a new value is scheduled for assertion at net2 at $10 + 5 = 15$ ns. Also, the scheduler completes the assignment to out1 assuming that it has not been preempted. The scheduler advances the current simulation time to 15 ns and completes the assignment to net2. The entity body is re-initiated and the statement at label L3 is executed. A new value is scheduled to be asserted at out1 at time $15 + 5 = 20$ ns. A pictorial view of the activities as a function of simulation time is shown in Figure 7.5(b). The fact that the sequence of activities in Figure 7.5(b) is more detailed than in Figure 7.5(a) reflects the fact that the Dataflow entity body represents more details about the input–output function in contrast to the entity body Behavior.

Figure 7.4 presents a structural entity body where a top-level entity may be expressed in terms of the lower-level constituent components. Since the components are in essence entities, they may, in turn, be expressed through behavior, dataflow, or structural descriptions. Thus, the structural is the most general and versatile description and closely resembles the instantiation scheme inherent in ADLIB-SABLE and Ada as an HDL. In Figure 7.4, the Simple_circuit is expressed through three gates, G1, G2, and G3, that relate to two types of components—and2 and or2. Both and2 and or2 are entities with their respective entity bodies. In the structural description, first the entity declarations are re-stated and then the specific instances of the components are enumerated. The statements in the structural description are concurrent, i.e. the entity bodies corresponding to gates G1 through G3 may execute simultaneously in principle. In practice, however, the execution semantics is similar to that discussed earlier for Figure 7.3.

Corresponding to an entity name, the user may specify an entity body that is desired to be used in the simulation through a "configuration" language construct. The binding of the exact entity body to the entity name must be completed statically, i.e. prior to initiating simulation. In Figure 7.6, the Dataflow body is selected for use in simulation corresponding to the entity Simple_circuit.

```
Configuration abc of Simple_circuit is
  for Dataflow
  end for;
end abc;
```

Figure 7.6 The use of configuration construct.

VHDL's capability to express complex behavior lies in its use of powerful language constructs such as if then else, case, loop, for, while, exit, next, etc. VHDL permits the declaration of complex variables and assignments to them. Along with the in, out, and inout ports, the definition of signals of any complex data type and signal assignments reflect VHDL's expressive power.

In VHDL, the specification of multiple entity bodies corresponding to a single entity declaration and the static binding of a specific entity body for use in the simulation through the "configuration" construct is conceptually not unique. It merely reflects a user convenience. In ADLIB-SABLE, corresponding to a NAND gate, one may specify a number of different comptypes—one to reflect a TTL NAND gate, another representing a CMOS version, a third that utilizes rise and fall delays, etc. The use of different unique comptype names is no different from the multiple tuples, {entity name, entity body identifier}, in VHDL, each of which uniquely refers to a specific entity body.

7.1.2 Connectivity

A given hardware system is represented in VHDL at the highest level through an entity. Where the description of the corresponding entity body is either a dataflow or a sequential process, no further lower-level composition details are available, implying that the use of connectivity information is irrelevant. In contrast, when the body of the entity is expressed through structure, i.e. through components, that are, in turn, defined as entities, the binding of the "formal" ports of the components to the "actual" ports of the higher-level entity constitutes connectivity. In VHDL, the connectivity is achieved implicitly, i.e. by matching the formal with the actual parameters, unlike in ADLIB-SABLE where the user develops an explicit SDL program or in Ada as an HDL where the user synthesizes an explicit interconnection database. Thus, VHDL serves as a vehicle to both represent hardware behavior and express connectivity.

Consider the Simple_circuit in Figure 4.4 and contrast the connectivity description in ADLIB-SABLE as shown in Figure 4.6 with that in VHDL as presented in Figure 7.4. The highest level of abstraction in ADLIB-SABLE is "Test" while that in VHDL is the tuple {Simple_circuit, Structure}. While the instantiation of the two AND gates and the single OR gate is similar, the mechanism for binding the port names is different. In Figure 4.6, net1 and net2 explicitly connect the output ports of G1 and G2, respectively, to the input ports of G3. In Figure 7.4, the fact that the port identifiers "net1" and "net2" occur in the port maps of G1, G2, and G3 constitutes an implicit connection between G1 and G3, and G2 and G3. The primary input and output ports deserve different treatment in ADLIB-SABLE, Ada as an HDL, and VHDL.

7.1.3 Timing

To achieve the timing description of both synchronous and asynchronous hardware systems, VHDL utilizes signal assignment statements, wait statements for synchronizing between signals, and timing assertions.

SIGNAL ASSIGNMENTS. A signal assignment statement assumes the form, "S <= 1 after 10 ns;" where S represents a signal of a specific data type, integer in this case, and, upon execution at the current simulation time, T, a 1 is scheduled to be asserted at S at time $(T + 10)$ ns. Unlike ADLIB-SABLE which implements a simple anticipatory scheduling, the current VHDL IEEE Standard 1076-1993 proposes two different types of delays—inertial and transport—to address inertial delays found in digital devices and line delays associated with buses in digital systems. For accuracy in the simulation results, the VHDL simulator also implements the preemptive semantics for the inertial delays, which has been discussed earlier in Section 3.4.4. Thus, VHDL achieves the timing accuracy described in Ada as an HDL

```
L1: S1 <= 1 after 10 ns;
L2: S2 <= 1 transport after 10 ns;
```

Figure 7.7 Delays in VHDL.

while utilizing explicit timing constructs as in ADLIB-SABLE to facilitate ease of comprehension among designers.

The statements corresponding to the labels L1 and L2 in Figure 7.7 reflect inertial and transport delays. Under inertial delays, the assignment of 1 to signal S1 may not ultimately be realized due to preemption by a more recent assignment. In contrast, nothing can prevent the assertion of 1 to the signal S2 at 10 ns beyond the current simulation time.

It is pointed out that assignments to variables in VHDL may be viewed as a special case of signal assignments in that the delay involved is exactly zero and there can be no preemption.

WAIT STATEMENTS. A key to VHDL's capability of describing asynchronous interactions lies in its elegant design of the "wait" construct. While inspired by the "waitfor" construct, VHDL's wait construct eliminates the difficulty in ADLIB-SABLE, as elaborated in Section 4.4.6. Clearly, the goal of wait is to suspend the sequential execution of an entity until a specific timing condition is satisfied.

```
L1: wait on clock;
L2: wait until clock = '1';
L3: wait for 10 ns;
```

Figure 7.8 Syntactical usages of wait in VHDL.

Figure 7.8 presents three different syntactical uses of wait. Under label L1, the statement waits until the signal clock is subject to an event. The statement at label L2 pauses until the value of the clock signal changes to 1. Last, the statement at label L3 suspends the sequential execution for 10 ns from the current simulation time and resumes the execution of the subsequent sequential statements thereafter.

TIMING ASSERTIONS. Unlike ADLIB-SABLE and Ada as an HDL, where the synthesis of the timing assertions is complex and non-intuitive to the designer, the VHDL design includes elegant language constructs to check for timing assertions. While the timing assertions in DABL, shown in Figure 3.14, are perhaps more straightforward and easily noticed in that they are declared upfront within the behavior model, the ability to access the history of signals in VHDL and to comparatively evaluate their relative timing provides greater flexibility to the designer, not only to verify the timing assertions but also to describe the timing behavior more intuitively.

The key attributes of signals include 'EVENT, 'LAST_VALUE, 'LAST_EVENT, 'STABLE(T), and 'DELAYED(T). For a given signal, S, the operation S'EVENT returns true if S is subject to a transition at the current simulation time. The operation S'LAST_EVENT returns the elapsed simulation time, relative to the current simulation time, since the most recent transition of S. Even where S'EVENT is true, the S'LAST_EVENT returns the time interval between the current simulation time and the time of the previous transition of S. The operation S'LAST_VALUE returns the previous signal value prior to the current value of S. The operation S'STABLE(T) is designed to accept a time interval, T, as an argument and determine whether S incurs no transitions in the interval {NOW, NOW – T}, where NOW refers to the current simulation time. The operation returns a synthesized signal of duration *T*, starting at the current simulation time and extending towards the past. It has a value true when S lacks transitions in the interval *T* and false otherwise. The operation S'DELAYED(T) accepts a time interval *T* as an argument and returns a new signal that is delayed by *T*, relative to the original signal, S. Since S is known with certainty from the origin, i.e. 0 simulation time, up to the current simulation time, the new signal is defined from $0 + T = T$ time units up to (current simulation time $+ T$) time units. Clearly, the aim of this operation is to permit the user to view and manipulate the past behavior of S, beyond the last event, information on which may be obtained through the S'LAST_EVENT and S'LAST_VALUE attributes. It is critical to bear in mind that, at the current simulation time, the newly synthesized signal cannot offer the user a view into the future, i.e. beyond the current simulation time. It is also important to note that, if the new signal is synthesized solely for the purpose of verifying timing relationships, it may not correspond to reality and the VHDL description may be viewed as unnatural and non-intuitive.

```
Process (Clk)
L1: if (Clk='0' and Clk'EVENT)
L2:     ASSERT (Clk'LAST_EVENT >= clock_high_width)
        REPORT "Minimum clock high width violated"
        SEVERITY ERROR;

L3: if (Clk = '1' and Clk'EVENT) THEN
L4:     Qout <= Din after propagation_delay;

L5:     ASSERT (Clk'LAST_EVENT >= clock_low_width)
        REPORT "Minimum clock low width violated"
        SEVERITY ERROR;

L6:     ASSERT (not Din'EVENT) and (Din'LAST_EVENT >= setup_delay)
        REPORT "Setup violation"
        SEVERITY ERROR;
    end if;

L7: wait for hold_delay;

L8:     ASSERT (Din'Stable(hold_delay))
        REPORT "Hold violation"
        SEVERITY ERROR;
...
end process;
```

Figure 7.9 Synthesizing timing assertions in VHDL.

Figure 7.9 presents the synthesis of timing assertions in VHDL for the four types of timing checks usually encountered in flip-flops and are representative of timing checks in hardware systems. The statements at labels L2 and L5 verify that the high and low durations of the clock pulse meet the minimum requirements. The statement at L4 reflects the actual function of the flip-flop. The statement at label L6 verifies the setup requirement. The current simulation time (or NOW) is advanced to the positive clock edge and the state of the Din input between NOW and setup_delay time units prior to NOW is examined. The verification of hold time cannot be accomplished at NOW since it requires the examination of the state Din hold_delay time units into the future. The statement at label L7 suspends the entity body for hold_delay time units and allows the simulation scheduler to advance the current simulation time. When the entity body is re-initiated, the statement at label L8 examines the state of Din signal to determine whether the hold condition is satisfied.

7.1.4 Hierarchy

The structural style of entity body description is ideally suited for describing a hierarchical digital design in VHDL. The description is expressed in terms of lower-level component instances and components, in turn, are entities with the corresponding entity bodies assuming either dataflow, behavioral, or structural descriptions. Consider the example scenario described in the requirement labeled (4a) in Section 4.5 of the VHDL DRFP. In VHDL, the highest level of abstraction is expressed as an entity named microprocessor. The entity body of microprocessor is described in structure style that includes a declaration of the components—register, memory, alu, control, and program counter—and their instantiations along with the appropriate port map. Associated with each of register, memory, alu, control, and program counter, in turn, are entity declarations and the corresponding entity bodies. The entity body of alu is again expressed as a structure that includes a declaration of the component, adder, and its instantiation with the appropriate port map. Associated with adder is an entity declaration and the corresponding entity body is again described through a structure that includes a declaration of the components—primitive gates—and their instantiation accompanied by port maps. Associated with the primitive gates are entities. Since the gates lack further structural decomposition, the corresponding entity bodies must be expressed either as dataflow or behavior descriptions. Thus, the VHDL language design successfully meets the DRFP requirements.

In Figure 7.10, entity X represents the highest-level abstraction of the hardware system and the configuration alpha commits the final binding of the architecture Y to X prior to initiating simulation.

```
entity X is
  port (P1: in BIT; P2: out BIT);
  ...
end entity X;

architecture Y of X is
 ...
end architecture Y;

configuration alpha of X is
  for Y
  ...
  end for;
end configuration alpha;
```

Figure 7.10 Highest level of design abstraction.

7.1.5 Naturalness and Consistency with Underlying Hardware

One of the most significant achievements of VHDL is its endorsement of the idea of entity as the microcosmic representation of hardware at any level of abstraction, starting at the highest level and down to the gate level. The inclusion of dataflow, behavior, and structure descriptions for entity bodies enables VHDL to realize a hierarchical organization of a hardware system that satisfies the VHDL DRFP. The use of transport and inertial delays to describe bus lines and digital devices coupled with the implementation of the preemptive semantics of inertial delays in the scheduler enables designers to write simpler code and causes VHDL to be viewed as a natural and user-friendly HDL. The design of the synchronizing construct, wait, is motivated by ADLIB-SABLE, intuitive, and is a key in describing asynchronous interactions. The language constructs to facilitate timing assertions are realized accurately. A key contribution of VHDL is the discipline that a bus resolution function must be defined for every inout port and bidirectional bus signal. In ADLIB-SABLE and Ada as an HDL, the burden to model a bus is placed squarely on the designer. While VHDL does not relieve the designer of that task, the discipline forces the awareness and superior HDL code is realized.

7.2 INADEQUACIES OF VHDL AND INCONSISTENCIES WITH THE UNDERLYING HARDWARE DESIGN

This section critically examines the VHDL language syntax and semantics, relative to the fundamental requirements of a behavior-level HDL. It is noted that the aim here is not to solely criticize the language architects. The architects are to be commended for a splendid effort. The aim is to understand it from first principles to enable us to have a deeper understanding and appreciation of the HDL discipline.

7.2.1 The "VHDL Entity" as an Incomplete Entity and Difficulties with Concurrency

Although VHDL employs the word entity to describe the basic hardware abstraction, its understanding of the underlying meaning of entity is incomplete and, in part, erroneous. An entity, by definition, must be self-contained, independent, and concurrent. Yet, a number of VHDL's syntactical constructs and semantics violate this basic premise. The IEEE Standard VHDL Language Reference Manual [33] does not detail the characteristics of an entity. Even the article [36], where the original VHDL architects introduce the core philosophy underlying VHDL, lacks any mention of the key characteristics of an entity. They defined a design entity to consist of an interface and one or more bodies, with the interface specifying the external characteristics of the entity and each body representing an alternate design approach consistent with those characteristics.

For instance, the inclusion of global signals in a structure description implies that there is an implicit connection of every component instantiated in the entity body. Since every component, in turn, is an entity, the notion of global signals poses a difficulty with the self-contained characteristic. By virtue of being a global signal, any update to it by a component instance must be visible to all the component instances. This requires either every instance to have explicit communication with every other instance or a centralized agent to control the activities of all instances. Upon examination of a real hardware system, it is clear that the behavior of a hardware system is controlled only by those signals that are explicitly connected to it. In fact, there is no known way to design hardware without complete knowledge of all signals, whether input, output, or both input and output. The issue of global signals also opposes a fundamental requirement, (1c), of the VHDL DRFP, namely to represent asynchronous systems. Unfortunately, the DRFP itself appears to be inconsistent, with the requirements (1c) and (1d) conflicting in principle.

The issue of global variables in dataflow and structure descriptions of entity bodies, labeled "shared" variables, pose the same problems as global signals. An added difficulty is that global variables not only require an immediate assignment but that the updated fluorescence must be visible instantaneously to every entity or every concurrent statement. There is no known physical system, let alone hardware, that can achieve this semantics. In essence, global variables do not reflect any aspect of hardware and do not belong in an HDL, even if they imply programming convenience. It must be remembered that VHDL is a hardware description language, not a programming language.

Because it violates a basic characteristic of entity, unlike the actual operation of the target hardware system, VHDL descriptions of hardware modules, in their current form, may not be executed concurrently on a real parallel processor. Thus, the requirement (6b) of DRFP [55], in theory, may not be satisfied.

The VHDL architects view VHDL as a total concurrent language and needed to invent a special mechanism, process, to allow sequential statements. Any statement, unless encapsulated within a process, is viewed as a concurrent statement. Although the purpose is to reflect hardware that is basically a concurrent process, the purpose is easily achieved through the structure description where the lower-level constituent components are, in turn, also entities. Conceptually, it is unclear whether there is a unique role for dataflow descriptions. Any concurrent statement in a dataflow description must reflect a hardware process at some level of abstraction and, if so, it may easily be represented as an entity. At the pragmatic level, the individual concurrent statements in a dataflow description cannot be executed simultaneously since they are too fine grained, they are not self contained, i.e. they may depend on global variables and signals, and they do not necessarily have their own notion of time. The added fact that VHDL is controlled by a centralized scheduler renders the effort of designing individual concurrent statements even more unnecessary.

Such redundancies and unnecessary constructs, especially without clear justification of their purpose, tend to make VHDL bulky, slow in compilation and especially execution, unnecessarily complex, and difficult to learn.

There is yet another reason underlying the problem of executing VHDL models in parallel and it lies in the difficulty with algorithms for accurate, distributed event driven simulation. This issue will be addressed in detail in Chapter 9.

In the Ada [66] programming language, whose task concept constitutes the basis for the entity concept in VHDL, a task specification is first declared and then followed by a task body. While the declaration ties the task identifier name with the task, the task body contains the execution semantics. Ada requires that a task specification and the corresponding task body share the same identifier. Ada does permit the creation of instances of a task type, all of which share the exact same task body. In contrast, in VHDL, any number of entity bodies, each different, may be associated with a single entity declaration. This is a violation of the requirement (5) in DRFP which mandates that any Ada syntax, if borrowed, must be adopted exactly. Furthermore, upon analysis, it is unclear as to the precise purpose and role of the entity declaration. In essence, an entity declaration enumerates the input, output, and bidirectional ports of an abstract hardware system. Clearly, the ports, by themselves, do not reflect the hardware system since a completely different hardware system may have identical input, output, and bidirectional ports. For instance, consider a two-input AND gate and a two-input OR gate. Except for the different entity names, the corresponding entity declarations are indistinguishable and, essentially, convey nothing important.

Consider an alternate strategy where we eliminate the entity declarations of VHDL and redefine an entity to include the port descriptions, etc., as well as the body of the architecture. This newly defined entity is self-contained and one can synthesize multiple instances of it, as necessary, similar to comptypes in ADLIB-SABLE. We will even introduce a convention of naming the new entities as X_Y, where X refers to the entity declaration and Y the architecture in the current VHDL standard. Thus, the designer can immediately recognize that two entities ADDER_GATES and ADDER_BEHAVIOR both refer to the same abstract hardware system—an adder. The designer would also understand that the entities are two distinct realizations of the basic adder. In the current VHDL, both the name in the entity declaration and the architecture identifier are required during static elaboration time. The result of this simplification would imply a simple yet strongly justified VHDL that would greatly facilitate quick and precise learning without sacrificing functionality.

7.2.2 The Difficulty with Connectivity in VHDL

Unlike in ADLIB-SABLE and Ada as an HDL where the nets interconnecting the hardware modules are explicitly stated, in VHDL the nets are implied. A net is a physical wire(s) and the explicit style is a natural way of specifying the connectivity. In addition, a potential advantage of the explicit style may be in allowing the designer to specify its electrical characteristics and other relevant properties.

In virtually every HDL, the ports are specified as either input, output, or bidirectional. This is logical, natural, and, based on our current knowledge, one cannot design hardware with any different kind of port. Yet VHDL proposes the use of buffer as a port type without carefully justifying the need. Consider a RS latch at two levels of abstraction, behavior and structure, as shown in Figures 7.11(a) and 7.11(b), respectively.

As required in VHDL, an entity declaration must be synthesized for the RS latch. While the ports Set and Reset are unmistakably of the input or "in" mode, there is confusion with the ports Q and Qb. Although, logically, they must be of mode "out," in the book titled "VHDL", the author [67] states that Q and Qb must be of mode "buffer" and argues the following. The port Q of the entity declaration must be mapped on to the structure level, where Q is both an input to one of the two NAND gates and an output from the other NAND gate. Since labeling them as "inout" is likely to be viewed as an outright contradiction of common knowledge, Q and Qb are labeled as mode buffer. This is consistent with the VHDL semantics that mandates that a formal parameter (in this case, the ports of the NAND gates) of mode "in" may be mapped only onto an actual parameter (in this case, the ports of the entity declaration latch) of modes "in," "inout," or "buffer." Figure 7.12 presents the code segment. In another book titled "A Guide to VHDL", the authors Mazor and Longstraat [68] characterize the ports Q and Qb of mode "inout," as shown in Figure 7.13.

While the use of mode "inout" for Q and Qb in Figure 7.13 is erroneous, the use of buffer in Figure 7.12 is without proper justification. A likely source of this confusion and ambiguity is the original VHDL language design strategy. A cardinal rule of language design is never to permit ambiguity. The VHDL DRFP had reiterated this in its mandate (1). A number of issues are raised through this scenario. First, the ports Q and Qb of a RS latch are

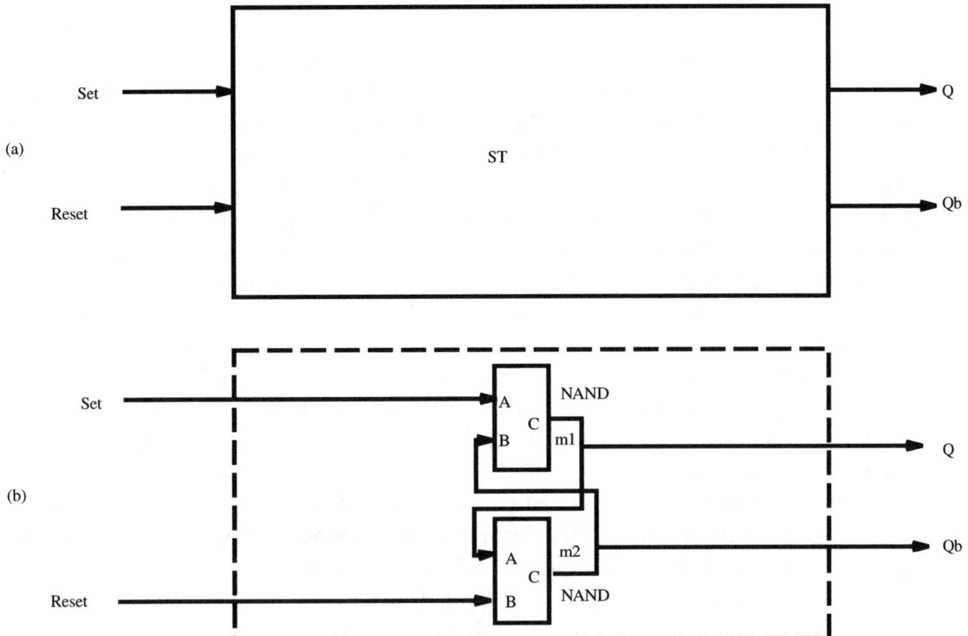

Figure 7.11 Port declarations in a VHDL entity.

```
entity rsff is
  port (set, reset: in BIT; q, qb: buffer BIT);
end rsff;

architecture netlist of rsff is
  component nand2
    port (a, b: in BIT; c: out BIT);
  end component;
  Begin
    U1: nand2 port map (set, qb, q);
    U2: nand2 port map (reset, q, qb);
End netlist;
```

Figure 7.12 Entity declarations for a RS latch in the book "VHDL" [67].

```
entity rsff is
  port (set, reset: in BIT; q, qb: inout BIT);
end rsff;
```

Figure 7.13 Entity declarations for a RS latch in the book "A Guide to VHDL" [68].

clearly outputs and there should be no doubts about it. Regardless of its reasons, VHDL cannot require designers to specify Q and Qb as anything but of mode "out."

Second, an entity declaration, according to VHDL, is expected to define a high-level interface that is not only applicable to any architecture body but is independent of the lower-level details. While the issue of "buffer" vs. "inout" is raised when the architecture is of type structure, it would not have been raised for an architecture of type behavior. In the behavior architecture, the state vector of the latch would be encapsulated through a variable. Thus, VHDL's objectives are self conflicting. Third, this scenario further strengthens the criticism that the notion of entity declaration has been borrowed from Ada without fully understanding its role and purpose in VHDL. Fourth, VHDL's proposal of "buffer" as a mode is unnatural in actual hardware systems and leaves the designers vulnerable and confused.

The VHDL language reference manual forbids a formal "in" port to be connected to an actual "out" port. This is understandable since such a mapping will leave the input port of the lower-level component undefined. However, such a restriction is unnecessary and harmful since one may encounter a formal "in" port connected to a formal "out" port and an actual "out" port, as is the case with the latch. Such types of connections are absolutely correct and ubiquitous in hardware and VHDL cannot treat them as illegal. The VHDL compiler must check to verify that whenever a formal "in" port is connected to an actual "out" port, there is also a connection with a formal "out" port. Ironically, according to the language reference manual, such a checking is done in VHDL anyway whenever "buffer" mode is used to ensure that there is a single driving source. Therefore, at any given level of abstraction in the design hierarchy, the ports must be defined uniquely and correctly, independent of the lower-level implementation.

7.2.3 Timing Inadequacies in VHDL

CONCEPTUAL DIFFICULTIES WITH CONLANS BCL TIME MODEL. According to the original VHDL architects [36], VHDL's model of time is similar to that of Conlan [35] in that it has a macro- and a micro-time scale. While the macro-time scale utilizes real time such as nanosecond, microsecond, etc., the micro-time scale represents unit delay and is not measurable. Any number of micro-units of time may exist between any two macro-time units. Conlan [35] is possibly the first HDL to propose the notion of dual time scales. It provides a discrete time model that breaks real time into discrete instants separated by a single time unit. Between any two successive time units there are an indefinite number of computation "steps" identified with integers greater than zero. The discrete instants and computation steps correspond to the macro- and a micro-time scale in VHDL.

Clearly, according to VHDL, in theory, the micro-time scale is not measurable. Thus, even if one were to invent hardware devices in the future to measure femtoseconds or an even finer granularity of time, the micro-time scale would still defy measurement. Under these circumstances, unless there are compelling reasons, it may be neither logical nor scientific to include the concept of micro-time scale in VHDL. The micro-time scale simply bears no correspondence to actual hardware. Furthermore, given that one is able to design hardware, precisely and accurately today, the need for the micro-time scale is brought into question.

The BCL time model poses several conceptual difficulties. First, given that a host computer is a discrete digital system, it cannot accommodate an indefinite number of steps within a finite time unit. Second, although the individual computation "steps" must imply some hardware operation, they do not correspond to discrete time instants which are utilized by the underlying discrete event simulator to schedule and execute the hardware operations.

Thus, the computation "steps" may not be executed by the simulator and, as a result, they may not serve any useful purpose. It is also noted that, fundamentally, in any discrete event simulation, the timestep or the smallest unit through which the simulation proceeds, is determined by the fastest sub-system or process. For accuracy, this requirement is absolute. Assume that this timestep is *T*m. If, instead of *T*m, a timestep *T* is used deliberately ($T > T$m), the contributions of the fastest sub-system or process cannot be captured in the simulation, leading to errors in interactions, and eventually incorrect results. The choice of *T*m as the timestep also refutes any claim that the dual time scales may yield faster simulation speed. Third, the dual time scales implied by the BCL model oppose the concept of universal time, explained earlier in Chapter 3.

Other difficulties with the BCL time model may be expressed as follows. First, the reasons and justifications underlying this timing model are missing in the literature. Conlan and VHDL has presented no scientific evidence to prove that there exists at least one real hardware system that may be designed and built yet cannot be described through the existing timing model in ADLIB-SABLE or in Ada as an HDL. Second, although Conlan [35] does not state any reasons, the original VHDL [36] architects state that the dual time scales enable the designer to simulate the clocked registers utilizing the real timing data while the intervening combinational logic may be simulated utilizing unit delays. The claim is logically incorrect in that ignoring the delays of the combinational logic may lead to erroneous results. Third, the VHDL architects overlooked a key requirement that VHDL enables the description and simulation of asynchronous designs. Their choice of the clocked registers example in defense of the macro- and micro-time scales constitutes a synchronous design. if the two registers were to belong to two asynchronous modules that are driven by two independent clocks, then ignoring the delays of the intervening combinational logic may yield erroneous results.

The use of zero- and unit-delay assignments in VHDL descriptions is discouraged for two reasons. First, fundamentally, no hardware devices are known to us that operate in zero delay. Second, we still don't know how to design two pieces of hardware with exactly the same delay. Thus, zero- and unit-delay devices are non-realistic. Third, it is well known from the digital design literature that zero-delay and unit-delay simulations, though easy to implement, are highly error prone. While the use of either zero or unit delays with combinational circuits fails to detect hazards, for sequential circuits, zero delays lead to inconsistency and unit-delay usage fails to identify potential races. In asynchronous systems, the absence of actual delay values relegates any effort to describe and simulate hardware pointless since relative timing is the key.

DIFFICULTIES WITH DELTA DELAYS. A manifestation of the macro- and micro-time scales in VHDL is the notion of delta delays. In theory, VHDL allows signal assignments with zero delays, i.e. the value is assigned to the signal in zero macro-time units but some finite, delta, micro-time units. The actual value of delta is inserted by the VHDL compiler, transparent to the user.

Before we critique delta delays, let us understand why the VHDL architects felt compelled to introduce the notion. Consider the RS latch in Figure 7.11(b). Assume that both NANDs are 0 ns delay gates and that they execute concurrently in a structural description. Assume that the initial value at Q and Qb are both 1 and that the value at set and reset input ports are both 1. At simulation time 0, both gates execute and generate the following assignments: (1) a value 0 is assigned at Q at time $0 + 0 = 0$ ns and (2) a value 0 is assigned at Qb at time $0 + 0 = 0$ ns. Assume that there is a race between (1) and (2) and that (1) is executed infinitesimally earlier. As a result, the lower NAND gate is stimulated and it generates an assignment: (3) a value 1 is assigned at Qb at $0 + 0 = 0$ ns. Upon examining (2) and (3), both assignments are scheduled to affect the same port, Qb, at the same exact time, 0 ns, one armed with a "0" and the other armed with a value "1." To alleviate this ambiguity, VHDL attempts to order the executions in a micro-time scale, and proposes the delta delay concept.

The first difficulty with delta delay is that the VHDL language reference manual [33] does not state how a value for the delta is selected. This is an important question since VHDL may return different results corresponding to different choices of the delta, as illustrated through Figure 7.14.

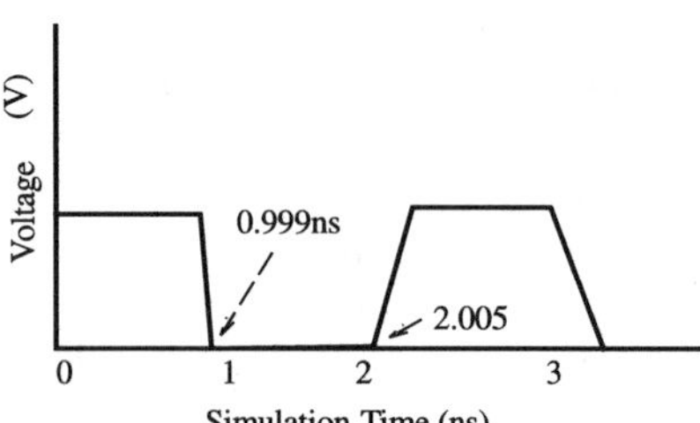

Figure 7.14 Impact of the choice of delta value on simulation results.

Figure 7.14 presents a signal waveform. Assume that the value of delta is 1 ps. When the current simulation time is either 1 ns or 2 ns, VHDL safely returns the value 0 for the signal value. However, where the delta value is 5 ps, VHDL will return the value 0 corresponding to the current simulation time of 2 ns but fail to return a definite value corresponding to the current simulation time of 1 ns. Since the signal waveform is realized at runtime, i.e. as the entities execute during simulation, and as the VHDL compiler must select a value for the delta delay at compile time, it is difficult to ensure the absence of ambiguous results.

The second difficulty is that one can construct any number of example scenarios in VHDL where the result is inconsistency and error. Consider the process, PROC1, shown in Figure 7.15. While not critical to this discussion, it is pointed out that the process PROC1 does not include a sensitivity list which is permitted by the VHDL language [33]. As an example of usage of a process without a sensitivity list, the reader is referred to page 57 of [33].

The statements S1 and S2 are both zero delay signal assignments. While S1 updates the signal "a" using the value of signal "b" and the variable "c", the statement S2 updates the signal "b" using the value of the signal "a" and the variable "c". To prevent ambiguity of assignments to the signals "a" and "b", the VHDL compiler inserts, at compile time, a delta delay of value delta1 say, to each of S1 and S2. Thus, S1 is modified to: a <= b + c after delta1, and S2 is modified to b <= a + c after delta1. For every iteration, the subsequent assignments to "a" and "b" are realized in increments of delta1. That is, first (NOW + delta1), then (NOW + delta1 + delta1), and so on. These are the micro-time steps in the micro-time scale and we will refer to them as delta points. Between the two consecutive macro-time steps, the VHDL scheduler may only allocate a maximum but finite number of delta points which is a compile time decision. Conceivably, the designer may choose a value for the number of iterations such that, eventually, the VHDL scheduler runs out of delta points.

```
architecture X of Y is
 signal a, b: resolve BIT ..;
begin
 PROC1: process
 variable c: int;
 begin
 for i in 0 to 1000 loop
S1: a <= b + c;
S2: b <= a + c;
  end loop;
 end process;

 PROC2: process
 variable d: int;
 begin
  for i in 0 to 7 loop
S3: a <= b + d;
S4: b <= a + d;
  end loop;
 end process;
end X;
```

Figure 7.15 An inconsistency with delta delays in VHDL.

Under these circumstances, VHDL will fail. Thus, the idea of signal deltas, transparent to the user, is not implementable.

Third, the notion of delta delays, in its current form, poses a serious difficulty with VHDL's design philosophy of concurrency. Consider Figure 7.15, where the processes PROC1 and PROC2, by definition, are concurrent with respect to one other. The two sets of statements—{S1,S2} in PROC1 and {S3,S4} in PROC2—both affect the signals "a" and "b" and "resolve" constitutes the resolution function, as required by VHDL. The statements S1, S2, S3, and S4 are all zero delay signal assignments, so delta delays must be invoked by the VHDL compiler. Since the dynamic execution behavior of processes are unknown a priori, the VHDL compiler may face difficulty in choosing appropriate values for the delta delay in each of the processes. In this example, however, logically, the VHDL compiler is likely to assign a very small value for the delta delay (say delta1) in PROC1, given that it has to accommodate 1001 delta points. In contrast, the VHDL compiler may assign a modest value for the delta delay (say, delta2 where delta2 $\gg$ delta1) in PROC2, given that only 8 delta points need to be accommodated. As stated earlier, assignments to the signals "a" and "b" will occur from within PROC1 at (NOW + delta1), (NOW + delta1 + delta1), and so on. From within PROC2, assignments to the signals "a" and "b" will occur at (NOW + delta2), (NOW + delta2 + delta2), etc. By definition, a resolution function resolves the values assigned to a signal by two or more drivers at the same instant. Therefore, here, "resolve" will be invoked only when (NOW + m $\times$ delta1) = (NOW + n $\times$ delta2), for some integer values "m" and "n." In all other cases, assignments to the signals "a" and "b" will occur either from within PROC1 or PROC2. Thus, the values of the signals "a" and "b," from the perspectives of processes PROC1 and PROC2 are uncoordinated, implying ambiguity and error.

In essence, the desire to use zero delay assignments necessitates the inclusion of delta delays in VHDL. In turn, delta delays foster ambiguity, uncertainty, and provide a false sense of accuracy under zero delay assignment usage. It is recommended that delta delays be eliminated from VHDL and that designers first determine the universal time for a given hardware system and then utilize realistic inertial and transport delay values, as appropriate, in the VHDL description of the hardware modules. The VHDL scheduler has been carefully designed to ensure the accurate description of realistic hardware descriptions. Last, the synthesis tools need to be enhanced to accept realistic VHDL descriptions.

INTERNAL SIGNALS. In addition to every "in," "out," and "inout" port being viewed as signals, VHDL permits the designer to define internal signals within the architecture body of an entity. The DRFP, under item (1b), recognizes the value of internal signals as an artifact to simplify descriptions and cautions that it does not correspond to a real signal. Unfortunately, in VHDL, the use of signal definitions in an architecture description of type structure assumes the role of a global signal to all of the constituent components and conflicts with the fundamental characteristics of an entity—independence, asynchronicity, and concurrency. Therefore, the question that becomes important is whether the need to incorporate internal signals to describe real hardware is so great that the violation of the entity concept must be tolerated, as is the route chosen by VHDL? The answer turns out to be negative. To understand the rationale, consider three principal uses of internal signals—A, B, and C.

SCENARIO A. Within a structure description, an internal signal may be defined to interconnect two ports of two components. Consider again the RS latch in Figure 7.11(b) for which the VHDL description is presented in Figure 7.16.

```
entity RS is
  port (set, reset: in BIT; q, qb: out BIT);
end RS;

architecture RS_gate of RS is
  signal m1, m2: BIT;
  ...
  U1: nand (set, m2, m1);
  U2: nand (reset, m1, m2);
  q <= m1;
  qb <= m2;
end RS_gate;
```

Figure 7.16 The use of internal signals to interconnect components.

In Figure 7.16, m1 and m2 connect the outputs of the NAND gates to the respective input ports and are useful. However, upon reflection, clearly these signals represent unidirectional interconnecting wires. Since a wire is as much a digital device as a NAND gate, it is natural and logical that the two wires be defined as unique components, exactly as the NAND gates. Each wire component may be labeled with explicit input and output ports and its behavior expressed through the use of transport delays. Furthermore, a wire has no memory associated with it, and the use of signals to represent such wires, as shown in Figure 7.16, reflects inaccurate modeling. Where two or more components are connected through an internal bus, using similar arguments, one would need to define a bus component with appropriate "inout" ports within the structure description.

SCENARIO B. A designer may also choose to use signals, in a structure description, to store data along with the timing information, as shown in Figure 7.17, since variables do not carry any notion of timing. While the statement S1 implies immediate information storage, statement S2 implies information storage delayed by 5 ns. In an actual hardware system, registers are used to store information. Therefore, under these circumstances, it is natural and logical to model registers as components within the structure and utilize inertial delays, just as other components such as NAND gates, etc.

```
architecture X of Y is
  signal p1, p2: BIT;
  ...
S1: p1 <= '1' after 0 ns;
S2: p2 <= '0' after 5 ns;
  ...
end X;
```

Figure 7.17 The use of internal signals to store information along with timing.

It is also evident from scenarios A and B that the semantics of signals may correspond to a memory-less wire at one time while reflecting that of a register at a different time.

SCENARIO C. A number of designers feel that signals may be used profitably to represent internal buses, to store a value internally at a specific time in the future, or to project an event into the future to serve as a reminder, in a behavior description. However, as we will see, all of these functions may be easily achieved within a behavior description without resorting to signals. An internal bus or a state vector of a hardware system may be emulated through a variable(s) in VHDL. Consider that a designer wishes an internal assignment to take effect *T* time units in the future and employs an internal signal. The assignment process must continuously check the simulation time to compare whether it is time to commit the assignment. Since the behavior description executes sequentially, clearly there must be a process to handle the assignment and it must execute concurrently. This is a contradiction since the behavior description is, by definition, purely sequential. The designer, however, does possess the capability to suspend the behavior execution by executing a "wait" clause for *T* time units, and realize the assignment upon completion of the "wait." Last, a designer may use a signal to project an event into the future to serve as a reminder. This resembles a timer, and utilizing the argument above, a timer is a concurrent concept for which there is no room in a sequential behavior description. Thus, the use of internal signals is either observed to oppose the HDL principles or may be addressed successfully through other VHDL language constructs.

WEAKNESS OF THE INERTIAL TIMING DELAY. Although the inertial timing delay construct of VHDL correctly detects and deletes inconsistent events, it does not share this information with the user. In reality, such events within a component may cause spikes in its output waveform, thereby upsetting the correct functioning of other digital devices connected to it. It is therefore important for the inertial timing delay construct to trigger the propagation of this information to the user.

INADEQUACY OF THE TRANSPORT TIMING DELAY. The difficulty with the transport timing delay semantics is a serious one and is addressed in Chapter 10 in detail.

LIMITATIONS OF SIGNAL ATTRIBUTES. The primary goal of the signal attributes is to facilitate specifying the timing assertions in hardware systems accurately, unambiguously, and easily. If one or more of the attributes assist in simplifying timing descriptions, that would be an added bonus. Timing assertions may involve examining the high or low pulse width of a single signal or the relative timing between two or more signals. In either case, what is required is access to the values and the assertion times of the values of the signals. VHDL does provide the value of a signal at the current simulation time. It is pointed out that a signal contains only the history of values and times from the origin to the current simulation time. What would have been ideal is for VHDL to have designed an attribute function that would accept a past time value, *T*, as an argument and yield the logical value of the signal under question corresponding to *T*. To a designer, such an attribute would have constituted a natural scheme to look into the history of signals.

According to the VHDL language reference manual, the operation S'STABLE(T) involves the STABLE attribute, accepts a time value, *T*, and returns a signal whose value is either true or false, depending on whether any events have occurred on signal S in the past *T* time units starting at the current simulation time. In essence, S'STABLE(T) yields a single value. Therefore, to label it a signal appears to be unnecessary.

In essence, the S'DELAYED(T) operation, as discussed earlier, allows the designer to look into the past behavior of the signal, for an interval of *T* time units, starting at the current simulation time. Clearly, the synthesis of the new signal S'DELAYED(T) occurs at runtime and is very time consuming since it requires examining the history of the signal from the origin simulation time to the current simulation time and their manipulation. For the purpose of checking the representative timing assertions, the DELAYED operator appears to be redundant. As discussed earlier, the operators STABLE(T), LAST_EVENT, and LAST_VALUE are adequate. The DELAYED operator does possess a unique capability in that it allows the designer to look into a signal's history at any point in the past, up to the origin. Perry [67] proposes a technique to verify the hold time of a D-type flip-flop, utilizing the DELAYED attribute. He delays the clock signal by hold time and utilizes it to activate a process hold time units into the future, relative to the current simulation time. The process verifies whether the value on the D input is stable for hold time units following the active clock edge. While perfectly legitimate, the solution requires the compute- and memory-intensive operation involving the DELAYED operator. In essence, the DELAYED operator allows for projecting a reference point in the future. When the simulation time had advanced in the future to the reference point, target signals, etc., may be examined. This scheme is unnatural for the following reason. If one wants to project a reference point into the future, one uses a timer in a hardware system. A sequential behavior description cannot model a timer since, by definition, a timer is concurrent. The goal, however, is correctly and easily realized by simply suspending the behavior description by executing a wait for *T* time units, at the end of which the description is reawakened.

A number of signal attributes—'ACTIVE, 'LAST_ACTIVE, 'QUIET(T), 'TRANSACTION(T), etc.—are included in VHDL to allow a designer to view the transactions. By definition, transactions are uncertain for they may be preempted. Therefore, it appears unnatural for VHDL to include such attributes. Also, one has to be very careful when interpreting such results since they are subject to change due to preemption. Since the total number of transactions may be significantly larger than the total number of events, operations such as S'QUIET(T) and S'TRANSACTION(T) are expected to be even more compute- and memory-intensive. Clearly, knowledge of the transactions may help greatly in debugging VHDL programs and in verifying the correctness of the VHDL environment. VHDL would be better served if such attributes were either invoked only under a special compilation and execution mode, say debugging, or removed from the language and placed under the VHDL simulation environment.

As a cautionary note, it is pointed out that the notion of bus resolution function, while useful, does not absolve the designer of the responsibility of correct design. The occurrence of a bus contention reflects a serious design error or flaw in the bus arbitration algorithm that should be addressed immediately and not brushed aside to be taken care of by the resolution function.

7.2.4 Limitations of Hierarchy in VHDL

Since VHDL is designed on top of Ada, and as the constructs in Ada are rich enough to support selective elaboration of subsets of the hierarchy at runtime, as illustrated in Chapter 6, it is logical to expect the state-of-the-art VHDL to address the concept of zooming.

7.2.5 Need for Naturalness and Consistency with Underlying Hardware

While critically evaluating VHDL, several of the language constructs were cited for lack of naturalness and consistency with the underlying hardware. They include the issues of shared variables, global clocks, global signals, internal signals, entity, transport delay, and a few of the signal attributes. The difficulty with shared variables has finally surfaced in the multi-threaded implementations of VHDL [69].

8

Case Studies: Developing Hardware Descriptions for Real-world Digital Systems

Prior to launching an effort in developing a hardware description of a digital system, it is critical to understand (1) the operation of the hardware, (2) the purpose of the exercise, and (3) the limits of the underlying hardware description language. In this chapter, we will develop hardware descriptions for three increasingly complex real-world digital systems, starting from first principles as outlined in Chapter 3. We will use a combination of VHDL and pseudo English, utilizing VHDL constructs where their use is meaningful and instructive, and pseudo English elsewhere to facilitate conceptual understanding.

Given a design—either developed already or to be fabricated—first it must be analyzed. The analysis must yield the key entities that execute independently and concurrently. The decision to identify specific units as entities must be based on the desired resolution of precision, i.e. which units whose detailed behavior, when understood, will help immensely in the understanding of the entire system. In general, when a unit, X, is lumped with other units, the collective behavior masks the finer aspects of the behavior of X. However, is X is a key element, i.e. it exerts influence over other units, it is crucial to focus on its behavior.

The identification of entities must also take into account the issue of concurrency. For efficiency in simulation, entities may be executed in parallel on multiple processors. However, if the amount of computational effort in an entity is limited, it may not be wise to execute it concurrently as the overhead is usually high. For example, if a simple circuit consists of three gates, while it is true that they are concurrent, it may not be efficient to identify them as entities solely for the purpose of executing them concurrently on three separate processors. Thus, if one were to describe the circuit in a dataflow manner in VHDL, i.e. consisting of three concurrent statements, the advantage is primarily theoretical. There are exceptions, of course. For instance, during fault simulation where every gate is executed many times, it may be advantageous to declare them as concurrent entities. The greatest benefit is realized when the entities are asynchronous and represent a substantial computational effort.

Once the entities have been identified, next the universal time must be determined. The universal time is the common denominator of time that permits entities with different internal clocks to interact meaningfully and accurately. Consider a system consisting of a CPU and a RAM. The CPU's timing behavior may be described in integrals of 8 μs while that for the

RAM is 9 μs. Clearly, the universal time must be set at 1 μs to permit meaningful interaction at this level. Imagine that the designer plans to utilize zooming, i.e. dynamically elaborate into the lower-level units of the CPU such as ALU, registers, multiplexors, etc. Assume further that the timing behavior of these units may be described in integrals of 200 ns. Under these circumstances, the universal time must be modified to 200 ns as the basic unit of time and let us name it a nittasec. The CPU's timing behavior will be expressed as $1\ \mu s \div 200\ ns \times 8 = 40$ nittasec. Similarly, for the RAM, the value is 45 nittasec. Of course, we can choose to utilize 1 ns as the universal time instead of defining the nittasec.

Where the system being modeled is synchronous, i.e. a single clock controls the timing of all units, the complete behavior—functional and timing—of the system may be accurately captured through a single sequential behavior description. In VHDL, this would correspond to a single process in a single entity. However, the designer is at liberty to either use multiple process statements or multiple entities, one for each unit.

Where the system being modeled is asynchronous, i.e. each sub-system maintains its own unique clock, accurate representation requires the sub-systems to be modeled through entities. Under these circumstances, the entities must not contain any shared variables. The behavior of every entity must be completely unknown to all other entities and any information that is needed to be shared between entity X and Y must be expressed through an interconnection between X and Y. The progress of execution of every entity will be controlled solely by a corresponding unique scheduler, subject to synchronization with other entities as necessitated.

Next, the interconnection of the entities is completely specified. Where possible, the timing characteristics of the interconnecting wires may be specified to permit accurate simulation.

Then the logical behavior of the entities must be developed in the hardware description language. It is noted that the ability to describe the non-timing, logical behavior of digital systems accurately is inherent in most hardware description languages. To represent accurate timing behavior, the designer should utilize inertial delays as supported by VHDL and transport delays as described in Chapter 10. The detection and preemption of inconsistent events is critical to accuracy and its absence in Verilog HDL is crippling. In developing the timing descriptions, zero delays should be categorically avoided. We are not yet capable of manufacturing hardware with zero delays. If the propagation delay of a unit is significantly lower, relative to other units, one must use appropriate units. Thus, even where every module has propagation delays of μs and a lonely gate has a delay of 20 ns, the actual gate delay value should be used. In event driven simulation, every legitimate event must be simulated and the use of zero delays does not make the simulation go any faster. In the absence of zero delays, the issue of delta delay of VHDL is a moot point. To describe the timing including asynchronous behavior, one may utilize the relevant timing control and timing check constructs of the hardware description language as explained in the previous chapters.

8.1 AM2903 BIT-SLICE PROCESSOR—A SYNCHRONOUS DESIGN

The Advanced Micro Devices Inc. manufactured AM2903 [70] bit-slice processor, shown in Figure 8.1, is a synchronous system. That is, a single clock controls every sequential sub-unit and the timing characteristics of the combinational blocks influence the choice of the high and low duration widths of the clock. The simplified AM2903 consists of a RAM, two latches A and B, two multiplexors A and B, an ALU, and a shifter. The Q register, Q shifter, and output buffer of the original AM2903 are ignored here without any loss in generality, relative to the

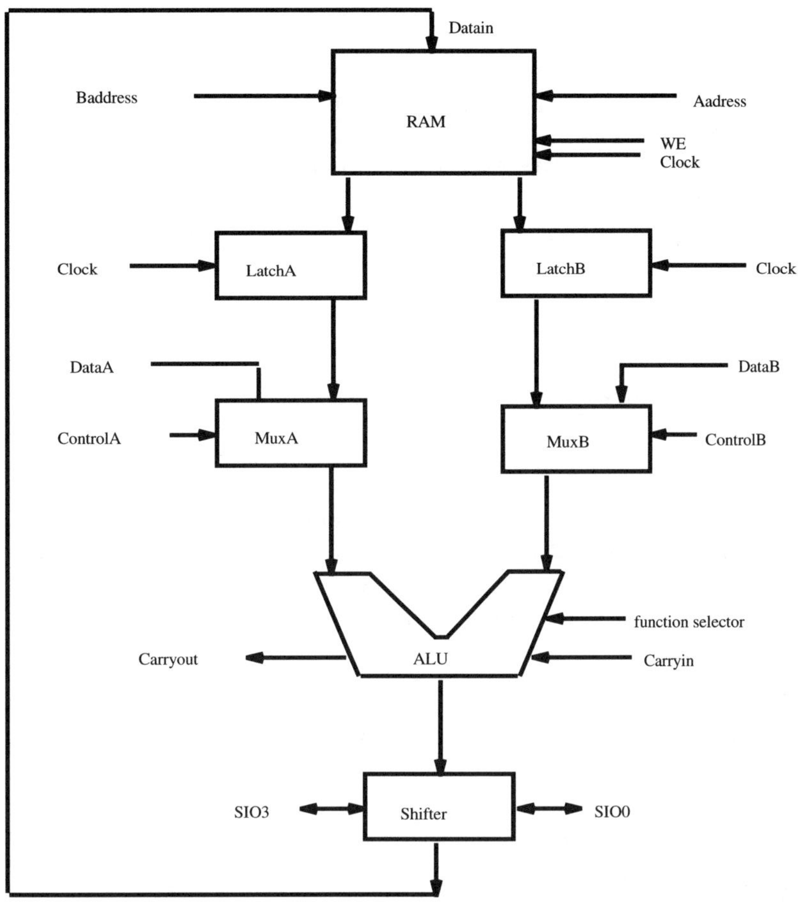

Figure 8.1 A simplified AM2903 bit-slice processor.

basic purpose of this chapter. The output of the shifter is connected to the datain port of the RAM. The operation of the AM2903 may be described as follows. When the clock signal is high and the RAM is read-enabled, i.e. WE is low, the RAM puts out its contents on the two output ports based on the addresses at Aaddress and Baddress, respectively. Also, the latches LatchA and LatchB are both open. When the clock signal incurs a transition to low, the data from the RAM are latched into LatchA and Latch B and are available at its outputs. The assumption is that, by the time of the clock transition to low, the Aaddress and Baddress are both stable as are the outputs of the RAM output ports. When the clock signal is low, the multiplexors select between the two inputs, and provide the operands to the ALU. The ALU generates the output and propagates it to the shifter. The shifter's output is propagated to the bus that connects to the datain port of the RAM. When the clock incurs a transition to high, the input value at the datain port of the RAM is read into the RAM, provided the WE signal is high. The assumption is that, by the time the clock transition to high occurs, the multiplexors, ALU, and shifter have completed their operations and the result may be safely written into the RAM.

Since all three sequential devices are controlled by a single clock, the AM2903 may be modeled accurately through a single VHDL process, as shown in Figure 8.2. One could also

```
entity AM2903 is
  port (
    Aaddress, Baddress, DataA, DataB: in bit_vector (3 downto 0);
    WE, Clock, Carryin, ControlA, ControB: in bit;
    function_selector: in bit_vector (2 downto 0);
    Carryout: out bit;
    Yout: out bit_vector (3 downto 0);
    SI00, SI03: inout bit;
  );
end AM2903;

architecture behavior of AM2903 is
  begin
   process (Clock)
     variable
       Amemout, Bmemout, Alatchout, Blatchout: integer range 0 to 3;
       MuxAout, MuxBout, ALUout, Shifterout: integer range 0 to 3;
       memdelay, latchdelay, muxdelay, aludelay, shifterdelay: integer;
       memsetup, memhold, latchsetup, latchhold: integer;
       final_result: integer;
    begin
       memdelay = ...;
       latchdelay = ...;
       muxdelay = ...;
       aludelay = ...;
       shifterdelay = ...;
       memsetup = ...;
       memhold = ...;
       latchsetup = ...;
      latchhold = ...;

L1:    if (Clock = '0') then
L2:       if (Aaddress'LAST_EVENT >= (memdelay + latchsetup))
              and (Baddress'LAST_EVENT >= (memdelay + latchsetup))
            then setup constraints for latches A and B are satisfied;

L3:       if (Aaddress'Stable(memdelay - latchhold)) and
             (Baddress'Stable(memdelay - latchhold))
              then hold constraint for latches A and B are satisfied;

L4:    Amemout = content of RAM at address Aaddress, defined at (memdelay
                 + latchsetup) time units prior to the current simulation time;
       Bmemout = content of RAM at address Baddress, defined at (memdelay
                 + latchsetup) time units prior to the current simulation time;
L5:    Alatchout = Amemout;
       Blatchout = Bmemout;
```

Figure 8.2 Hardware description of a simplified AM2903 in VHDL and pseudo English. (*continued on next page*)

```
L6:    wait for latchdelay ns;
L7:    if ControlA = '0' then MuxAout = Alatchout
                         else MuxAout = DataA;
       if ControlB = '0' then MuxBout = Blatchout
                         else MuxBout = DataB;
L8:    wait for muxdelay ns;

L9:    Carryout, ALUout = operation(MuxAout, MuxBout, function_selector,
                          Carryin);
L10:   wait for aludelay ns;

L11:   Shifterout = ALUout;
       SI00 = ...;
       SI03 = ...;

L12:   final_result = NOW;
L13:   Yout <= Shifterout after shifterdelay ns;
L14:   wait for shifterdelay;

L15:   wait for (WE = '1' or Clock = '1');

L16:   if (WE = '1' and Clock = '1') then
          RAM setup and hold constraints violated;
L17:   else if (WE = '1') then
L18:      wait for Clock = '1';
L19:   if (WE'LAST_EVENT >= memsetup) and (Aaddress'LAST_EVENT >=
                  memsetup) and (NOW - final_result) >= memsetup) then
          RAM setup constraint satisfied;
L20:      RAM gets Shifterout value at Aaddress address;
       end if;
    end if;

L21:   else if (Clock = '1') then
L22:     wait for memhold ns;
L23:     if WE'Stable(memhold) and Aaddress'Stable(memhold) then
            RAM hold constraint satisfied;
       end if;
    end process;
end behavior;
```

Figure 8.2 *(continued)*

view each of the RAM, latches A and B, multiplexors A and B, ALU, and shifter as components or individual VHDL processes and develop their corresponding descriptions similar to that in Figure 8.2. We will concentrate on the most difficult and challenging aspect of developing a hardware description—the timing. The non-timing characteristics such as decoding instructions, elaborating the arithmetic operation of the ALU, etc., are relatively straightforward.

In the description shown in Figure 8.2, the input, output, and inout ports are defined in the entity declaration and they correspond to the diagram of the AM2903 in Figure 8.1. The body of the entity is of type behavior and it contains a single process that is activated whenever the Clock signal incurs a transition. A number of variables are defined to hold the intermediate outputs that are generated by the different units. Thus, Amemout and Bmemout represent the outputs of the RAM through its two output ports, while Alatchout and Blatchout represent the outputs of the two latches A and B. The outputs of the two multiplexors, the ALU, and the shifter are represented through MuxAout, MuxBout, ALUout, and Shifterout, respectively. In addition, the variables memdelay, latchdelay, muxdelay, aludelay, and shifterdelay reflect the propagation delays of the units, while the variables memsetup, memhold, latchsetup, and latchhold encapsulate the timing constraints. The variable, final_result, is intended to store the simulation time at which the shifter's output is generated.

According to the AM2903, when the Clock is 1, the RAM is in read mode and the latches are open. Also, when the Clock is 0, the latches lock on to the input data and RAM permits a write into it. However, the principal activities appear to occur when the Clock signal incurs a transition. Thus, the process uses Clock as its parameter. The body of the process is organized into two major parts: the first part corresponding to the Clock being equal to 0, and the second part corresponding to the Clock being equal to 1.

In the second part of the process, when the Clock changes from a 0 to a 1 and just becomes equal to 1, assuming that WE = 0, i.e. read is enabled, Amemout and Bmemout may be determined using the knowledge of Aaddress and Baddress. However, such determination is pointless since the Aaddress and Baddress may not yet be stable. What is critical are the Aaddress and Baddress values just prior to the Clock incurring a transition from 1 to 0. The same argument applies for the latches. Although they are open while Clock = 1, i.e. the output is identical to the input data, the critically important issue is its input value from the RAM just prior to Clock transition from 1 to 0. Thus, the body of the code for the most part of Clock = 1 is trivial and its key part focuses on just prior to the transition from 1 to 0, as reflected in the code at label L21, which will be discussed subsequently. A relevant timing behavior that occurs close to the transition from 0 to 1 but within the period Clock = 1 is described in the first part of the process at label L20 and will be discussed subsequently.

In the first part of the process, as soon as the Clock = 0, following the transition of the Clock from 1 to 0, we need to know the output values of the two latches. In Figure 8.3, the point "c" marks the occurrence of the transition. To achieve this, we must first find out the input values to the latches at latchsetup ns just prior to the clock transition. This corresponds to the point "b" in Figure 8.3. But the input to the latch corresponds to the data output of the RAM. Thus, in turn, we require to know the value of Aaddress and Baddress at memdelay ns prior to the point "b," which is indicated by point "a." Thus, the temporal distance from "a" to "c" is (latchsetup+memdelay) ns. Clearly, if the Aaddress and Baddress have been unstable prior to point "a" but have been stable precisely at point "a" then the setup condition for the latch is satisfied. The code at label L2 achieves this goal. Next, we have to verify the hold constraints for the latches. The input to the latches must remain stable up to latchhold ns following point "c" which is represented by point "d" in Figure 8.3. This implies that the Aaddress and Baddress must remain stable at (memhold−latchhold) ns prior to the point "c" which is marked by point "e" in Figure 8.3. The code at label L3 achieves this check. It is pointed out that, although we need to check only that the Aaddress and Baddress are stable between the points "a" and "e," for safety, the codes at labels L2 and L3 ensure that Aaddress and Baddress are stable between points "a" and "c." The output of the RAM is evaluated based on the Aaddress and Baddress values as defined at point "a," as shown in the code at label L4. These outputs are contained in the variables Amemout and Bmemout. The output of the latches A and B, namely Alatchout and Blatchout, are identical to Amemout and Bmemout, as confirmed by the code at label L5. Now, the outputs of the latches are not necessarily available at point "c" but at point "f" in Figure 8.3 which is latchdelay ns away from point "c." So, before the values Alatchout and Blatchout may be used by the multiplexors, the behavior description must pause for latchdelay ns, which is achieved by the code at label L6. Note that the code between labels L1 and L5 executes corresponding to the current simulation time equal to the point "c" and the code at label L6 will alter the simulation time to a new value corresponding to which the statements at label L7 and beyond will execute.

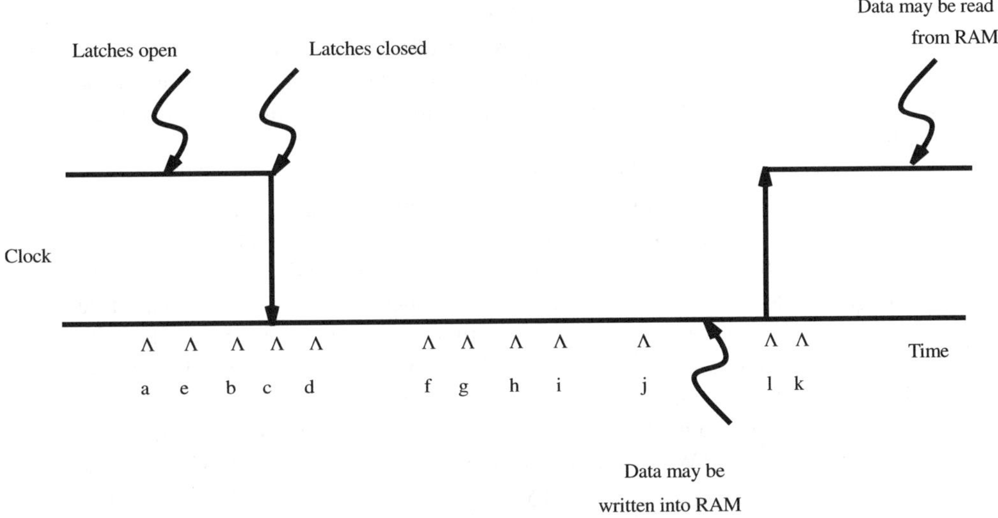

Figure 8.3 Understanding the operation of the AM29203 bit-slice processor.

The code at label L7 corresponds to the point "f" in Figure 8.3 and generates the outputs of the multiplexors, utilizing the control signal and the two input values. Similar to the latches, the outputs of the multiplexors will be available following a propagation delay value given by muxdelay. This is reflected by the code at label L8. Note that there is a reason why the wait statement at label L8 succeeds the statement at label L7. The presence of a wait statement implies that the subsequent code executes at a different simulation time. Since there is the possibility that the values of signals and variables may undergo change at a different simulation time, it is prudent to execute an assignment statement prior to a statement containing a wait clause. The statement at label L9 relates to the ALU and corresponds to the point "g" in Figure 8.3. The codes describing the details of the non-timing aspects of the ALU's behavior are straightforward and are not shown here. The code at label L10 reflects the

ALU's propagation delay. The code at label L11 corresponds to the point "h" and relates to the shifter. The output of the shifter is assigned to the external output port, Yout, following a delay of shifterdelay ns. The execution of the assignment statement at label L13 does not alter the timing of the behavior code, and since there is activity left to model in the process, the wait clause is included at label L14, which when completely executed will correspond to point "i" in Figure 8.3.

At this point, all that remains to do is to write the output of the shifter into the RAM at the Aaddress location, provided WE is 1. Within the VHDL process, we are at point "i," and one may be tempted to write the shifterout value into RAM if WE = 1. Such an action may be incorrect since the value of WE = 1 may not have yet been assigned and the Aaddress may not yet be stable. The difficulty is that, within the VHDL process, we are not aware of where we are precisely in relation to the subsequent transition of the Clock from 0 to 1, denoted by point "l" in Figure 8.3. We come up with the following plan. We will suspend execution and wait until either WE is observed equal to 1 or the Clock incurs a transition and becomes equal to 1. The code at label L15 achieves this objective. The need to check for Clock = 1 is important for the following reason. If WE does not equal 1, i.e. the designer does not care to write the shifterout value into RAM, within the current interval of Clock = 0, the suspension of the process will continue until WE becomes equal to 1 in some future Clock pulse. Even though the VHDL process has required that changes to Clock be monitored, the suspension initiated by the statement at label L15 will predominate. The simulator will not trigger the execution of the process corresponding to the subsequent transition of the Clock. Clearly, this will lead to an error in that the hardware description will no longer correspond to the actual hardware.

There is an important lesson here, relative to the semantics of HDL constructs. For any wait clause, utilized within a process statement, the timing condition that is specified, including the duration, must be consistent with the high-level objectives of the process. The presence of conflicts will very likely cause errors and it is probably impossible for a compiler to detect all possible inconsistencies, such as the one discussed earlier.

Where WE changes to 1 at the same time the Clock incurs a transition from 0 to 1, the setup and hold constraints of the RAM will certainly be violated as encapsulated by the code at label L16. In contrast, should WE equal 1 before the Clock transitions from 0 to 1, as encapsulated by the code at label L17, then there is a chance that the shifterout value may be written into the RAM. While we may be tempted to write into the RAM at this point, we must remember that the Aaddress may not yet be stable. The eventual RAM location to which the shifterout value will be written will be governed by the Aaddress value just prior to the Clock transition from 0 to 1. In addition, we have to verify that the setup and hold constraints for the RAM are satisfied. To achieve this goal, the code at label L18 executes a wait clause until the Clock just becomes equal to 1. Then, the code at label L19 verifies that WE has remained at 1 at point "j" in Figure 8.3 which is memsetup ns away from the point "l." The code at label L19 also verifies that both the Aaddress has been stable at point "j" and the shifterout value has been available at point "j." If the setup constraints are satisfied, the code at label L20 ensures that the Aaddress defined at point "j" is used to write into the RAM. The statement at label L20 concludes the first part of the VHDL process. However, we still have not verified the hold time for the RAM and this is accomplished in the second part of the VHDL process, starting at label L21. It is noted that the codes at label L18 and label L21 both correspond to events at point "l" in Figure 8.3 and, in theory, may be executed by the simulator in any order. However, in either case, the results must be identical. The code at label L22 pauses the execution of the VHDL process for memhold ns, allowing the simulation time to progress to point "k" in Figure 8.3. At this point, the WE and Aaddress are verified to be stable between the points "l" and "k." Where the hold constraints are satisfied, the write into the RAM is valid. Otherwise, an error is signaled.

The above exercise reflects only one way to develop a hardware description. A designer may synthesize a very different, yet equivalent description, and there is virtually no end to human creativity.

8.2 ASYNCHRONOUS VME BUS—PROCESSOR–MEMORY INTERACTION

Synchronous digital systems are relatively easy to design, fabricate, and debug. However, by definition, the width of the synchronous clock pulse must exceed the maximum of all possible delays due to the combinational sub-parts and the delays are generally dependent on the inputs. As a result, synchronous systems are generally unable to offer the highest performance. In contrast, asynchronous systems exploit the maximal concurrency, inherent in the hardware design, to yield the highest achievable performance. There are other advantages of asynchronous designs. Consider the following anecdote. At Bell Labs, the initial design of the 32-bit CMOS processor, Bellmac32, utilized the synchronous design philosophy. When the chip was tested, following fabrication, it experienced failure. By nature, CMOS circuitry draws negligible current at either one of the stable states, but consumes significant power during a state transition. The synchronous design style enabled all transitions at the clock edges and, as a result, the processor's current consumption increased dramatically at every clock transition, causing both the power supply and the processor to fail. The Bellmac32 was subsequently redesigned into modules that maintained their own clocks but interacted asynchronously. Despite its inherent advantages, the principal difficulty with asynchronous designs is the difficulty in designing, fabricating, and debugging. Clearly, the task of developing hardware descriptions for such systems is not likely to be easy.

The popular VME [71] bus permits asynchronous reading and writing operations between a processor and a memory module. It is pointed out that the Motorola M680XX processors also exploit asynchronous memory interactions. Figure 8.4 presents a schematic diagram consisting of a CPU, a RAM module, and a bus structure that is organized into a control bus, a data bus, and an address bus. For this case study, we will focus on the key elements of the control bus, namely the address strobe (AS), the read/write (R/W) control, the data strobe (DS), and the data acknowledge (DTACK) signal. Each of the CPU and RAM units may be internally synchronous or asynchronous designs. Assume that the CPU is a

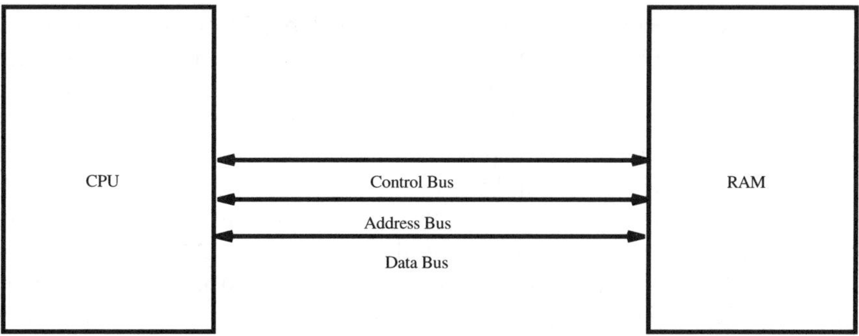

Figure 8.4 Asynchronous VME bus.

synchronous design and driven by a 10 MHz clock which yields a clock time period of 100 ns. The RAM unit is assumed asynchronous. The operation of the bus is organized into read and write cycles and the corresponding timing diagrams are shown in Figures 8.5 and 8.6.

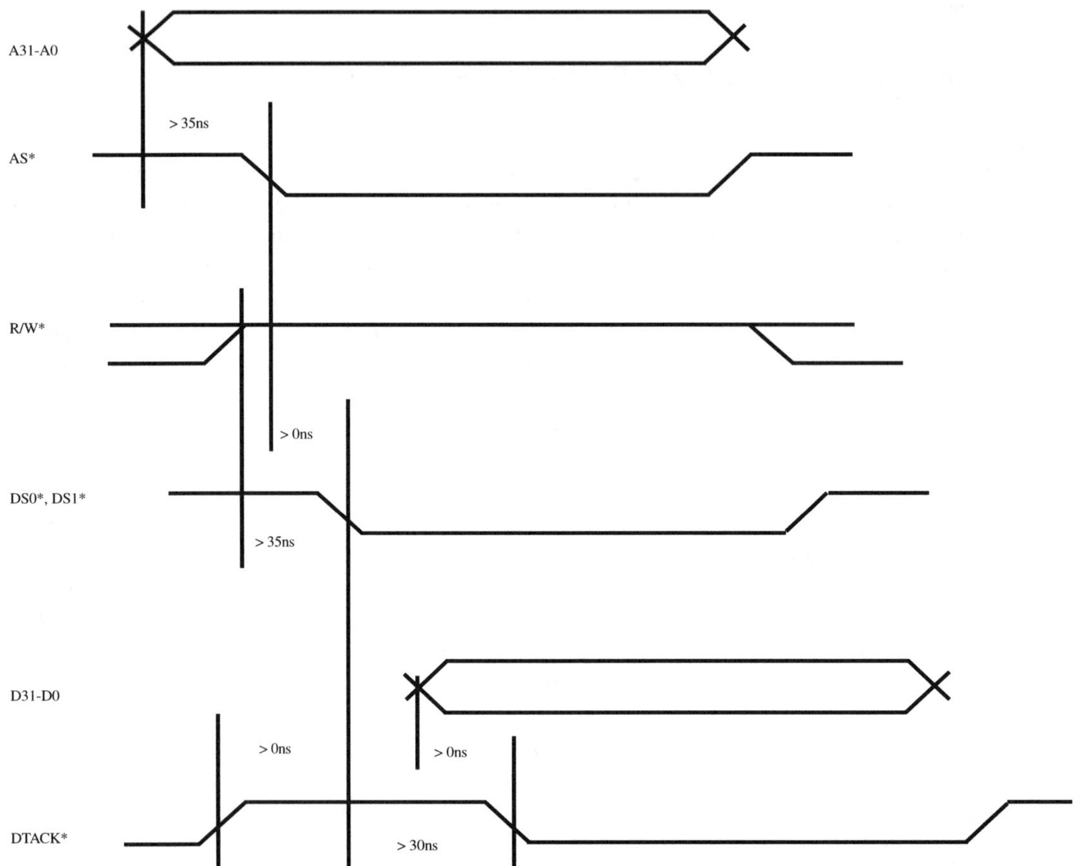

Figure 8.5 Timing diagram for asynchronous memory read.

The read cycle may be described as follows. First, the processor asserts the address and then sets R/W = 1. The AS signal is asserted at least 35 ns after the address. The DS is the last signal asserted, i.e. DS is set to 0, by the CPU only after it detects that the DTACK signal is 1. The DTACK signal is driven by the RAM module. The DS signal must succeed the AS signal by at least 0 ns. At this point, the processor awaits the response from the RAM module and continues to scan for changes in the DTACK signal from its current value of 1. After it has received the address and R/W control, the RAM module prepares to access the appropriate contents. After intercepting the AS followed by the DS, the RAM module first asserts the data on the databus and then asserts a 0 onto DTACK at least 0 ns after it asserts the data and at least 30 ns after it had intercepted the DS. Upon recognizing the change in DTACK to 0, the processor reads the data off of the databus and deactivates DS, AS, and the address. Upon

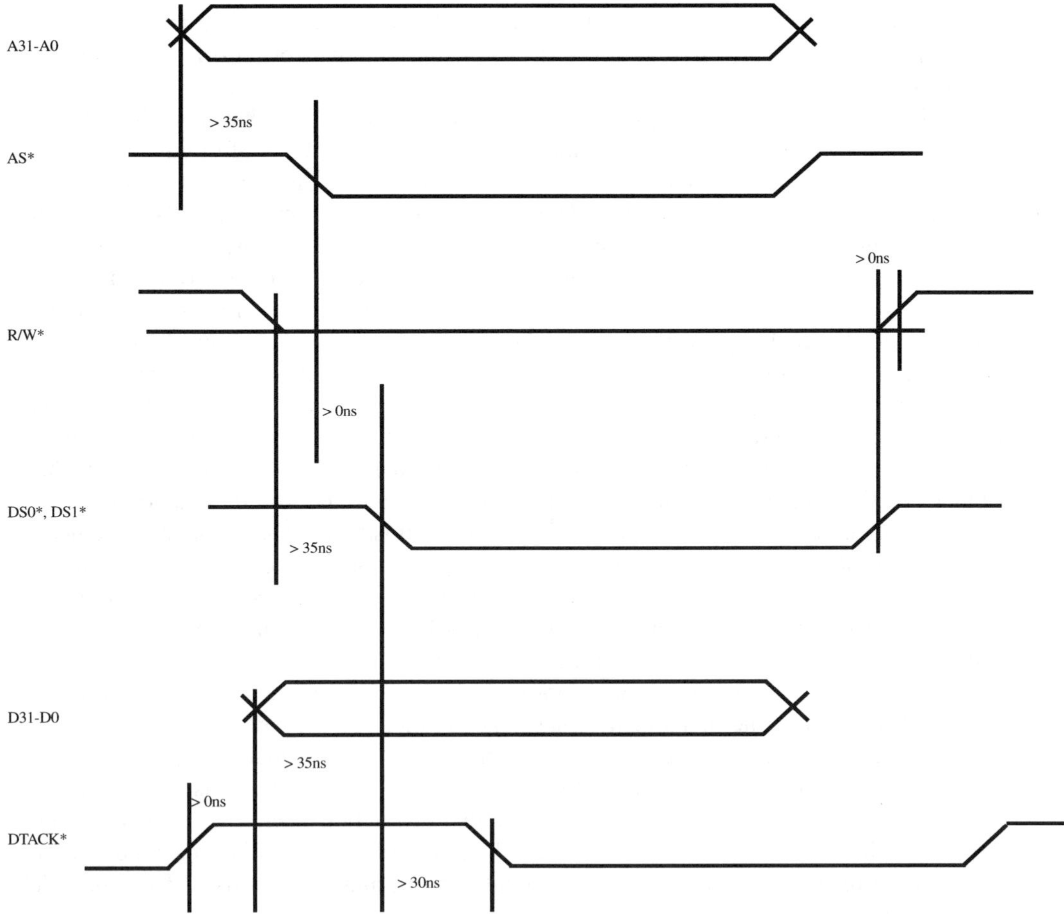

Figure 8.6 Timing diagram for asynchronous memory write.

detecting that DS is at 1, the RAM module releases the databus and deactivates the DTACK signal, i.e. resets it to 1.

In the write cycle, the processor asserts the address and then sets R/W = 0. The AS signal is asserted at least 35 ns after the address. The CPU asserts the data onto the databus when it detects DTACK = 1. The CPU asserts a 0 onto the DS signal at least 35 ns after the R/W control and only after it has detected DTACK = 1. At this point, the CPU awaits the response from the RAM module and examines the DTACK for a transition to 0. After it has received the address, R/W control, and data, the RAM module prepares to write the data into the appropriate location. After intercepting the AS followed by the DS, it writes the data and forces a 0 onto the DTACK signal at least 30 ns after it had detected DS. Upon detecting DTACK = 0, the CPU withdraws that data off of the databus, address, and AS, and the DS is reset to 1. Upon detecting that DS is at 1, the RAM module deactivates the DTACK signal, i.e. resets it to 1.

Thus, the CPU drives the address, AS, R/W control, and the DS which are read by the RAM unit. In turn, the RAM module drives the DTACK signal which the CPU reads. While the CPU drives the databus in the write cycle, the RAM unit drives the databus in the read cycle. However, under no circumstances is the databus simultaneously driven by both CPU and RAM.

The VME bus system represents asynchronous interactions between the CPU and RAM, both of which are independent units with their unique notions of timing. To represent their behavior accurately, the CPU and RAM must be expressed through entities, as discussed in Chapter 3. Therefore, each of the CPU and RAM descriptions must be self-contained and contain no shared variables. The notion of universal time is nanoseconds, reflecting the lowest common denominator of timing among the CPU and RAM. The hardware description interprets the specification, ">0 ns" as "≥ 1 ns." Internally, the CPU is a synchronous machine with its clock represented by the Clock signal. The RAM description in Figure 8.7 reflects an asynchronous design. The CPU drives the address, AS, R/W, and DS signals and these are its output ports. The RAM drives the DTACK signal and is its output port. The data signal is declared bidirectional since the CPU drives it during the write operation while the RAM drives it during the read operation. As in VHDL, the description in Figure 8.7 includes a bus resolution function, "resolve", for the databus to yield the resulting value in the event that it is inadvertently driven by more than one source. It is noted that, while the specification of the bus resolution function constitutes good practice, it should not be misinterpreted as a solution to an underlying problem of bus conflicts. The occurrence of a bus conflict, in general, reflects a serious flaw in the design—both logical and timing—of the bus and the interacting entities, and it must be carefully analyzed.

The CPU is a high-performance, complex unit and the reading and writing of memory over the VME bus is one of its many tasks. The architecture body of the CPU is described through a process that is activated corresponding to every Clock cycle. The variable "m_state" reflects the current, principal state of the CPU, i.e. whether it is reading, writing, executing register operations, fetching and decoding instructions, etc. The variable "s_state" represents the sub-state within the principal state. As a sequential machine, the key specification of the CPU must include its input, output, and inout ports and the state. The state is expressed through a combination of the "m_state" and "s_state" variables which may be viewed, from the perspective of hardware, as "state containing registers" within the CPU. These registers are initialized with the initial state, prior to initiating the CPU for execution, they may be modified during the operation of the CPU, and are continuous between the successive executions of the process, driven by the clock pulses. The registers are not re-initialized during every successive execution of the process. Given the speed and cost of the CPU and the consequent desire to maximize its usage, the goal here is to engage the CPU in performing local activities, such as register to register operations, while simultaneously waiting for an asynchronous memory operation to complete.

The description presented in Figure 8.7 is illegal in VHDL since the grammar [33] disallows the inclusion of variables within an architecture body preceding the specification of a process. VHDL permits shared variables in entity declarations and only signals and shared variables in the corresponding architecture body. The probable reason for this restriction is that, logically, the structural and dataflow styles only contain instantiations and interconnection signals between the instances. Furthermore, a process in the behavioral style is permitted to define variables within itself. Thus, VHDL fails to realize that it may be necessary, for a sequential machine, to define and use variables, outside of a process, to encapsulate the state of the underlying machine and that such state variables may be emulated in hardware through registers. These variables are associated with a specific process and not shared between multiple processes. While alternate schemes in VHDL may circumvent this problem, the

```
Package fourpack is
  type fourval is (X,L,H,Z);
  type fourvalvector is array (natural range <>) of fourval;
  type mainstate is (read, write, execute_reg_instruction, other);
  type substrate is (s0, s1, s2, s3, s4, s5, s6, s7);

  function resolve (...) return fourvalvector;
end fourpack;

Use work.fourpack.all

Entity CPU is
  port (
    Clock: in bit;
    DTACK: in bit;
    address: out bit_vector (31 downto 0);
    AS, R/W, DS: out bit;
    data: inout fourvalvector;
  );
end CPU;

Architecture Behavior of CPU is
  variable m_state: mainstate := other;
           s_state: substrate := s0;
  begin
   process ;(Clock)
     begin
       case m_state is
         when other =>
           - fetch instructions, decode, sets m_state to read or write, etc.
         when execute_reg_instruction =>
           execute register, i.e. non-memory, instruction;
           if (s_state = 's1' or 's2') then m_state = read;
           else if (s_state = 's4' or 's5') then m_state = write;
L1:      when read =>
           case s_state is
             when s0 =>
               address <= address-lines-from-instruction after 1 ns;
               R/W <= 1 after 2 ns;
               AS <= 0 after 36 ns;
               wait for 36 ns;
               if DTACK = '1' then DS <= 0 after 1 ns; s_state = s2;
                               else s_state = s1; m_state =
                               execute_reg_instruction;
               end if;
```

Figure 8.7 Hardware descriptions of a CUP–RAM–VME bus system in VHDL and pseudo English. (*continued on next page*)

```
L2:       when s1 =>
            if DTACK = '1' then DS <= 0 after 1 ns; s_state =s";
                               else s_state = s1; m_state =
                               execute_reg_instruction;
L3:       when s2 =>
            if DTACK = '1' then
                   s_state =s2;
                   m_state = execute_reg_instruction;
            else if DTACK = '0' then
                   accumulator or register <= data after 1 ns;
                   address <= inactive after 1 ns;
                   R/W <= inactive after 1 ns;
                   AS <= 1 after 1 ns;
                   DS <= 1 after 2 ns;
                   m_state = other;
                   s_state = s0;
            end if;
          end case;
L4:     when write =>
          case s_state is
            when s3 =>
              address <= address-lines-from-instruction after 1 ns;
              R/W <= 1 after 2 ns;
L5:           AS <= 0 after 36 ns;
              if DTACK = '1' then
                                data <= data from accumulator after 1 ns;
                                DS <= 0 after 36 ns;
                                s_state = s5;
                              else
                                s_state = s4;
                                m_state = execute_reg_instruction;
              end if;
L6:         when s4 =>
              if DTACK = '1' then
                                data <= data from accumulator after 1 ns;
                                DS <= 0 after 36 ns;
                                s_state =s5;
                              else
                                s_state =s4;
                                m_state = execute_reg_instruction;
              end if;
L7:         when s5 =>
              if DTACK = '1' then
                     s_state =s5;
                     m_state = execute_reg_instruction;
              else if DTACK = '0' then
```

Figure 8.7 *(continued)*

```
                    address <= inactive after 1 ns;
                    DS <= 1 after 1 ns;
                    AS <= 1 after 1 ns;
                    R/W <= inactive after 2 ns;
                    data <= inactive after 1 ns;
                    m_state = other;
                    s_state = s3;
    end case;
  end process;
...

Entity RAM is
  port (
      address: in bit_vector (31 downto 0);
      AS, R/W, DS: in bit;
      DTACK: out bit;
      data: inout fourvalvector;
  );
      constant memreaddelay: int = ...;
      constant memwritedelay: int = ...;
end RAM;

Architecture Behavior of RAM is
  begin
    process (AS)
      variable addr, ram_data: ...;
               l_state: mainstate;
      begin
        if (AS = '0') then
          addr = address;
          if R/W = 1 then l_state = read;
                     else l_state = write;
          end if;
          case l_state is
            when read =>
              DTACK <= 1 after 1 ns;
              wait until DS = '0';
              data <= memory contents of addr after memreaddelay ns;
              wait for larger of (30, memreaddelay) ns;
              DTACK <= 0 after 1 ns;
              wait until DS = '1';
              release data;
              DTACK <= 1 after 1 ns;
            when write =>
              DTACK <= 1 after 1 ns;
              wait until DS ='0';
```

Figure 8.7 *(continued)*

```
              ram_data = data;
               wait for larger (30, memwritedelay) ns;
               DTACK <= 0 after 1 ns;
               wait until DS = '1';
               DTACK <= 1 after 1 ns;
      end case;
    end if;
  end process;
end Behavior;

Entity System is
  port (Clock1: in bit);
end System;

Architecture Structural of System is
  component CPU
      port (Clock: in bit;
            DTACK: in bit; address: out bit_vector (31 downto 0);
            AS, R/W, DS: out bit;
            data: inout fourvalvector;
  end component
  component RAM
      port (address: in bit_vector (31 downto 0);
            AS, R/W, DS: in bit;
            DTACK: out bit;
            data: inout fourvalvector;
  end component;
  signal DTACK, AS, R/W, DS: bit;
  signal address: bit_vector (31 downto 0);
  signal data_bus : resolve fourvalvector := ...;
  begin
L8: U1: CPU port map (Clock1, DTACK, address, AS, R/W, DS, data_bus);
L9: U2: RAM port map (address, AS, R/W, DS, DTACK, data_bus);
  end Structural;
```

Figure 8.7 *(continued)*

approach to developing hardware descriptions presented in Figure 8.7 is legitimate and logical.

In the architecture body for the CPU in Figure 8.7, the principal states include "other", that represents instruction fetch, decode, etc., "execute_reg_instruction" that refers to local operations which do not require memory accesses, "read", and "write." Execution during the other state may set the subsequent state to read or write. During a read or write operation, if an asynchronous waiting is encountered, the "m_state" is set to "execute_reg_instruction" so that during the subsequent clock pulse, control is returned to executing the local operations. Following the completion of the local operation, the state is set to read or write, as appropriate, so that during the subsequent clock pulse, the status of the previously initiated asynchronous operation may be examined.

The read operation is represented by the code statement at label L1. The sub-states correspond to s0, s1, and s2. In sub-state s0, first the address is asserted 1 ns following the current simulation time, second a 1 is asserted on the R/W signal 1 ns beyond the assertion of the address, and then a 0 is asserted onto the AS 35 ns following the assertion of the address. The CPU must wait for at least 35 ns following the assertion of R/W. Therefore, the description contains a wait for $(35 + 1) = 36$ ns. It is noted that the CPU is assumed rated for 10 MHz, giving a width of 100 ns for the Clock, and so the duration of 36 ns for the wait statement is acceptable. If, however, the clock pulse had been 10 ns wide, the duration of wait would be considered too large and the efficiency requirements would have demanded a different strategy. After the completion of the wait, however, the CPU cannot yet assert the DS signal until the DTACK signal is high. If DTACK is high, the CPU asserts the DS 1 ns beyond the current simulation time. It now waits for the RAM module to respond with a 0 value at DTACK and so sets "s_state" to s2. If DTACK is not yet high, since DTACK is controlled by the asynchronous RAM unit, the description is suspended and the m_state is set to "execute_reg_instruction" and "s_state" is set to s1. When control returns to the read operation in a future clock pulse, execution resumes at s1 (label L2) if DTACK was not high earlier or s2 (label L3) if the CPU had successfully asserted DS and is waiting for the RAM to respond. The control jumps beyond label L2 when the CPU observes a high value at DTACK and is able to successfully assert the DS. If DTACK continues to be at 1, the code at label L3 executes, the control exits the description, and execution is resumed at L3 again in a future clock pulse. Otherwise, control jumps successfully forward and the CPU description reads the data from the databus. Then, the address, R/W, AS, and DS are deactivated. The read operation is complete for the CPU and it resets s_state to s0 and m_state to other.

The write operation is represented by the code at label L4. The operation consists of three sub-states: s3, s4, and s5. In sub-state s3, first the CPU asserts the address lines 1 ns following the current simulation time. Second, a 0 is asserted on R/W 1 ns after the address lines are asserted. The CPU asserts a 0 on AS 35 ns following the assertion on the address lines. The CPU then examines the state of DTACK. If the DTACK is at 1, the CPU first asserts the data on the databus 1 ns following the current simulation time and then the DS signal 36 ns following the assertion of data. The CPU must wait for the RAM module to lower the DTACK signal asynchronously and it sets "s_state" to s5 and suspends execution. Where DTACK is not at 1, the CPU cannot assert the data and DS. It must wait until the RAM module resets DTACK to 1 asynchronously which is realized by setting "s_state" to s4. The hardware description does not contain a "wait for 35 ns" statement following the code at label L5, unlike in the read counterpart, since the CPU is eager to assert the data and allow the RAM sufficient time to read it. The state s4 at label L6 corresponds to the situation that the DTACK is not yet 1 and the CPU has not successfully asserted the data and DS. Under state s4, the CPU examines the state of DTACK and proceeds forward to state s5 only after it detects a 1 at DTACK. Otherwise, it sets "s_state" to s4 and "m_state" to execute_reg_instruction so that the CPU may focus on executing local operations in the subsequent clock cycle. The control is returned to the write operation following the completion of the local operation in the subsequent clock cycle. Under state s5, at label L7, the CPU awaits a 0 at DTACK and remain at s5 until the CPU detects a 0 at DTACK. Then, it deactivates the address, DS, AS, and data, and finally the R/W signal. The "m_state" is set to other to allow the CPU to continue with its operations. The architecture body of the CPU entity is complete.

For the entity RAM, representing a memory module, the input ports are address, AS, R/W, and DS. The data represents a bus that it drives under VME read operation and DTACK is an output port. The read access and write recovery delays are represented through the two constants memreaddelay and memwritedelay, respectively. Unlike the CPU, the RAM module lacks a clock signal since it is focused solely on the mutually exclusive read and write operations. Under either read or write operations, when the RAM is idly awaiting a response from the CPU asynchronously, there is no need to suspend the current operation and attend to a different activity. Once initiated, a read or write operation must execute through completion. Thus, the RAM contains only two states: read and write. The body of the architecture corresponding to the entity RAM is expressed as a process that is activated corresponding to changes in AS. A new read or write cycle is implied only when AS incurs a transition from 1 to 0.

First, the RAM module reads the address and stores it into its local register variable, addr. It then reads the R/W signal. The reader may recall that the CPU asserts the address and R/W signal long before asserting the AS. Utilizing the value at R/W, the RAM determines whether the current VME cycle corresponds to read or write.

Under read operation, the RAM module asserts a 1 at DTACK. This may be redundant since the RAM module asserts a 1 at DTACK following the completion of the previous read and write operations. It then waits until it detects a 0 at DS which is driven asynchronously by the CPU. At this point, the RAM module transfers the contents of the memory, corresponding to addr, onto the data port following a delay of memreaddelay ns. The RAM module executes a wait for the longer of memreaddelay ns and 30 ns, where the latter is recommended by the VME bus design. Next, the RAM module asserts a 0 at DTACK 1 ns following the current simulation time. The description then awaits for the CPU to reset DS to 1, asynchronously. Upon detecting a 1 at DS, the RAM releases the data port and raises DTACK to inactive, 1 ns following the current simulation time.

Under write operation, first the RAM description asserts a 1 at DTACK. As with the read operation, this assertion may be redundant. It then waits until it detects a 0 at DS which is driven asynchronously by the CPU. At this point, the RAM reads the data into its local register variable ram_data and executes a wait for the longer of memwritedelay ns or 30 ns where the latter is recommended by the VME bus design. This ensures the safe storage of ram-data into the RAM at the location specified by addr. The RAM then asserts a 0 at DTACK 1 ns following the current simulation time and then awaits for the CPU to reset DS, asynchronously, to 1. When it detects a 1 at DS, it resets the DTACK to inactive, 1 ns after the current simulation time. The description for the RAM is complete.

The complete VME system is represented in Figure 8.7 through the entity "system", and its corresponding architecture body labeled "Structural". The architecture body contains the specifications of CPU and RAM as components and are instantiated as U1 and U2, respectively, at labels L8 and L9. The entity, "System", is characterized by a single input, Clock1, which constitutes the clock of the CPU. The address, AS, R/W, data, DS, and DTACK are internal to "System" and are defined as signals.

8.3 MODELING A VME BUS SYSTEM WITH MULTIPLE RAM MODULES

Figure 8.8 represents a VME bus system with a single CPU and three RAM modules—RAM1, RAM2, and RAM3—with mutually exclusive address ranges. Thus, when the CPU asserts an address, only one of RAM1, RAM2, and RAM3 will respond. Similar to the description in Figure 8.7 in the previous Section 8.2, the address, AS, R/W, and DS are driven by the CPU and constitute output ports of the CPU. Thus, they will serve as input ports of RAM1, RAM2, and RAM3. Also, the data lines will be viewed as inout ports within each of

CPU, RAM1, RAM2, and RAM3. In contrast to the hardware description in Figure 8.7, DTACK may be driven exclusively by RAM1, RAM2, or RAM3 and it must be expressed as inout port within RAM1, RAM2, and RAM3. The bus resolution function will yield a 0 or low whenever any of the signals is a 0 and a 1 or high when all of the signals are 1. The CPU will view DTACK as an input port.

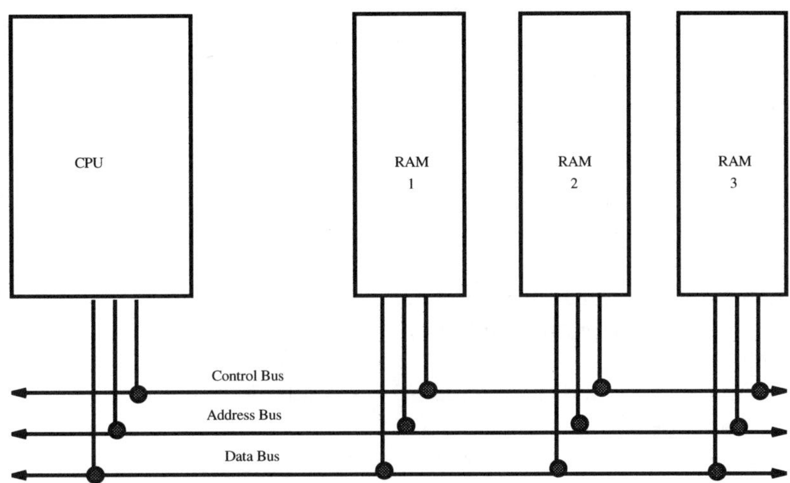

Figure 8.8 A VME bus with a CPU and multiple RAM modules.

Figure 8.9 presents the hardware description of the VME bus with the CPU and three RAM modules. The description for the CPU is identical to that in Figure 8.7 and is not shown here. The description for each of the three RAM modules is encapsulated through entity RAMX and it differs slightly from that in Figure 8.7. The entity RAMX includes a low_address and high_address value that specifies the valid address range. Also, the DTACK is defined as an inout port. The body of the architecture corresponding to RAMX is similar to that in Figure 8.7 except that at label L1, the description verifies whether the input address lies within the valid range specified through low_address and high_address. If the input address lies outside the valid range, the RAM module in question is not selected, it does not assert a 0 on the DTACK signal, and it waits for the subsequent transition in AS from 1 to 0.

The extended VME system is represented by the entity, SystemX, which defines the CPU and RAMX entities as components. It includes an instance U1 corresponding to the CPU and instances RAM1, RAM2, and RAM3 that emulate the three RAM modules which differ only in their valid addresses.

```
Entity RAMX is
  generic (low_address, high_address: integer);
  port (
      address: in bit_vector (31 downto 0);
      AS, R/W, DS: in bit;
      DTACK: inout fourval;
      data: inout fourvalvector;
  );
      constant memreaddelay: int = ...;
      constant memwritedelay: int = ...;
end RAMX;

Architecture Behavior of RAMX is
  begin
    process (AS)
      variable addr, ram_data: ...;
               l_state: mainstate;
      begin
L1:    if (AS = '0' and low_address =< address =< high_address) then
        addr = address;
        if R/W = 1 then l_state = read;
                   else l_state = write;
        end if;
        case l_state is
          when read =>
            DTACK <= 1 after 1 ns;
            wait until DS = '0';
            data <= memory contents of addr after memreaddelay ns;
            wait for larger of (30, memreaddelay) ns;
            DTACK <= 0 after 1 ns;
            wait until DS = '1';
            release data;
            DTACK <= 1 after 1 ns;
          when write =>
            DTACK <= 1 after 1 ns;
            wait until DS ='0';
            ram_data = data;
            wait for larger (30, memwritedelay) ns;
            DTACK <= 0 after 1 ns;
            wait until DS = '1';
            DTACK <= 1 after 1 ns;
      end case;
    end if;
  end process;
end Behavior;
```

Figure 8.9 Hardware descriptions of a VME bus with multiple RAM modules in VHDL and pseudo English.

```
Entity SystemX is
  port (Clock1: in bit);
end SystemX;

Architecture Structural of SystemX is
  component CPU
      port (Clock: in bit;
            DTACK: in bit; address: out bit_vector (31 downto 0);
            AS, R/W, DS: out bit;
            data: inout fourvalvector;
  end component
  component RAMX
      generic (low_address: integer := ...; high_address: integer := ...);
      port (address: in bit_vector (31 downto 0);
            AS, R/W, DS: in bit;
            DTACK: inout bit;
            data: inout fourvalvector;
end component;
signal DTACK, AS, R/W, DS: bit;
signal address: bit_vector (31 downto 0);
signal data_bus : resolve fourvalvector := ...;
begin
  U1: CPU port map (Clock1, DTACK, address, AS, R/W, DS, data_bus);
  RAM1: RAMX
      generic map (..., ...);
      port map (address, AS, R/W, DS, DTACK, data_bus);
  RAM2: RAMX
      generic map (..., ...);
      port map (address, AS, R/W, DS, DTACK, data_bus);
  RAM3: RAMX
      generic map (..., ...);
      port map (address, AS, R/W, DS, DTACK, data_bus);
end Structural;
```

Figure 8.9 *(continued)*

9

Simulation Algorithms for Concurrent Execution of HDLs on Loosely-coupled Parallel Processors

Simulation algorithms play a key role in enabling the distributed execution of hardware descriptions on actual parallel processors and this chapter presents two successful algorithms. The problem of distributed hardware description execution is particularly unique in the literature in that, unlike queueing networks that require a simple constant delay for every unit, each hardware module generally requires multiple, input dependent delays. The input dependent delays cause the problem of inconsistent events which must be detected and preempted for accuracy. Of the two algorithms presented here, the first is simpler and the second one complex and more efficient.

9.1 REVIEW OF THE LITERATURE

The principal objective [55] of the DoD VHDL [33] hardware description language has been to allow digital hardware designs to be described, simulated, and validated accurately and efficiently in the computer, prior to building a prototype. In addition, an important objective of VHDL [55] has been to allow the execution of VHDL processes in parallel on multi-computers and multiprocessors. To date, the execution of VHDL models continues to be limited to uniprocessors and is excruciatingly slow. The literature relative to deadlock-free, null-message-free, and conservative algorithms for execution of VHDL models is nil. To achieve concurrent and efficient simulation of VHDL descriptions requires an efficient algorithm for distributed, discrete event simulation.

The literature reports three principal techniques for distributed discrete event simulation, namely synchronous, rollback, and asynchronous. The "time first evaluation" algorithm, proposed by Ishiura, Yasuura, and Yajima [72], is potentially inappropriate for distributed execution on parallel processors. The reasons are as follows. In this approach, communication between the gates, located on different processors, is reduced. However, only a very few gates may execute at any time instant while others await their input waveforms. Consequently, much of the execution will be sequential, yielding very low concurrency on parallel processors. The most significant drawback of this approach is with feedback loops [72], where the input signal waveform to a gate at a time instant is dependent on the waveform at a previous time instant as well as the behavior of other gates. This, in turn, severely limits the concurrency on a parallel processor. Fujimoto [73] reports a state-of-the-art survey on the execution of simulation models on parallel processors. In the synchronous approach [74], a processor is designated as a centralized controller (master processor), and it is responsible for allocating all other entities to the processors (slaves) of the parallel processing system and for initiating their executions. In addition, the controller resynchronizes all processors at the end of every activity and maintains the simulation clock, i.e. global simulation time. Chamberlain and Franklin [75] describe a variation of the synchronous approach wherein a distributed algorithm is used to execute the tasks associated with the master processor. While this approach permits the concurrent execution of entities corresponding to two or more events at the simulation time given by $t = t_1$, its limitations include the following. The processors must resynchronize at the end of every activity even in the absence of data dependency, and an uncertainty is associated with the completion of message communication at the end of an activity. Consequently, the synchronous approach is unable to utilize the maximal inherent parallelism. Soule and Blank [76] present an implementation of the synchronous algorithm. Although they eliminate the problem of the centralized queue by placing a queue on every processor for round robin event insertions, they report limited parallelism. In another variation of the synchronous approach, two events with different timestamps may be executed concurrently as long as they are within a small delta of each other.

In the rollback mechanism [77], also known as virtual-time algorithm or optimistic asynchronous algorithm, the states of individual models need to be saved from time to time, as events occur, and a list of messages received must be saved [78] such that the simulation system may be permitted to rollback to its previous state in the event of an error caused by processing messages out of order. In the absence of information regarding a signal at an input port, a model assumes that the signal value at that input port has remained unchanged and the results of execution based on the assumption are propagated to subsequent models. If a subsequent message is received by the component that contradicts the previous assumption, new results in the form of anti-messages are propagated to subsequent models. The theoretical limitations of the rollback mechanism include the significant storage requirement [79] and the uncertainty constituted by a combination of messages and anti-messages propagating throughout the simulation system. In general, the actual performance of virtual-time algorithm is unpredictable and will depend on the example circuit being simulated, the partitioning and allocation of the components of the circuit to processors, and relative speeds of the processors and message communications. In extreme cases, domino rollbacks may occur [79], devastating the progress of simulation, and processes may spend more time on rollbacks than useful computation [78]. Sissler [80] notes that a concurrent simulator using virtual-time algorithm may impose one to two orders of magnitude of processing overhead. Chawla *et al.* [81] propose the design of a 256-node parallel computer that utilizes virtual-time algorithm to synchronize the execution of the concurrent processes. While their approach requires the computation of global simulation time which, in turn, requires all processes to synchronize, they acknowledge that the effects of rollback on the total simulation execution time is difficult to predict. A number of improvements to the virtual-time algorithm and implementations have been reported in the literature [79, 82–88]. Steinman [87] defines an event horizon as the timestamp of the earliest new event generated within a cycle and utilizes it to process events. This approach combines elements of the virtual-time and synchronous algorithms and estimates speedups of 32 on the 64-processor Caltech/JPL Mark III Hypercube, relative to a uniprocessor. West [88] proposed the use of lazy reevaluation but its implementation [73] was observed to be excessively expensive. Akyildiz *et al.* [83] propose a new, "cancelback", protocol to reclaim storage when the rollback system runs out of memory and develop an analytic model. The model is validated assuming that the timestamps of the processed uncommitted events in each processor follow a Poisson distribution and that the distribution of timestamps for messages and anti-messages are independent, which may not be true in reality. Nicol and Heidelberger [84] describe the use of uniformization for virtual-time algorithm to reduce the frequency of state saving and eliminate the Global Virtual Time (GVT) computation. They note good speedups of queueing network simulations. Som and Sargent [85] propose a new criterion to assign logical processes to processors for virtual-time algorithm wherein logical processes which may experience rollbacks caused by a common logical process are assigned to the same processor. This approach is limited in that rollbacks are unpredictable, i.e. their a priori prediction is difficult. During a simulation interval of 30,000 units of simulated clock ticks for example models, a total of 13,656 messages needed to be rolled back. Wen and Yelick [86] report good speedups for transistor-level timing simulations but their failure to include the cost of rollbacks based on the presence of an "oracle" to predict the presence of events is unrealistic. Furthermore, the reported parallelisms for benchmark combinational circuits is unnecessary since the conservative asynchronous algorithm suffers from deadlock only for circuits with feedback loops. Briner *et al.* [79] report a speedup of 23 on a 32-processor system over a good sequential algorithm, through the use of lazy cancelation and virtual-time algorithm at the transistor level. They note that, while lower bounding window sizes reduce speedup, higher bounding window sizes generate speedup but suffer inefficiencies from excess event handling, rollback, and wasted model evaluations. It is noted that the simulation frequently proceeds to a large simulation time and then rolls back to an earlier time value, thereby requiring large amounts of storage memory and producing limited results. Krishnaswamy and Banerjee [89] utilize a parallel time warp simulator for VHDL but lack discussion on the issue of inconsistent event preemption. Furthermore, the global virtual time management—a key component of their parallel algorithm—utilizes a centralized scheme, a clear inconsistency.

Unlike the synchronous approach, the asynchronous algorithm has the potential of utilizing the maximum parallelism inherent in the simulation system. Bailey [90] reports that, for logic-level circuit simulation, the execution time of conservative asynchronous strategy is a lower bound over the synchronous strategy for variable delay, assuming an unlimited number of processors and without taking into account the overheads. In this approach, a model may execute as soon as it receives its signal transitions at its input ports, subject to certain principles that are detailed subsequently, that guarantee accuracy. In contrast to the rollback approach, the asynchronous algorithm is conservative and always generates accurate results. The term "accurate" refers to preciseness and logical correctness. However, the asynchronous algorithm has the potential of executing into a deadlock. The literature contains a number of approaches—Chandy and Misra [91], Chandy *et al.* [92], and Peacock *et al.* [93]—to asynchronous distributed discrete event simulation in the context of queueing networks. The approach in [91] suggests a deadlock avoidance method through the propagation of "null" messages whenever a model executes but fails to produce a new output transition. Thus, this approach violates the basic premise of discrete event simulation that a message is sent only when it implies a change compared to its previous value. Conceivably, this technique leads to inefficiency [94]. Furthermore, the technique fails to address the issue of deadlocks in the case of networks with feedback loops. Misra [74] presents an approach for distributed simulation of networks with feedback loops. Peacock *et al.* [93] propose a method of detecting deadlock through the use of "probe" messages that is also, conceivably, inefficient. The approach in [92] suggests a distributed technique to detect and recover from deadlocks in networks with feedback loops. However, implementations [95] of this approach have been observed to be non-linear and highly inefficient. Soule and Gupta [154] report speedups of up to 16 on a 64-node processor without memory interconnect or operating system contention. Given that the Chandy–Misra [91] approach lacks the ability to handle asymmetric rise and fall delays, Soule and Gupta predict a 50% degradation in speed due to glitch suppression and event removal, thereby implying a speedup of only 9.4 on a 64-node processor. Chandy and Sherman [96] present a new, synchronous, conservative approach that utilizes the notion of conditional events. Simulation experiments of a network of switches yield speedups ranging from 2 to 8 on a 24-processor Intel iPSC/1. DeVries [94] proposes methods to reduce the number of null messages in the approach presented in [74] for a few limited networks. Furthermore, performance measurements are absent in [94]. The Chandy–Misra [91] approach incurs two fundamental limitations. First, it fails to recognize the occurrence of inconsistent events and is unable to preempt them, thereby implying incorrect results. Second, it violates the basic principle of event driven simulation and incurs the penalty of unnecessary and repeated executions of the models even when no new input stimuli are received.

Debenedictis, Ghosh, and Yu successfully introduced a novel algorithm for asynchronous, distributed discrete event simulation, YADDES [97], that is mathematically proved to be deadlock-free and accurate. In discrete event simulation, a behavior model C^1, referred to as component or entity, executes when stimulated by an input event or cause. Upon execution, C^1 may generate an output event which is then propagated to subsequent components, $C^i, \ldots, C^j$, connected to the output of C^1. Subsequently, one or more of the components, $C^i, \ldots, C^j$, may be executed. The process continues until the number of outstanding events is nil. In general, execution of the components, $C^i, \ldots, C^j$, connected to the output of C^1, may be initiated by the output event generated from the execution of C^1 corresponding to an earlier event concurrent with the execution of C^1 corresponding to the subsequent event. To increase concurrency in the simulation and thereby improve efficiency, YADDES utilizes a scheme, referred to as the data-flow network, that quickly generates the "earliest time of next event" at the output of C^1 prior to the generation of the output event of C^1 following the completion of its execution. The "earliest time of next event," also referred to as the predicted event time, represents a conservative estimate on the time of the subsequent event at the output of C^1. Thus, components $C^1, \ldots, C^j$ may concurrently execute, without violating any data dependency, any and all input events with times less than "earliest time of next event." The underlying assumption is that the computation cost of predicting the earliest time when an event E^j may be generated as a result of the execution of C^1, corresponding to some input event E^i, is less than the computation cost of executing C^1. For simulations with complex VHDL models, this assumption is, in general, true. An added problem when simulating digital designs with feedback loops, utilizing the traditional distributed discrete event simulation algorithm, is that, often, an output event may not be generated following the execution of a component, C^1. Thus, other components connected to its output cannot execute. If the output of one of these components is, in turn, connected to the input of C^1, then deadlock occurs leading to an abrupt and indefinite halt of simulation. The data-flow network of YADDES successfully addresses this problem.

In YADDES [97], corresponding to every component in the actual simulation, the data-flow network consists of pseudo-primed and pseudo-unprimed components. These pseudo components are purely mathematical entities that execute specific functions to generate W (or W') values at their outputs. They are interconnected in an acyclic manner that is detailed subsequently in this chapter. The data-flow network executes concurrently with the actual simulation. The W (or W') value asserted at the input of a pseudo component from the output of another pseudo component reflects an estimate of the earliest time when an event may be propagated along the path. For complete definitions of W (or W'), the reader is referred to [97].

Consider a behavior model C^n with inputs, $\{1, 2, \ldots, i\}$. The "window of temporal independence", t^n_{win}, is defined as the minimum over all W (or W') values over all inputs $i \in \{1, \ldots, I\}$ of the pseudo component(s) corresponding to the model C^n. Any and all input events with time t on an input of C^n may be executed accurately, without violation of data dependency, provided t is less than t^n_{win}. In YADDES, t^n_{win} is represented by K^n. Correctness is guaranteed in YADDES through accurate values for W (or W'). Formally,

$$t^n_{win} = minimum(W_i \text{ } or \text{ } W'_i) \, \forall i \tag{9.1}$$

While successful as a novel, asynchronous, distributed discrete event simulation algorithm, YADDES is inappropriate for simulating behavior models such as VHDL models since it is limited to a single, input-independent, propagation delay between the input and output of a component. Unlike queueing networks, most digital components require the use of unique high-to-low (t_{phl}) and low-to-high (t_{plh}) propagation delays to accurately represent the timing behavior. Consider a component C and a signal asserted at an input port at $t = t_1$. Where a high to low transition is caused at the output port of C, the output is asserted at a time given by $t = t_1 + t_{phl}$. In the event that a low to high transition is caused at the output port of C, the output is asserted at $t = t_1 + t_{plh}$. For many digital components, the values of t_{plh} and t_{phl} may differ from one another by a factor as high as 3, leading to the generation of inconsistent events. For correctness in the simulation, preemption or descheduling of inconsistent events is necessitated. The YADDES algorithms lacks any mechanism to achieve inconsistent event descheduling.

9.2 AN ASYNCHRONOUS APPROACH FOR SIMULATING BEHAVIOR LEVEL MODELS ON PARALLEL PROCESSORS

9.2.1 A Novel Asynchronous Distributed Approach to Simulation

In this approach, a digital system, consisting of deterministic digital components, is partitioned by the user and models corresponding to the components in each partition are assigned to a unique processor. The system is assumed to possess well-defined external inputs and outputs, referred to as primary input and primary output ports respectively. Furthermore, the simulation is confined to a well-defined logical value system including the three-valued (0,1,X) and five-valued (0,1,X,S,CH) systems, where 0, 1, X, S, and CH refer to logical low, logical high, unknown, stable, and changing. A model is an executable entity that represents the logical and timing behavior of the component. The input and outputs correspond respectively to the input and output ports of the component. The electrical paths between the components are represented through communication protocols through which models exchange messages. Each model is an asynchronous, concurrent entity and, thus, at any instant of time, more than one model may execute concurrently. Additionally, a model is completely unaware of the presence of other models in the system. A model's only relationship to the world outside itself is through the input and output protocols. In order that the models execute correctly, i.e. the order of execution of the events is preserved, in this approach, a necessary and sufficient condition is the absence of the global simulation time and the encoding of time in the messages. That is, whenever a model either receives or propagates the logical value of a signal transition as a message, the assertion time of the signal is encoded in the message. Assume, on the contrary, that the asynchronous distributed approach utilizes the notion of global simulation time and lacks time encoding. Thus, every model must have access to the global simulation time. Since a model's only means of communication with the external world is through the input and output protocols, every model must either connect (i) to a special, single device that maintains the global simulation time or (ii) to every other model. The alternative (i) reduces to the synchronous approach. In alternative (ii), every model maintains an identical copy of the global simulation time by accessing the results of execution of all other models in the system. The network of protocols in alternative (ii) requires synchronization between all the models even when every component may not be physically connected to every other. Thus, the simulation system is not asynchronous and, as a result, the assumption must be fallacious. In the absence of global simulation time, a model must receive, at its input ports, the logical values $(v_1, v_2, \ldots, v_n)$ and the corresponding assertion times $(t_1, t_2, \ldots, t_n)$ of the transitions from either other components or the external world. Then, it computes the minimum, M, of all the assertion times through the equation $M = minimum(t_1,\ t_2,\ \ldots,\ t_n)$. The quantity M signifies that the model has complete knowledge of the signals at every input port $\{1, 2, \ldots, N\}$ between $t = 0$ ns and $t = M$ ns. Thus, the model may accurately determine the output signal value, O_p, and output assertion time $T = M + d$, where the propagation delay, d, is equal to either t_{phl} or t_{plh}, depending on the nature of the output transition.

An output transition generated following the execution of a model is not immediately asserted at the output port. It is stored within the model. At a later time when the model is subsequently scheduled for execution at $t = M_2$, the output transition is asserted at the output port subject to the conditions that $M_2 \geq T$ and that the output transition has not already been preempted.

When a model attempts to receive or propagate a message, it utilizes non-blocking primitives. That is, corresponding to a request by a model to propagate an output transition, a response—either success or failure—is immediately propagated by the operating system to the model. When the buffer, associated with a protocol, is partially filled, a request to transmit a message may succeed in placing the new message in the buffer. When the buffer is full, the request fails and the model must attempt to retransmit the message in the future. The operating system guarantees the accurate delivery of the messages already located in the buffer. Furthermore, the operating system also guarantees that the order in which messages are delivered to a recipient model is consistent with the order in which messages were sent by the sender model. When the buffer corresponding to an inward protocol of a model is empty, a request by a model to receive a message fails immediately. Such a request would have succeeded immediately if the buffer was non-empty. The algorithm is presented first for combinational digital designs and then for sequential systems.

9.2.1.1 Simulation of Combinational Digital Designs. Consider the simulation of a system shown in Figure 9.1. Although a simple digital system is chosen for simplicity of explanation, the distributed approach applies equally to complex behavior models. The output ports of components A and B are connected to the inputs of the two-input AND gate C and the signal transitions generated by each of A and B between $t = 0$ ns and $t = 30$ ns are shown in Figure 9.1. Assuming models A and B are allocated arbitrarily to processors 2 and 3, A and B are executed asynchronously and, as a result, the real time during simulation at which the signal transitions are propagated from A to C and B to C may not relate to each other. The

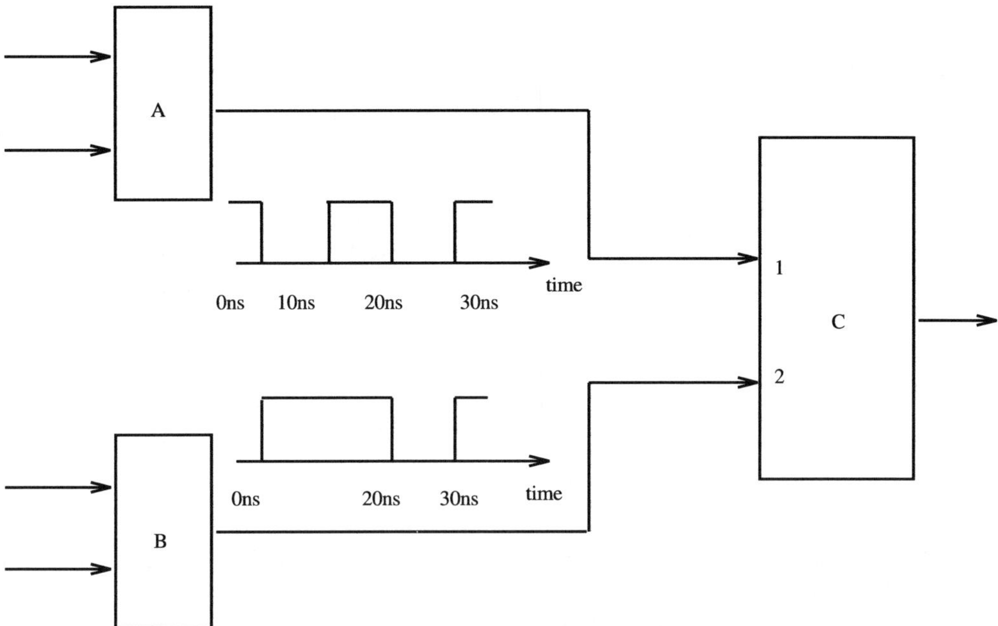

Figure 9.1 An example design.

individual transitions from A to C—n_1: 0 at $t = 0$ ns, n_2: 1 at $t = 10$ ns, n_3: 0 at $t = 20$ ns, and n_4: 1 at $t = 30$ ns—are guaranteed to be asserted in the same order as represented in logical time [98] shown in Figure 9.2(a). Figure 9.2(b) represents a similar ordering of the transitions from B to C—m_1: 1 at $t = 0$ ns, m_2: 0 at $t = 20$ ns, and m_3: 1 at $t = 30$ ns in logical time. For the purpose of explanation, assume that the ordering of the transitions in real time is represented by Figure 9.3. The correctness of the distributed approach is invariant to the ordering of the events in real time given that the logical ordering specified in each of Figures 9.2(a) and 9.2(b) is preserved. In Figure 9.3, assume m_1, n_1, n_2, n_3, m_2, m_3, and n_4 are asserted at C at real times $T = s_1$, $T = s_2$, $T = s_3$, $T = s_4$, $T = s_5$, $T = s_6$, and

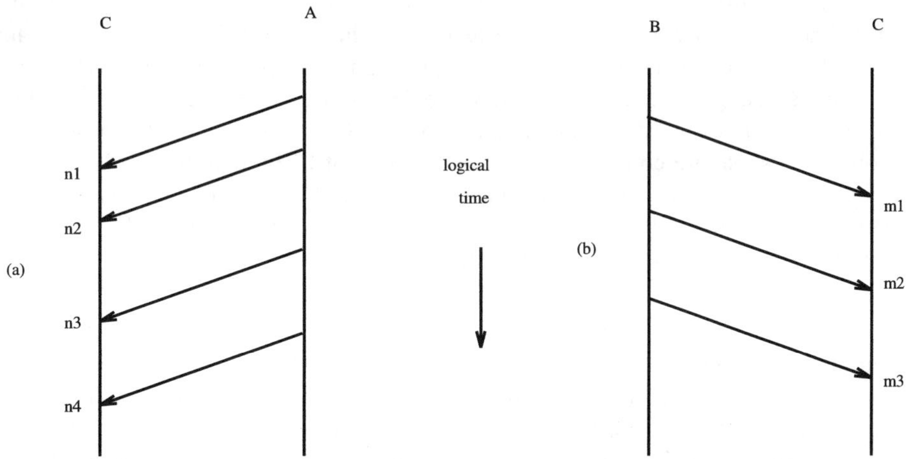

Figure 9.2 Logical ordering of events in an asynchronous distributed simulation.

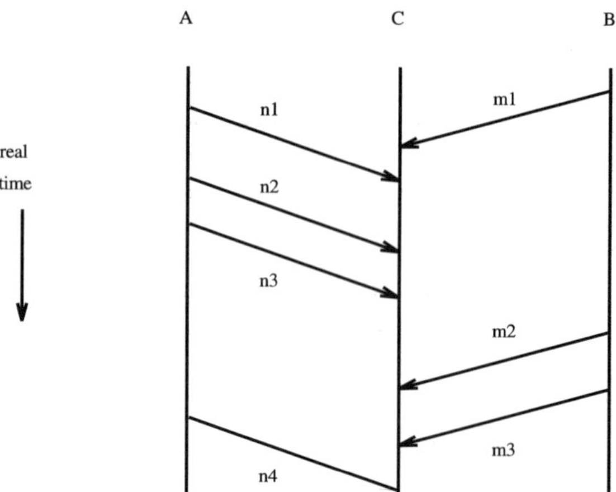

Figure 9.3 Real-time ordering of events in an asynchronous distributed simulation.

$T = s_7$, respectively, where T represents the progress of real time during simulation on the parallel processor.

Corresponding to the assertion of an input transition at $T = s_1$, C is unable to schedule for execution as the signal transition at port 2 is yet to be specified for $t = 0$ ns where t represents the progress of simulation time and corresponds to the hardware execution. At $T = s_2$, C schedules itself for execution and an output transition l_1: 0 at $t = 0 + 5 = 5$ ns is determined. Given that the previous value of the output was 0, l_1 does not imply any new information and is ignored. Corresponding to each of n_2 and n_3 at $T = s_3$ and $T = s_4$ respectively, the behavior description of C is not executed as the signal transition at port 2 has not been asserted beyond $t = 0$ ns. At $T = s_5$, signal transitions have been specified at $t = 20$ ns at both ports 1 and 2 and C is scheduled for execution. Output transitions l_2: 1 at $t = 10 + 16 = 26$ ns and l_3: 0 at $t = 20 + 5 = 25$ ns are generated but l_2 is observed to be inconsistent with l_3 and consequently discarded. Assertion of the output transition l_3: 0 at $t = 25$ ns is deferred as the transitions at the input port of C are defined at $t = 20$ ns and the model is yet unable to conclude with certainty that l_3 may not be discarded in the future. At $T = s_6$, transition m_3 is asserted at an input port of C but the input signal at port 1 is yet undefined beyond $t = 20$ ns. Consequently neither C may be executed nor any decision regarding l_3 be finalized. At $T = s_7$, input signals at ports 1 and 2 are both defined at $t = 30$ ns and given that l_3 has not yet been shown inconsistent and $30 > 25$, it may be asserted, with certainty, at the output port of C and consequently propagated to other components that are connected to the output port of C. In addition, the behavior description of C is executed and an output transition l_4: 1 at $t = 30 + 16 = 46$ ns is determined and stored within the model. The mechanism, described above, is referred to as the principle of deferred assertion of output assignments.

THE NOTION OF STARVATION. The number of active components during gate-level simulation has generally been observed to be between 5% and 20% [99] and this may be assumed to hold true for behavior-level simulation. Such a low activity may cause the following scenario during asynchronous distributed simulation. For example, in Figure 9.1 assume component A executes a number of times due to signal transitions at its input ports between $t = 0$ ns and $t = 1000$ ns (say) and asserts a number of transitions at port 1 of C.

Also assume B executes infrequently due to a limited set of transitions at its input port between $t = 0$ ns and $t = 1000$ ns with the consequence that only one transition at $t = 2$ ns (say) is asserted at port 2 of C. The value of the signal at port 2 remains essentially unchanged between $t = 2$ ns and $t = 1000$ ns. Consequently, C may not execute beyond $t = 2$ ns and this situation constitutes "starvation" with the cause being component B. In essence, the execution of model C suffers starvation due to a limited number of transitions at one or more input ports. This phenomenon has been referred to as blocking in the literature. In addition, other components, if any, that are connected to the output port of C either directly or indirectly will also starve implying a possibility of very low overall activity during simulation.

Starvation does not correspond to a physical process in hardware execution and its cause may be explained as follows. Event driven simulation requires, for efficiency, that only changes in the logical value of a signal be propagated. Consequently, in an uniprocessor environment, the value of the signal between two consecutive transitions e_1 and e_2 is identical to the value indicated in e_1. Such an assumption, however, is dangerous in the distributed asynchronous simulation on a parallel processor system for a message, sent to the input port of a component, may be delayed asynchronously and the behavior model may erroneously interpret the absence of message to imply "no change" in the logical value at that port. Consequently, a component must execute based on signals at input ports at $t = t_1$ such that transitions have been asserted at all of its input ports at $t \geq t_1$. Such a mechanism, as well as the principle of deferred assertion of output assignments, stated earlier, may increase the possibility of the occurrence of starvation.

In the event that starvation occurs during simulation of a design, first, it is detected in the following manner. When the number of input assignments at an input port of a component, that have not yet been used to generate output events, exceed a threshold, the component raises an exception. As a consequence of the exception, the execution mode of every processor is set to "exception-mode". The execution mode of the processors is reset from exception-mode when the cause of starvation is removed, i.e. the number of outstanding input assignments at the input port of the component falls below the threshold. The actual value of the threshold is empirically determined based on the amount of memory available in the processor for simulation and it influences the relative durations of normal- and exception-modes during a simulation. The characteristics of the exception-mode may be expressed as follows. Signal values are asserted at all input ports of components including the primary input ports, even when the logical values are unchanged with respect to their previous values. In addition, when a model is executed at $t = t_1$, either a previously generated correct signal transition at $t = t_x$ ($\leq t_1$), that was not yet asserted at the output port, is propagated to the output or the most recent logical value at the output is asserted at $t = t_1$ plus the minimum of the high-to-low and low-to-high propagation delays of the component.

The rationale for the strategy to resolve starvation may be expressed as follows. The causes of starvation experienced by a model include (i) signals at the primary input ports that are unchanged over relatively long periods of simulation time, (ii) models that generate a few transitions at their output ports in spite of frequency executions, and (iii) other starved models whose output ports are connected to one or more input ports of the model in question. When starvation occurs in a simulation system, it is logical to first determine the model responsible for causing starvation and then force that model to generate a special message indicating that its output has remained unchanged. In an asynchronous distributed simulation environment, the detection of the cause of starvation may be extremely difficult. Peacock *et al.* [93] suggest the use of probes which require complex message handling. Chandy and Misra [91] avoid the use of probes through the use of null messages. However, a system wherein every model propagates a null message whenever its execution fails to generate an output transition is conceivably inefficient. It may be argued that the ultimate reason for an incidence of starvation lies either with the signals at the primary inputs or with the nature of the models. A model may receive a large number of transitions defined up to time T_1 at all but one or a few of the input ports. At these few input ports, the most recent signal transition has an assertion time of T_2 ($\ll T_1$). Thus, the model is unable to execute beyond T_2 implying starvation. In reality, the signals at the few input ports may actually be unchanged. Therefore, it is logical to temporarily relax the constraint—propagation of changes only—when starvation is detected. This temporary action will cause the propagation of adequate information at the input ports of the starved models with the consequence that they may execute and consume the transitions at their respective input ports. At this stage, a large number of the outstanding unprocessed transitions at the input ports of the starved models may be utilized and the simulation system may be permitted to resume its "normal" mode of operation.

As an example, assume that the components A and B in Figure 9.1 are executed on processor I of a parallel processor system while model C is executed on another processor II of the system. Assume further that a significant number of signal transitions are asserted at the input ports of A and that the signals at the input ports of B are unchanged in their logical values throughout the duration of simulation. Consequently, A is executed frequently. On the other hand, B is executed very infrequently and very few output transitions are asserted at the input port 2 of C. The model C is unable to execute in the absence of signal transitions at the input port 2 and the number of outstanding input entries at the input port 1 of C may exceed the threshold. Consequently, C raises an exception and the execution mode of all the processors is set to exception-mode. In this mode, signal transitions are asserted at the input ports of B even though they are unchanged in their logical values. Consequently, B is executed more frequently and a modest number of output transitions are asserted at the port 2 of C. The model C is executed and the outstanding entries at the port 1 are utilized to generate output assignments and the cause of starvation is removed.

The precise functionality of each processor that contains one or more components of a partition is expressed through the pseudo algorithm in Figure 9.4. The pseudo algorithm applies only to the combinational components.

PROOF OF CORRECTNESS. The correctness of the algorithm requires that (i) every model generate accurate output transitions following its execution, (ii) the algorithm never deadlocks, and (iii) the execution of the algorithm terminates in finite time. The accuracy of a model is defined by the faithfulness with which the model description represents the activity of the original component. The accuracy of timing is ensured though the principle of deferred assertion of output assignments. Thus, a model is certain that a transition propagated at an output port may not be preempted.

For combinational designs, information flows in a single direction, i.e. from the primary input ports to the primary output ports. Additionally, communication is implemented through non-blocking primitives. When a model fails in its attempt to propagate a message, it continues to attempt the sending process until success. The success is guaranteed given that adequate memory is available in the processor and that it is continuously reused. When a transition is utilized, the corresponding memory location is freed and it may be used to store subsequent transitions. Consequently, the execution of the algorithm will always progress for combinational designs.

```
read in transitions at input ports- from external ports or other components;
analyze the transitions and schedule components for execution;
for (each component scheduled) {
  determine minimum time up to which all input transitions are defined;
  if (the minimum time exceeds the previous value) {
     compute the logical value and assertion time of the output;
     if (output value differs from previous value) {
        store the output transition;
     }
     verify that all of the stored transitions are consistent;
     discard duplicate and inconsistent transitions;
     for (all stored transitions) {
       if (the assertion time is less than or equal to the minimum time) {
          propagate the output transition;
       }
     }
     if (no output transitions are propagated && exception mode) {
        temp time = minimum time + smaller of high-to-low and low-to-high
        propagation d
        propagate the previous logical value at assertion time equal to temp
        time;
     }
  }
}
compute total number of unprocessed input transitions for all components in
the processor if (total number exceeds threshold) {
  set exception mode;
}
else if (currently exception mode && total number falls below threshold) {
  reset to normal mode;
}
```

Figure 9.4 Algorithm for each processor (for combinational components).

The absence of permanent starvation may be proved by the method of contradiction. Assume that the execution of a model, C_p, is starved due to lack of adequate transitions at an input port p_k $in $\{p_1, \ldots, p_{k-1}, p_{k+1}, \ldots, p_n\}$. Thus, while signals at all other input ports are defined up to a large value of the simulation time, the signal at p_k is defined up to a significantly smaller value of simulation time. Clearly, the cause of starvation must lie with a preceding model, C_q, that fails to generate adequate output transitions. If one traces in this manner, one must arrive at a primary input port that is defined only up to a value of simulation time that is significantly smaller than the maximum simulation time. This is a contradiction and therefore the simulation must be free from permanent starvation.

To show that the execution terminates in finite time, assume the contrary. That is, the execution of the algorithm never terminates. Since the simulation never hangs, at least one model continues to execute forever. When a model executes either in normal- or exception-mode, it will utilize signal transitions at the input ports and generate output signals—

transitions or values that are unchanged—with assertion times that must progressively increase. Also, a component may propagate only a finite number of output signals. Since the model executes forever, eventually the assertion times of the signals at the input ports must approach infinity. The signal transitions at the primary input ports are defined up to a finite time and the propagation delays of the components are finite. Consequently, for the assertion times of signal transitions to approach infinity, the simulation system must consist of an infinite number of components. This conclusion is fallacious and, therefore, the execution of the algorithm must terminate in finite time.

The proofs hold true even in the presence of the principle of deferred assertion of output assignments. At every primary input port, following the assertion of the final signal transition, a special assertion is propagated. Where such special signals are asserted at every input port of a model, all previously generated and stored output assignments, if any, are propagated at the output port. Then, a special signal transition is propagated at the output port. When special signal transitions are asserted at all of the primary output ports of the digital system, simulation is considered completed.

9.2.1.2 Simulation of Sequential Digital Designs. The most important characteristic of the asynchronous approach is that it permits every simulation model to execute independently in the absence of data dependency. Consequently, the success of this approach for an example simulation system is a function of the degree of data dependency between the models. For a set of entities constituting a directed cyclic graph [100], it was shown that the output generated as a consequence of execution of a model may influence its input at a later time. Consequently, directed cyclic graphs tend to require more synchronization than systems with no cycles and the effectiveness of the asynchronous approach for digital systems with directed cyclic graphs may be diminished. In addition, directed cyclic graphs present a more serious difficulty to the asynchronous approach as explained through the example in Figure 9.5.

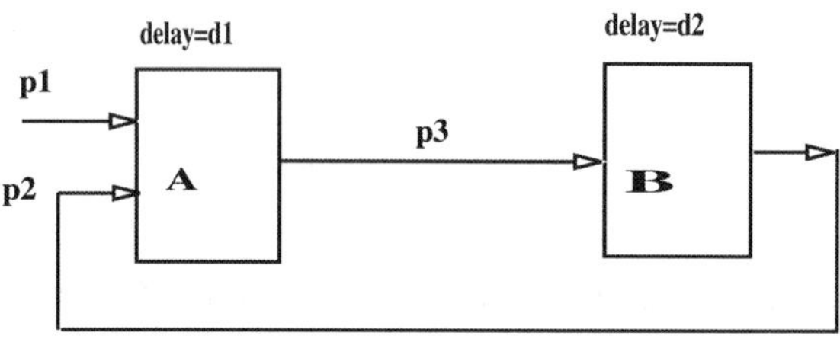

Figure 9.5 Asynchronous simulation of a system with directed cyclic graph.

In Figure 9.5, the input ports of model A are p1 and p2 and the input port of B is p3. The output of model B is connected to p2 of A and the output of A is connected to p3 of B. The propagation delays of A and B are d_1 and d_2, respectively. Assume distributed asynchronous discrete event simulation of a design where A and B are associated with two distinct processors of the parallel processor. Simulation is considered to have run to termination when all externally supplied usable signal transitions have been utilized to generate output transitions. Assume also that initial logical value 0 is associated with ports p3 and p2 at $t = 0$ ns. For a given signal transition 0 to 1 at port p1 of A at $t = 0$ ns, A may be executed. Assume that the logical value of 1 at p1 persists up to $t = T$ ns, where T is very large. Where the execution of A generates a transition at its output port at $t = 0 + d_1 = d_1$, the transition is propagated to B that causes B to be scheduled for execution at $t = d_1$. Where the execution of B at $t = d_1$ generates a transition at its output port at $t = d_1 + d_2$, the transition is propagated to port p2 of A. As a result, A may be executed again and the process will continue as long as models A and B may be correctly scheduled for execution. If the execution of B at $t = d_1$ does not generate a signal transition at its output port, no message is sent from B to A. Consequently, model A is unaware that the port p2 is at logical 0 up to $t = d_1 + d_2$ and is unable to execute beyond $t = 0$ ns, although the signal value of 1 at port p1 persists up to $t = T$ ns and, therefore, the simulation should execute until the transition $t = T$ ns is utilized. Consequently, execution of A cannot proceed in the absence of messages from B and the execution of B requires messages from A and this constitutes a deadlock. The cause of deadlock is due to the absence of information at an input port of a model and the lack of global knowledge that the signal value at that port has remained unchanged.

In the approach proposed in this section, a digital design is first analyzed and every component that is included in a feedback loop of the design is identified. The behavior models corresponding to these components are modified to execute permanently in the "exception-mode". That is, following the execution of a behavior model at $t = t_1$ an output signal is always asserted at the output port. Assume that t_{phl} and t_{plh} model the high-to-low and low-to-high propagation delays and $t_{plh} < t_{phl}$. When an output signal transition is generated at $t = t_1 + t_{plh}$, it is immediately propagated at the output port for it may never be preempted. If, on the other hand, an output signal transition is generated at $t = t_1 + t_{phl}$, it is stored within the model and a message containing the previous logical value and with an assertion time given by $t = t_1 + t_{plh}$ is propagated at the output. This transition may be propagated in the future when the model is subsequently executed for a transition at $t = t_2$ such that $t_2 > (t_1 + t_{phl})$ provided that the transition has not already been preempted. When no new transitions are generated at the output port, the model propagates a message containing the previous logical value and assertion time given by $t = t_1 + t_{plh}$. Thus, following the execution of a model, the propagation of a new message with a value of time larger than that of the previous one is guaranteed.

Consider the simulation of the digital design of Figure 9.5. First, model A is executed and then B is executed. Even though the execution of B generates no transition at its output, a message corresponding to the unchanged signal value at p2 at $t = d_1 + d_2$ is propagated to A. Then, model B executes again, generating an unchanged signal at its output at $t = 2d_1 + d_2$ which is subsequently propagated to B. The process continues until the external signal at p1 for $t = T$ ns is utilized and the output signals at ports p2 and p3 are appropriately determined.

Figure 9.6 presents the pseudo algorithm of each processor that contains one or more components of a partition. The pseudo algorithm applies only to the sequential components that are permanently executed in exception-mode.

PROOF OF CORRECTNESS. This approach guarantees the absence of deadlock for any digital design with feedback loops. This guarantee may be understood intuitively as follows. Whenever a model executes, an output signal is generated and propagated to other components connected to its output. The assertion time of the signal exceeds that corresponding to the previous output signal of the model. Thus, there is always progressive activity in the simulation system and it never deadlocks. For a subsystem containing one or more directed cyclic graphs or feedback loops, the first activity consists of a signal transition asserted at the external inputs of the subsystem.

The reasoning to show the absence of permanent starvation for sequential system simulation is straightforward. Starvation is caused either by models that do not generate new

```
read in transitions at input ports- from external ports or other components;
analyze the transitions and schedule components for execution;
for (each component scheduled) {
 determine minimum time up to which all input transitions are defined;
 if (the minimum time exceeds the previous value) {
    compute the logical value and assertion time of the output;
    if (output value differs from previous value) {
       store the output transition;
    }
    verify that all of the stored transitions are consistent;
    discard duplicate and inconsistent transitions;
    for (all stored transitions) {
        if (the assertion time is less than or equal to the minimum time) {
           propagate the output transition;
        }
    }
    if (no output transitions are propagated) {
       temp time = minimum time + smaller of high-to-low and low-to-high
       propagation d
       propagate the previous logical value at assertion time equal to temp
       time;
    }
 }
}
```

Figure 9.6 Algorithm for each processor (for sequential components).

transitions following their execution or due to the lack of transitions associated with the primary input ports. In the exception-mode, both of these conditions are absent and, as a result, starvation may not occur.

For a formal proof, consider a digital design wherein external signal transitions are asserted at every external input port up to $t \geq T$. Assume also that the system simulates into a deadlock. Choose a model M_1 that is affected by deadlock such that the smallest assertion time at an input port I_{M1} is given by $t = t_1$ ($t_1 < T$). The output of M_1 is defined at $t = t_1 + d_{M1}$, where d_{M1} is the propagation delay of M_1. Clearly, based on the assumption, input port I_{M1} cannot represent an external input port. Furthermore, it cannot be connected to the output of M_1, for the signal at the output of M_1 is defined at $t = t_1 + d_{M1}$ that exceeds t_1. Therefore, I_{M1} must be connected to the output port of another component, say M_2. Consider that the signal transition at the output port of M_2, defined at $t = t_1$, is caused by an input signal transition at the input port I_{M2} of M_2 that is defined up to $t = t_2$. Obviously, $t_2 = t_1 - d_{M2}$, where d_{M2} represents the propagation delay of M_2. Applying the same reasoning process as before, the input port I_{M2} must be connected to the output port of another component, say M_3. If one continues to trace in this manner, eventually one must arrive at a component M_n such that its input port I_{Mn} is connected to the output of one of the previously encountered components $\{M_1, M_2, \ldots, M_{n-1}\}$. The signal at I_{Mn} is defined at

$$t = t_1 - d_{M2} - \cdots - d_{Mn} \tag{9.2}$$

However, the signals at the output ports of M_1, M_2, ..., M_{n-1} are defined, respectively, at

$$t = t_1 \tag{9.3}$$

$$t = t_1 - d_{M1} \tag{9.4}$$

$$t = t_1 - d_{M1} - \cdots - d_{M(n-2)} \tag{9.5}$$

Clearly, equation (9.2) is inconsistent with any of equations (9.3) through (9.5). Hence the assumption must be fallacious and the simulation cannot deadlock. Ultimately, the signal transitions at the input port of the models must be asserted at a time greater than or equal to T, implying that all externally applied transitions have been utilized.

Since the simulation is free from deadlock, and as subsequent executions of a model generate output signals with progressively increasing assertion times, the simulation must eventually terminate given that the external signals are defined only up to finite time.

The principal limitation of this approach is its inefficiency with respect to the absolute standards. The inefficiency is most apparent in situations where the external signals to a design, that is cyclic yet non-oscillatory, remains unchanged for a long period of simulation time. For instance, assume that the example system in Figure 9.5 is non-oscillatory. Consequently, messages corresponding to unchanged signal values will be propagated from A to B and from B to A at intervals of $d_1 + d_2$. The total number of iterations around the cycle until simulation terminates is given by $(T/(d_1 + d_2))$ approximately. The total CPU time required for simulation is proportional to the number of iterations and where the ratio $(T/(d_1 + d_2))$ is large, a high degree of inefficiency is indicated.

9.2.2 Implementation Issues

The implementations of the algorithm on the Bell Labs hypercube [101] and ARMSTRONG system [102] at Brown University are both similar. However, they are complex and are described as follows. Given any complex system and a user-specified partition, the total number of processors required for the simulation equals $(N + 1)$ where N refers to the number of partitions. While the components of every partition execute on a processor, the primary input ports of the system are modeled through a single device, P0, and is executed on a unique processor. P0 asserts signal transitions at the primary inputs of the system. Also, when every transition has been asserted, P0 asserts a special message—termination vector—to signify the end of all input transitions. In the event that the system contains feedback loops, either the user must manually identify all components that are included in any of the feedback loops or one may resort to automatic programs [103]. This information is accepted by a pre-processor that generates input files for each partition. A significant component of the implementation consists of a kernel C description (approximately 1500 lines) that executes on every processor except the one that executes P0. Each processor accepts an unique input file that represents information on the models and their interconnections for the corresponding partition. The pre-processor assumes that the system is described in a hardware description language, ESL [104]. The descriptions of the activities of the components are represented through a combination of C language constructs and primitives that permit communication between the models. When a partition includes multiple models, an interconnection between two or more models on the same processor is implemented through a data structure. When the models are located on separate processors, an interprocessor protocol represents the connection.

Prior to initiating execution, every component that is included in a feedback loop is set to execute in "exception-mode". The flow of control during the execution of the algorithm may be described as follows. The main program is executed on a separate processor that serves as a host to and is external to the parallel processor system. It loads the program on each of the appropriate processors and initiates their executions through special operating system function calls. Initially, the processors are unable to initiate executions of the models given the lack of information at all of the input ports of the models. Following the assertion of signals at the primary input ports of P0, one or more models may receive transitions at all of their input ports. As a result, these models execute. A model assumes an initial logical value for $t = 0$ ns for all those input ports that are included in feedback loops. Whenever any model observes that the total number of unprocessed transitions at the input ports exceeds a predetermined threshold, a message is propagated to all processors, through a broadcast mechanism, requesting a change of the execution mode from "normal" to "exception". At a later stage, when the number of unprocessed input transitions fall short of the threshold, the execution mode is reset to "normal". Eventually, P0 propagates termination vectors at each of the primary inputs. When a component receives termination vectors at every input port, first it propagates every outstanding output signal transition stored in it and then it propagates a termination vector at every output port. The execution of the corresponding model is complete. Eventually, the executions of the models terminate and the processors report back to their host indicating the completion of the tasks.

9.2.3 Performance of the Algorithm

The efficiency of execution of a distributed asynchronous algorithm is dependent on the frequency of synchronization required between the concurrent and independent entities. Thus, the efficiency increases when the computational complexity of the models are high, the frequency of message communication between the models is low, and the time required for a message to propagate from one processor to another is relatively small. For the Bell Labs hypercube, the size of a message varies from 4 to 1000 bytes and it requires an average of 1 ms for propagation from one node to another.

It may be observed that in the discipline of computer-aided design of digital systems, there are no standard benchmark behavior-level systems that may be utilized to measure the performance of the algorithm. Also, the computational complexity of behavior-level models certainly exceeds those of gates by a significant margin. Furthermore, the execution times of the models such as MC6809 and MC6800 may differ from one another appreciably depending on the level of sophistication and representation details. In this research, the algorithm is executed for two behavior-level designs—one combinational and the second sequential—to study the performance issues.

9.2.3.1 Combinational Behavior-level Design. A combinational behavior-level digital design is constructed, based on the topology of a 2-bit adder (Figure 9.7), as follows. The gates are replaced by entities whose computational requirements, measured in terms of the CPU times, are similar to those behavior-level models of complex devices such as MC68040, MC88000, Intel 486, and Intel 860. Thus, the synthesized behavior-level system is analogous, for the purpose of studying the performance of the algorithm, to a network of complex digital devices. In these experiments, several parameters are controlled to generate data. First, the system is partitioned into 1, 2, 4, 8, and 16 partitions, thereby utilizing a total of 1, 3, 5, 9, and 17 processors of the hypercube. Second, for a few experiments, the computational complexities of all of the models are assumed to be identical and the value is varied from 0.34 ms to 3.4 ms, 34 ms, and 340 ms. The execution times are estimated for typical behavior models utilizing the lengths of their code descriptions, the MIPS rate for the 68000 processor, and the approximate number of machine instructions generated following the compilation of the high-level descriptions in languages such as C. The basic processor for the hypercube is the Motorola MC68000. It may be further observed that the algorithm proposed in this section may be used for other complex algorithms such as fault simulation where a behavior model may be executed not once corresponding to an input transition but, perhaps, hundreds of times, each corresponding to a fault in the system. Third, the number of external signal transitions, termed input vectors, are varied from 50 to 1000 respectively, reflecting realistic conditions under which simulation programs operate. The relatively large number of vectors pose a challenge to the issue of starvation, i.e. the suitability of current hypercube machines for behavior-level simulation. Finally, for a few experiments, the computational complexity of the models are assigned random values ranging from 0.34 ms to 3.4 ms, reflecting typical digital systems with assorted behavior-level models. The results are presented and analyzed as follows.

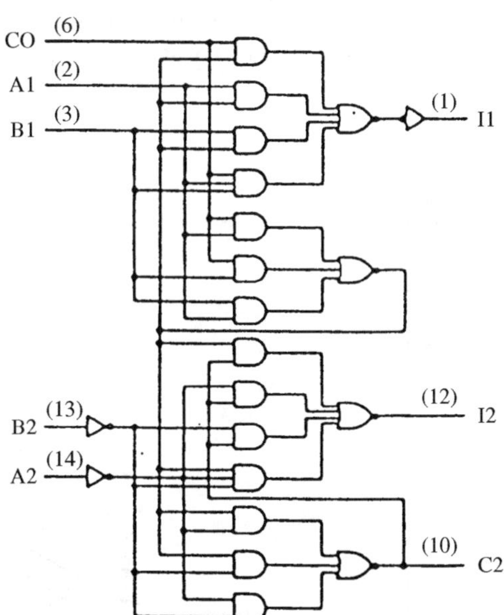

Figure 9.7 Combinational behavior-level design.

The results reported in Figures 9.8, 9.9, and 9.10 correspond to a set of experiments wherein every model is assumed to be characterized by the same computational complexity. Also, the data reported for the simulation utilizing a single partition implies that the distributed algorithm is executed, with all its overhead, on a single processor. Normally, a uniprocessor algorithm lacks such overhead and, thus, is more efficient on a single processor. In Figures 9.8(a) through 9.8(c), the graphs represent a logarithmic plot of the CPU time versus the number of vectors asserted at the primary inputs wherein the models' computational complexity is increased from 3.4 ms to 340 ms and the number of partitions is increased from 1 to 4 and then 8. It may be observed that the performance of the algorithm is linear with respect to increasing number of vectors. It may be further noted that the linearity is preserved

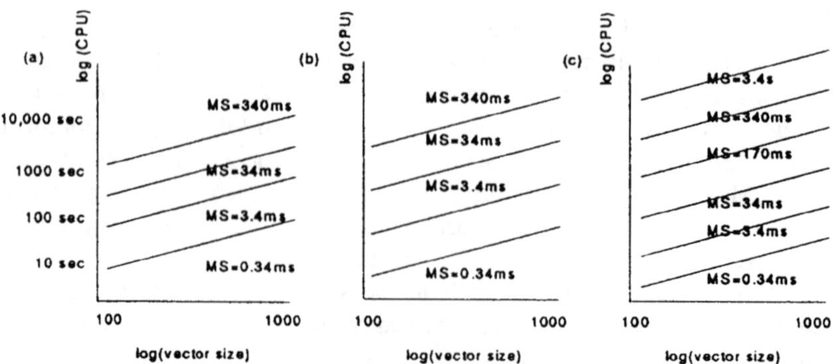

Figure 9.8 Graphs of CPU times vs. number of vectors. (a) 1 processor, (b) 4 processors, (c) 8 processors.

even when the simulation requires significant CPU times exceeding 10,000 s. The graphs in Figure 9.9 represent a logarithmic plot of the CPU time versus the models' computational complexity where the number of vectors is changed from 100 to 1000. Although the data in Figure 9.9 is obtained for simulation utilizing four partitions, similar graphs are observed for simulations on 4, 8, and 16 partitions. The performance of the algorithm is observed to be

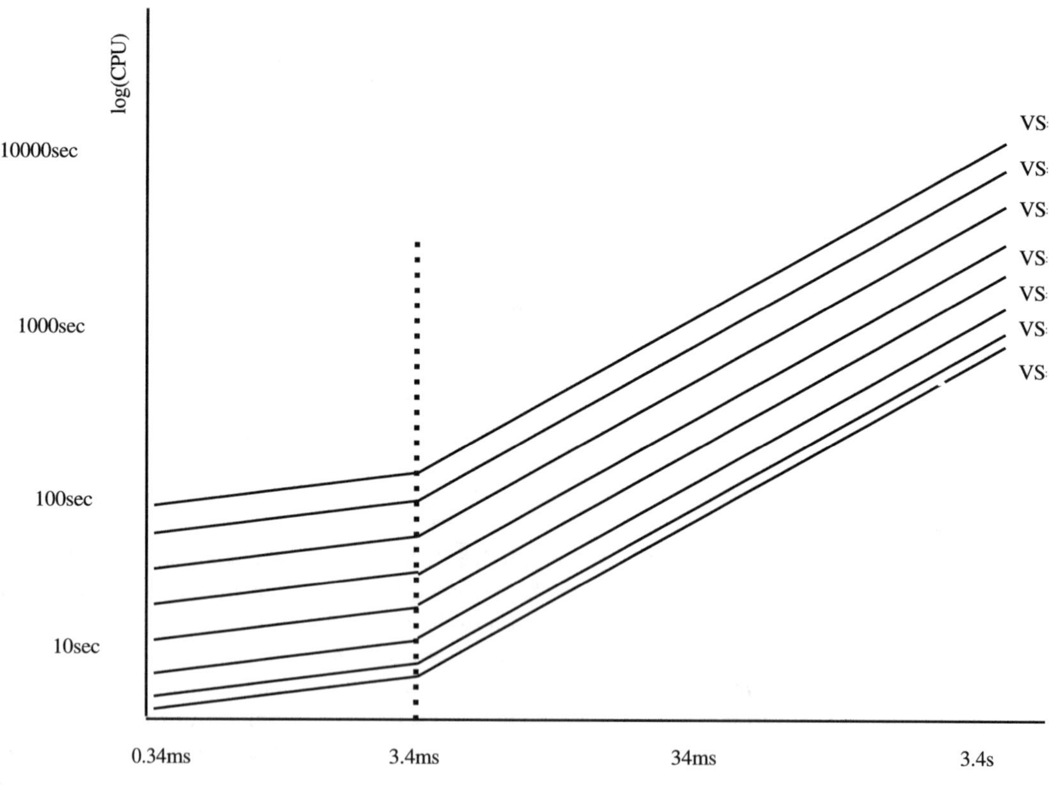

Figure 9.9 Graphs of CPU times vs. computational complexity of models (4 processors).

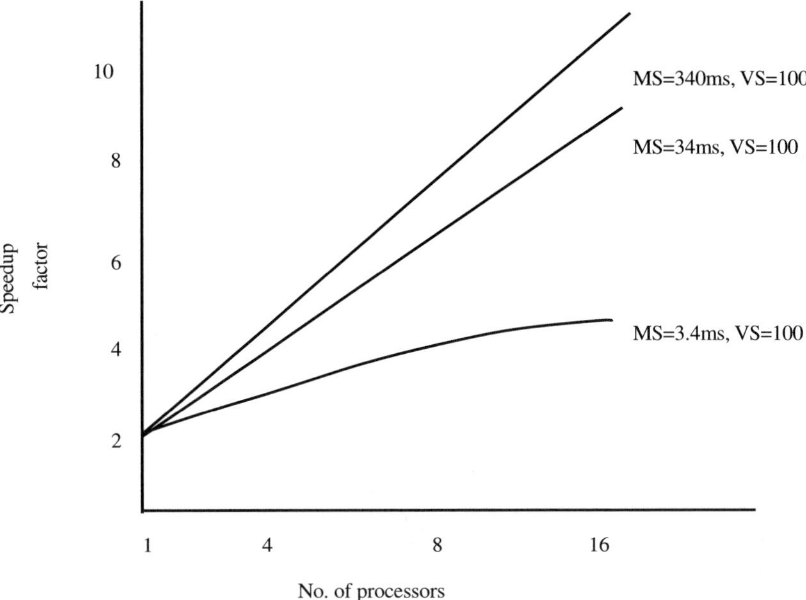

Figure 9.10 Graphs of speedup vs. number of processors.

linear with increasing models' computational complexity as one would expect. It may be further noted that the "knees" of the graphs corresponding to the models' computational complexity of 0.34 ms reflect the dominance of message communication in the hypercube over that of the computational complexity for the values of the latter below 0.34 ms. Where the computational complexity values exceed 0.34 ms, the effect of computation dominates over that of communication.

Figure 9.10 presents a plot of the speedup factor versus the number of processors for three specific pairs of model and vector sizes. The graph corresponding to the model size of 0.34 ms and vector size 100 resembles a saturation curve and reflects the dominance of the message communication over computation in the simulation. The other two graphs are both linear and indicate that the speedup factor increases linearly with increasing number of processors. The maximum speedup factor for this specific case is observed to be 12 when the design is partitioned and simulated with 17 processors. The slope differences of the graphs also indicate that increasing CPU time is spent in model computation as opposed to communication and other overheads for increasing models' computational complexity values.

The results in Figures 9.9 through 9.12 correspond to a set of experiments wherein the computational complexity values of the models are assigned random values in the range {0.34 ms, ..., 3.4 ms}. In Figure 9.11, the graph represents a plot of the CPU time versus the number of vectors for the case of four partitions where the simulation is permanently executed in normal-mode. The linearity of the graph implies that the performance of the algorithm is linear with respect to increasing number of vectors. The graph in Figure 9.12 corresponds to a simulation wherein all of the parameters are identical to that in Figure 9.11 except that the simulation executes permanently in exception-mode. While the algorithm's performance continues to be linear, the increased CPU times suggest that simulations, in exception-mode, are relatively slow. Figure 9.13 presents a plot of the speedup factor versus the number of processors for a total of 10 cases. Each case refers to a specific set of random

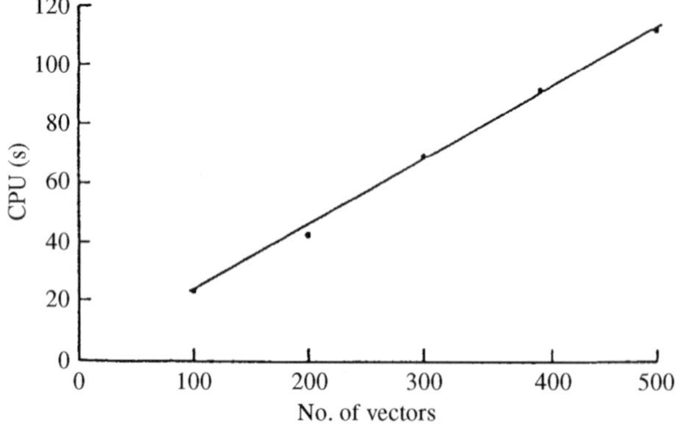

Figure 9.11 Plot of CPU time (s) vs. number of vectors, under normal-mode.

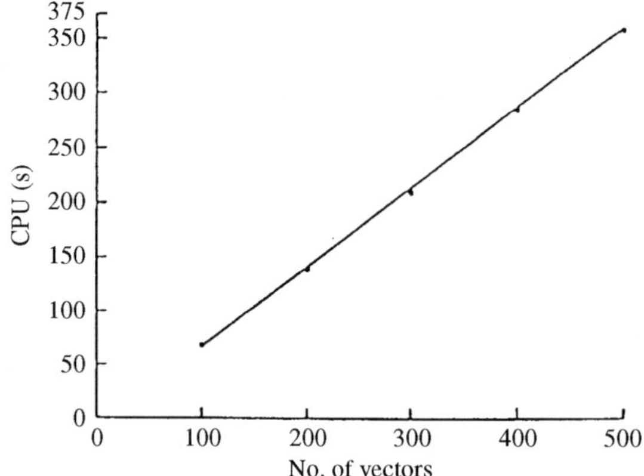

Figure 9.12 Plot of CPU time (s) vs. number of vectors, under exception-mode.

values assigned to the models as their computational complexity values. For each of these 10 cases, a total of 500 vectors are used for the execution of simulation. Also, the system is partitioned into 4, 8, 12, and 16 partitions for the purpose of simulation. In contrast to the previous set of experiments where the uniprocessor implementation is inefficient, for these experiments the uniprocessor simulation data refers to an industrial grade simulator, ESIM [104], at Bell Labs. Of the ten cases, while the following {1, 3, 5, 7, 9} are executed in

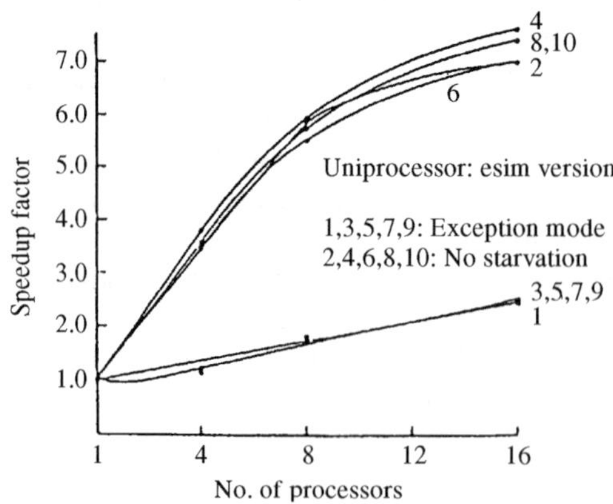

Figure 9.13 Plot of speedup factor vs. number of partitions.

exception-mode, for the set {2, 4, 6, 8, 10}, starvation never occurs. The onset of normal- and exception-modes are controlled through the use of significantly small and large values of the "threshold". It may be observed that, where starvation never occurs, a high speedup factor of 8.8 is observed for 16 partitions (i.e. 17 processors). The graphs resemble saturation curves implying that the efficiency of simulation increases with increasing number of processors but only up to a limit. Thereafter, simulation fails to offer additional computational load to the additional processors and the increase in efficiency is negligible. However, when the simulation executes permanently in exception-mode, a speedup factor of 3 is observed for 16 partitions.

Figure 9.14 presents the results of a study to understand the influence of threshold factors on the algorithm's performance. As expected, the execution mode switches from exception- to normal-mode when the value of threshold is increased. The threshold value, X, reflects the number of outstanding transitions associated with the models allocated to a processor. That is, it refers to $4 \times X$ bytes of memory that may be utilized to store the transitions at any instant during simulation. It may be concluded that, where memory limitation is severe, distributed simulation may be realized at a degraded level of performance under exception-mode. Also, where adequate memory is available, the simulation may never experience starvation. The critical value of the threshold factor increases with increasing number of vectors and, for the case of 500 vectors, the high value of threshold is 400. That is, if every processor is permitted the exclusive use of $400 \times 4 = 1.6$ Kbytes of memory for storing unprocessed entries, starvation may not occur. For typical hypercube processors with a low of 500 Kbytes of memory, this requirement may be tolerated easily. While the threshold factor will be clearly higher for larger numbers of vectors, it is pointed out that in the newer generation parallel processors, each processor may have access to 8 megabytes or more of memory. Under these conditions, it is highly probable that starvation may not pose a significant problem in the asynchronous distributed simulation of behavior-level digital systems.

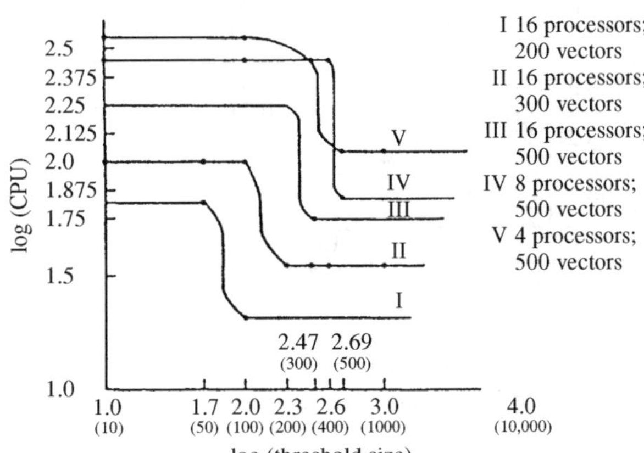

Figure 9.14 Logarithmic plot of CPU time vs. threshold value.

9.2.3.2 Sequential Behavior-level Design. The first design considered in this study is that of a simple cross-coupled NAND latch shown in Figure 9.15. Assume that both of the NAND gates, M_1 and M_2, have identical propagation delays expressed through $\Delta/2$ ns. The

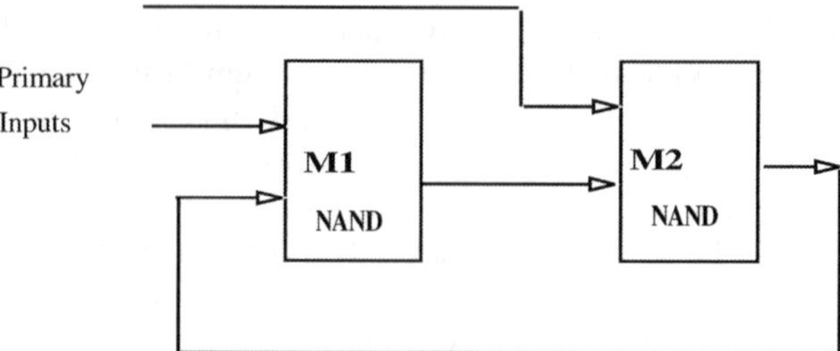

Figure 9.15 Simulation of a cross-coupled NAND latch.

computational complexity values for both M_1 and M_2 are assumed identical and the value is varied from 5 ms to 10 ms. External signal transitions are asserted at the respective primary input ports of M_1 and M_2 and assume that the average time period of the signals is 100 ns.

The graphs in Figure 9.16 show a plot of the CPU time (s) versus the ratio T/Δ, where Δ represents the cumulative propagation delay of the components M_1 and M_2 in the loop. The curves labeled I and II refer to the cases where the complexity values are 5 ms and 10 ms, respectively. It may be observed that with decreasing values of Δ from 100 ns to 10 ns, i.e. for increasing values of T/Δ from 1 to 10, the CPU time increases linearly. This is consistent behavior because smaller values of the propagation delays imply that M_1 and M_2 execute more frequently, thereby propagating messages at their output ports at an increased rate. This behavior continues to persist even when the model complexity, and as a result the CPU time, increases.

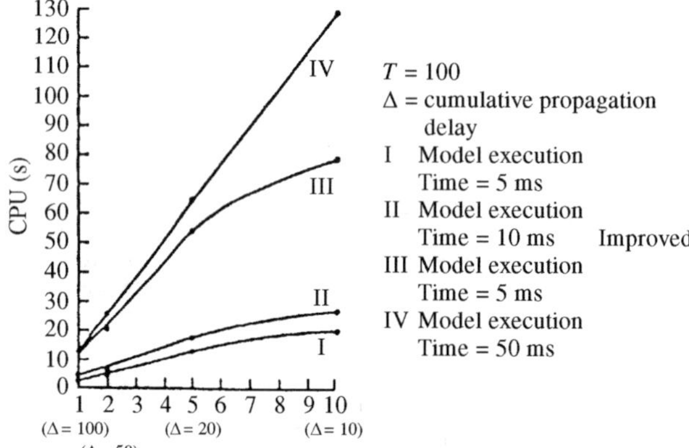

Figure 9.16 Plot of CPU time (s) vs. T/Δ.

Although the messages propagated in the simulation are essential towards preventing deadlock, the fact that most of them do not imply new logical values implies inefficiency. In general however, most sequential digital designs are defined for maximal clock speed that implies a ratio of T/Δ very close to unity. Thus, under these conditions, the inefficiency is minimal.

The graphs I through IV in Figure 9.17 represent the plot of CPU time versus the number of vectors for four sets of values for T/Δ: 1, 2, 5, and 10, respectively. The algorithm exhibits a linear performance with respect to increasing number of vectors, even for sequential systems.

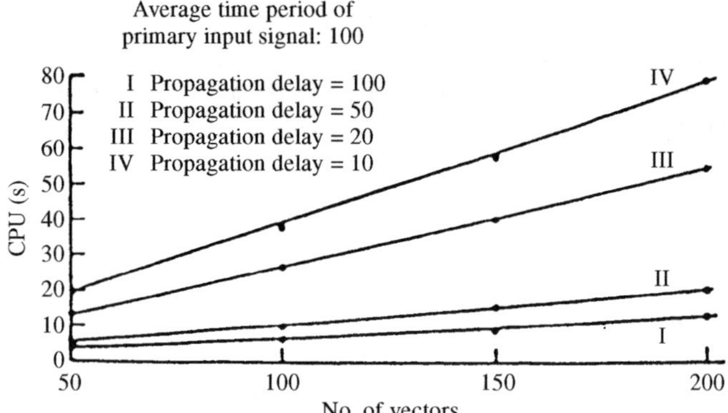

Figure 9.17 Plot of CPU time (s) vs. number of vectors.

A second sequential system, shown in Figure 9.18, is constructed by deliberately adding a feedback link to the combinational design of Figure 9.5. As with the combinational system, similar reasoning may be utilized to argue that the behavior of this sequential system is analogous, for the purpose of studying the performance of the algorithm, to a network of complex devices with feedback. The design is observed to consist of a purely combinational and a sequential subsystem. In this experiment, the design is partitioned into four partitions and executed on 5 processors. Additionally, random numbers between 0.34 and 3.4 are generated and assigned as the model's computational complexity values. The propagation delay for every gate is identical and equal to 20 ns. Also, three sets of external signal

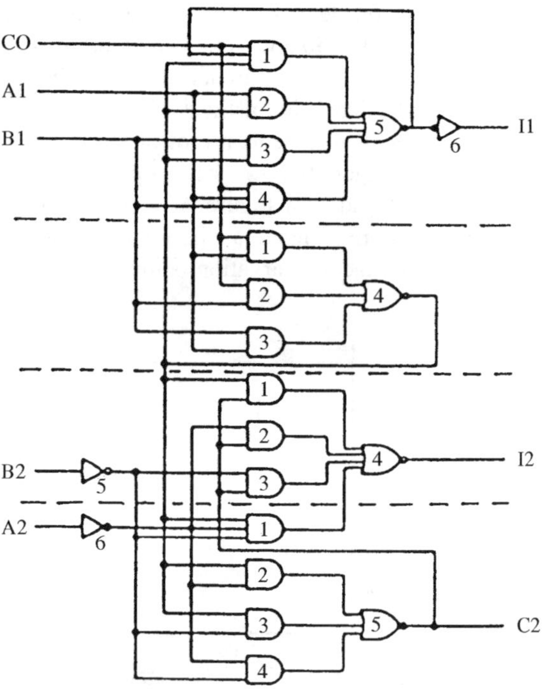

Figure 9.18 A sequential digital design.

transitions are utilized wherein the average time period varies from 50 ns to 100 ns and 200 ns. For each set, the number of vectors is varied from 50 to 100 and 300 vectors. Additionally, the value of the threshold factor is varied from 5 to 10, 50, 100, 200, 300, and 600. The graphs in Figure 9.19 represent a plot of CPU time (s) versus the factor T/Δ where Δ is the cumulative

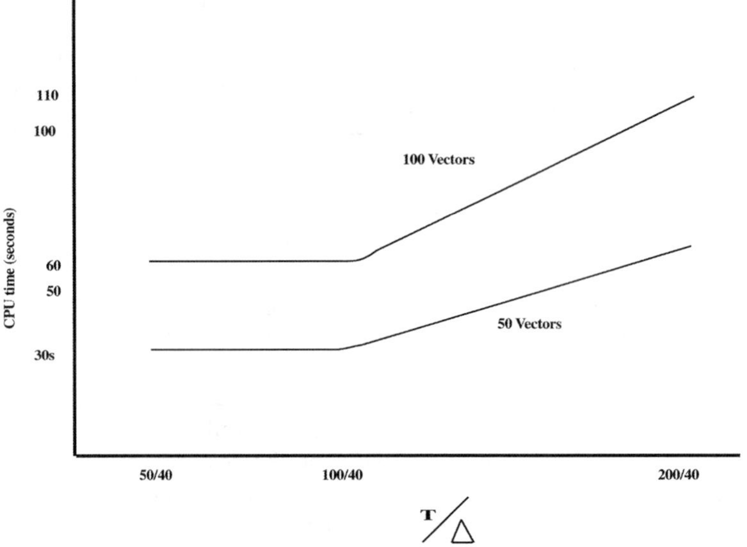

Figure 9.19 Plot of CPU time vs. T/Δ.

propagation delay of models 1 and 5 in Figure 9.18 and is equal to 40 ns. The two graphs correspond to 50 and 100 vectors and it is observed that the CPU time increases linearly with increasing value of the factor T/Δ. Although only models 1 and 5 in partition 1 are included in the feedback loop, their effect on the overall CPU time is unmistakable. The graph in Figure 9.20 represents a plot of the CPU time required for simulation versus the number of vectors. Figure 9.21 presents a plot of the CPU time versus the value of the threshold factor. Graphs I and II of Figure 9.21 correspond to simulations where two distinct sets of random numbers are assigned as the model computational complexity values. For lower values of the threshold factor, starvation is initiated in both cases. For case I, the decrease in the CPU time

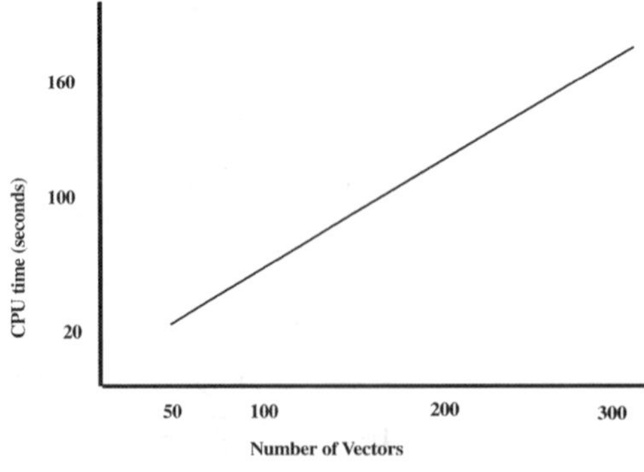

Figure 9.20 Plot of CPU time vs. number of vectors.

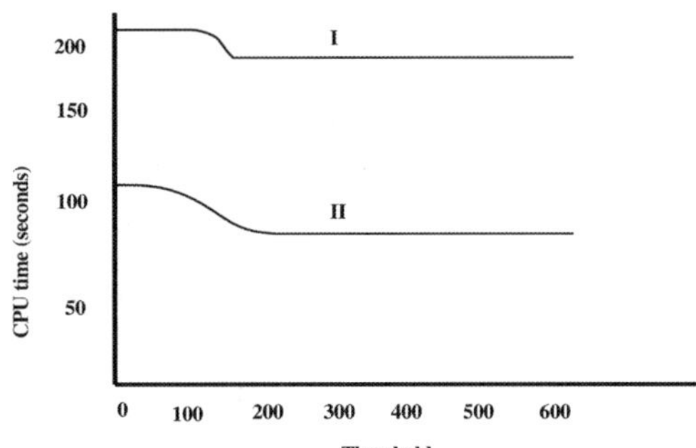

Figure 9.21 Plot of CPU time vs. threshold.

when the threshold factor is increased sufficiently to avoid starvation is relatively small. For case II, the corresponding decrease of CPU time is relatively large. This implies that the distribution of model complexity values for case I is such that the onset of starvation has little influence on the overall simulation. In contrast, for case II, the distribution of model complexity values is such that the onset of starvation casts a significant impact on the simulation performance. Additionally, the critical value of threshold is observed to be roughly equal to the number of test vectors. Therefore, the earlier discussion on the achievability of starvation-free simulation applies equally to sequential systems.

A limitation of distributed simulation of sequential systems consists in inefficient simulation and poor speedup for systems with either large number of feedback loops, a large value for the ratio T/Δ, or a combination of both of them. A large value of T/Δ implies the use of a large number of null messages which reduces the gain from distributed computational engines. Conceivably, under such circumstances, a uniprocessor algorithm, executing on a processor, may be faster than the distributed algorithm. Thus, the use of the distributed algorithm for the simulation of sequential systems requires a judicious analysis of the possible gains based on the value of the ratio T/Δ.

9.3 P^2 *EDAS*: AN ALGORITHM FOR ACCURATE EXECUTION OF VHDL DESCRIPTIONS ON PARALLEL PROCESSORS

This section presents a new approach, P^2*EDAS*, that is capable of asynchronous and concurrent simulation of VHDL models on parallel processors including the detection and preemption of inconsistent events. This section is organized as follows. Section 9.3.1 details the P^2*EDAS* algorithm while Section 9.3.2 presents an example to illustrate the approach. Section 9.3.3 provides mathematical proofs of correctness of P^2*EDAS*, freedom from deadlock, and termination in finite time. Section 9.3.4 reports on the implementation of P^2*EDAS* on both a multi-threaded shared memory multiprocessor and a network of work-stations configured as a loosely-coupled parallel processor and presents performance data for representative designs.

9.3.1 The P^2*EDAS* Approach

The fundamental advances of P^2*EDAS* over YADDES [97] enable it to simulate behavior descriptions, concurrently, efficiently, and accurately. They include:

1. In YADDES [97], each gate is characterized with a single, propagation delay, associated with all input to output paths. In contrast, P^2EDAS accommodates behavior descriptions, i.e. architectural bodies corresponding to VHDL entities, with any number of propagation delays for each input to output path of the model such as inertial, transport, t_{plh}, and t_{phl} propagation delays.
2. Unlike YADDES, the P^2EDAS algorithm acknowledges the generation of inconsistent output events, arising from the use of inertial delays. It possesses inherent mechanisms to preempt or deschedule such events, as defined by Ghosh and Yu [46] and imposed in VHDL [33].
3. The data-flow network, utilized by YADDES, is also enhanced in P^2EDAS to address output event preemption and reduce communications. P^2EDAS refers to the enhanced network as the "event prediction network."

P^2EDAS assumes that the behavior models constitute entities, similar to the VHDL entities, where the input, output, and bidirectional ports are clearly defined and the specification of the functional and timing behavior is complete and self-contained. Thus, P^2EDAS targets coarse-grain parallelism.

9.3.1.1 The Role of Event Prediction Network in P^2EDAS. A key concept in P^2EDAS is the event prediction network that is analogous to the data-flow network in YADDES and one that is responsible for asynchronous and concurrent yet accurate simulation of the behavior models.

The software simulation of a digital component's behavior description is, in essence, a mapping of the input to output behavior of the digital component corresponding to the given input stimulus. Following execution of a component, $C_{t,e}$, triggered by an input stimulus, an output response may be generated. In discrete event simulation, input stimulus and those output responses that are different from the previous values are termed events. Thus, an event corresponds to an ordered pair (t, e) that is propagated from the output of a model to the inputs of other models connected to it. The first parameter of the pair, e, refers to the logical value of the signal while t implies the assertion time of the signal.

Figure 9.22 shows a behavior model $C_{t,e}$ of a component with N input ports labeled i^1 through i^N, and M output ports labeled o^1 through o^M. Input signals, expressed through $i^1_{t,e}$ through $i^N_{t,e}$, are asserted at the input ports of $C_{t,e}$. The subscript set "t,e" reflects events with logical value "e" at assertion time "t" that trigger the execution of the model. The principles of event driven simulation require that where two events $i^1_{t1,e1}$ and $i^1_{t2,e2}$ are asserted in succession at the input port i^1,

$$t1 \neq t2 \quad \text{and} \quad e1 \neq e2 \tag{9.6}$$

In addition, the assertion times of events must increase monotonically, i.e. t2 must exceed t1. Following the execution of $C_{t,e}$, triggered by input events, output events represented by $o^1_{t,e}$ through $o^M_{t,e}$ are asserted at the corresponding output ports. As with input events, output events

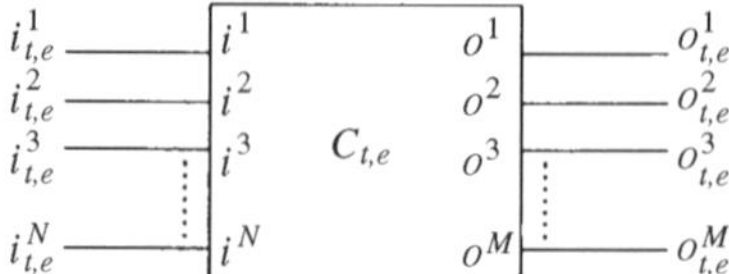

Figure 9.22 A behavior model of a digital component.

must also obey the conditions of equation (9.6) and the assertion times of output events at any output port must increase monotonically. The behavior of $C_{t,e}$ may be expressed through

$$o^{m}_{t,e} \leftarrow f^{m}(i^{1}_{t,e}, i^{2}_{t,e}, \ldots, i^{N}_{t,e}), \quad m \in \{1, \ldots, M\} \tag{9.7}$$

where M functions, f^1 through f^M, map the input signals to the output signal at the corresponding output port.

In uniprocessor-based discrete event simulation, the centralized scheduler maintains a queue of all events in the system. For accuracy, the event with the smallest time is first executed. That is, the event triggers the execution of the corresponding component. Then, the subsequent event, i.e. the next lowest assertion time, is executed. Corresponding to every component execution, new events may be generated and they must be appropriately sorted into the event queue. The process continues until all events are exhausted. Thus, correctness and accuracy are preserved by the strict chronological execution of the events.

To address this problem, the asynchronous distributed discrete event simulation algorithm P^2*EDAS* allocates each component of a digital system under simulation to a unique process on a processor of a parallel processor. Where an adequate number of processors are available, each component may be assigned to a unique processor. Components may be executed independently and concurrently by the underlying processors as soon as events are available at their input ports. Given the absence of the global event queue, to ensure correctness of simulation, P^2*EDAS* requires the following. A "simulation clock value," represented through $clock^N$, is computed for every component "N" and is defined as the time up to which the model has been simulated. A model may be executed accurately for all events with assertion times up to the minimum over all event times asserted at all input ports. Thus, the simulation clock value is identical to the minimum of all input events at all input ports. A difficulty in the simulation of digital systems is that, often, events are not generated at an output port of a component following its execution. Under such circumstances, events are not propagated over to the inputs of subsequent components that are connected to the output and the simulation clock value stagnates periodically.

A more serious problem occurs with sequential subcircuits, i.e. with feedback loops. Assume a simple circuit with two components C^1 and C^2 where the single output of C^1 is connected to the only input of C^2 and the output of C^2 is connected to the second input of C^1. The first input of C^1 is a primary input. Assume that C^1 and C^2 are allocated to two distinct processors. During the course of execution, assume that C^1 executes but generates no new output signal. Assume that there are no outstanding events at C^2. Thus, no event is propagated over to C^2. The simulation clock value does not advance and C^2 does not execute. Consequently, the second input of C^1 does not receive a new event and assume that this leads to no advance in C^1's simulation clock value. Clearly, neither C^1 nor C^2 executes since each expects events from the other. In reality, a large number of events may be asserted at the first input of C^1 but are not executed since C^1 is unaware that the output of C^2 has remained unchanged and it can safely execute the new events without causing any loss in simulation accuracy.

P^2*EDAS* addresses both problems through the use of the event prediction network. The network is synthesized for a given digital system under simulation and executes concurrently with the simulation of the behavior descriptions. It consists of mathematical entities, termed pseudo components, that generate "predicted event times." A predicted event time, defined at an output port of a pseudo component and associated with the corresponding output of the component, is the time at which an event is expected to be generated at that output and it is guaranteed that no events with assertion times less than the predicted event time will be generated. The predicted event time is computed separately, yet concurrently, from the actual simulation of the behavior descriptions. The simulation utilizes the predicted event times, generated by the event prediction network, to efficiently yet correctly schedule executions of the models for appropriate events.

Figure 9.23 presents a simple event prediction network for the behavior description in Figure 9.22. Although it is constituted by pseudo component(s), the network is conceptually represented through a single black box, EPN_t, with inputs $i^1, \ldots, i^N$ and outputs $o^1, \ldots, o^N$ that correspond to the inputs and outputs of the simulation model. The black box EPN_t receives predicted event times, represented by $pi_t^1, \ldots, pi_t^N$, at its input ports from the corresponding output ports of the preceding EPN_ts. Where an input port of the EPN_t is primary, the maximum predicted event time is defined by T_i. T_{max} is the maximum simulation time of interest and every primary input port is defined up to T_{max}. EPN_t generates two quantities: (1) the predicted event times at its outputs that are, in turn, propagated to the inputs of subsequent EPN_ts, and (2) t_{win} that defines which input events may trigger the execution of the corresponding behavior model. EPN_t does not require detailed knowledge of the behavior expressed in the model $C_{t,e}$. The behavior of EPN_t is complex and is detailed subsequently in this section.

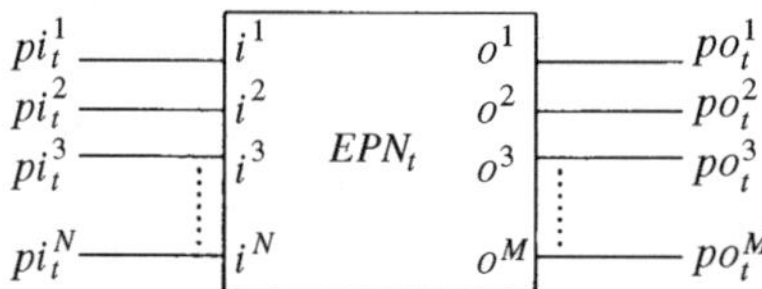

Figure 9.23 Event prediction network.

As an example, consider Figure 9.24 which shows a simple behavior model, C, with three inputs, i_1, i_2, and i_3. Assume that there are three events with assertion times 1 ns, 3 ns, and 5 ns associated with i_1, two events with assertion times 2 ns and 3 ns associated with i_2, and no events associated with input i_3. Clearly, the behavior model C may not be executed in the absence of input events at input i_3. Assume also that the event prediction network, EPN_t, receives the predicted event times, 7 ns, 3 ns, and 3 ns at input ports i_1, i_2, and i_3, respectively, prior to receiving the events at i_1, i_2, and i_3. In Figure 9.24, the predicted event times are associated with $*$ to distinguish them from regular event times. The EPN_t propagates the window of temporal independence to the model C and the latter safely executes all events at the input ports of C with assertion times less than t_{win} as per equation (9.1). The value of t_{win} is computed as the minimum over the predicted event times at all input ports, i.e. $minimum(7*, 3*, 3*) = 3$ ns. Thus, C is executed for the event with assertion time 1 at i_1 and the event with assertion time 2 at i_2 even before any event is asserted at input i_3 of C.

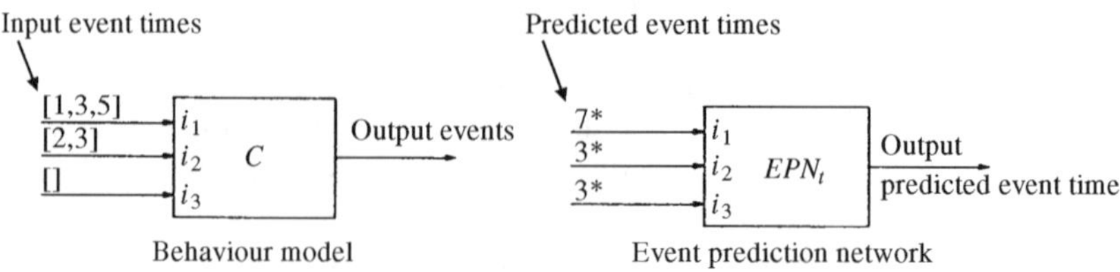

Figure 9.24 An example circuit to illustrate the role of event prediction network.

Therefore, the role of the event prediction network is to provide larger values for t_{win} and facilitate quick yet accurate simulation of the behavior models.

9.3.1.2 Synthesis of Event Prediction Network. In constructing the event prediction network for a digital system under simulation, first the feedback loops and the models included in the loops are identified.

Next, for a subcircuit containing feedback loops, a feedback arc set [100] S, given by $S = \{E_1, E_2, \ldots, E_n\}$, of a directed graph corresponding to the subcircuit is identified such that the graph may be rendered acyclic following the removal of all of the edges E_1 through E_n. The correctness of P^2EDAS is not affected by the identification of the minimal feedback arc set [100] which is difficult and time consuming. For each E_1 $\forall i \in \{1, 2, \ldots, n\}$, in the original directed graph, a new acyclic directed graph is constructed by replacing E_i with two unconnected edges E_i^{in} and E_i^{out} as illustrated in Figure 9.25. In Figure 9.25 a cyclic circuit consisting of a 2-input AND gate **A** whose output is connected through edge E_2 to the input of the inverter **B** is depicted. The output of the inverter **B** is connected through edge E_1 to an input of **A**. The other input port of **A** is edge E_3.

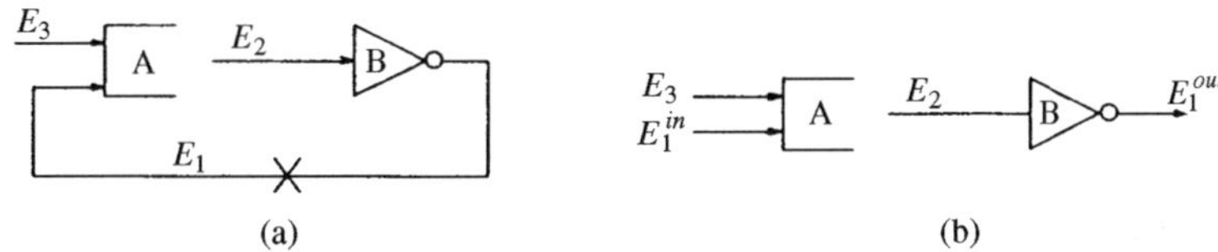

Figure 9.25 Reducing a cyclic directed graph to an acyclic directed graph.

Assume that the feedback arc set for the circuit is given by $S = \{E_1\}$. The graph is rendered acyclic in Figure 9.25(b) through the removal of E_1 and replacing it by E_1^{in} and E_1^{out} associated with the input of **A** and the output of **B**, respectively.

The event prediction network for the circuit is synthesized from connecting two identical copies of the acyclic circuit through a "crossbar switch." The crossbar switch represents a static connection of the output of a gate to the inputs of other gates that are affected. The two acyclic circuits to the left and right of the crossbar switch are referred to as the "tail network" and "head network" respectively. The entities in the event prediction network are termed pseudo components and identified individually as X_t and X_h respectively, where X refers to the corresponding simulation model. Pseudo components in the tail network are identified through X_t while those in the head network are expressed through X_h. Every input port of a pseudo component X_t that has a label of the form E_i^{in} is permanently held at an infinitely large number represented by the symbol ∞. Since inputs to pseudo components are times of events, the symbol ∞ represents the fact that no finite time values will be propagated through such inputs. An output port of every X_t that has a label of the form E_j^{out} is linked to the input port of any pseudo component Y_h in the head network that has a label of the form E_j^{in}. A link connecting the output port E_i^{out} of a pseudo component X_t in the tail network to an input port E_j^{in} of a component Y in the head network merely functions to propagate the predicted event times from the tail network to the head network. For the cyclic graph in Figure 9.25(a), the corresponding event prediction network is shown in Figure 9.26.

In Figure 9.26, the pseudo components A_t and B_t constitute the tail network where the input port E_1^{in} of A_t is permanently connected to ∞. A_t and B_t correspond to the AND and

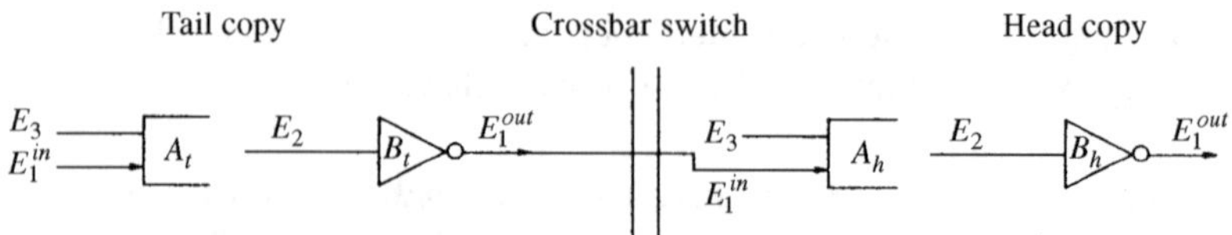

Figure 9.26 Event prediction network for the cyclic circuit in Figure 9.25(a).

inverter gates in the simulation system. Pseudo components A_h and B_h constitute the head network. The output port E_1^{out} of B_t is connected via the crossbar switch to the input port of E_1^{in} of A_h since the activities of the behavior models A and B may affect each other. The first input ports of both A_t and A_h are connected to the external path E_3.

Where an external input of a cyclic subcircuit X is, in turn, the output of another cyclic subcircuit Y, the two individual event prediction networks are linked together by connecting the output of the head pseudo component of Y to the corresponding external input port of the pseudo network for X. The final event prediction network corresponding to a digital design with three cyclic subcircuits, shown in Figure 9.27, is presented in Figure 9.28.

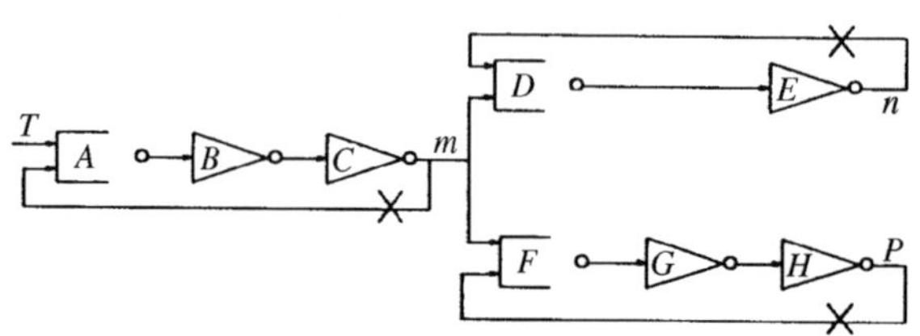

Figure 9.27 Digital circuit with cyclic subcircuits.

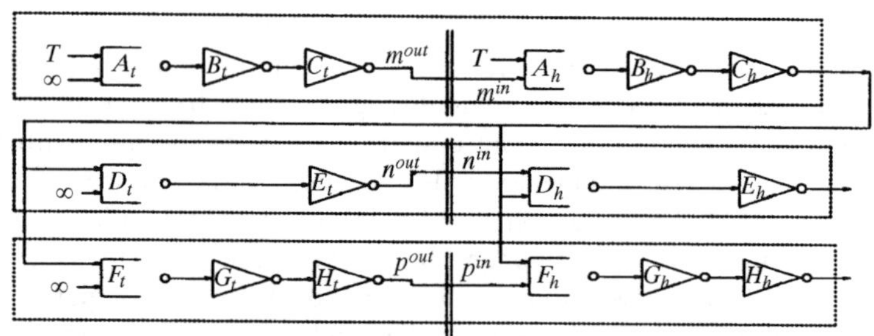

Figure 9.28 Event prediction network for Figure 9.27.

Corresponding to a combinational subcircuit, i.e. no feedback loops, the event prediction network consists only of the head network. Thus, for the digital system consisting of a combinational and two sequential subcircuits, as shown in Figure 9.29, the event prediction network is shown in Figure 9.30.

Within the event prediction network, the pseudo components compute the predicted event times and communicate them to other appropriate pseudo components through the interconnection links. For combinational subcircuits, the propagation of predicted event times terminate with the last head pseudo component. For sequential subcircuits, the process is more complex and explained subsequently.

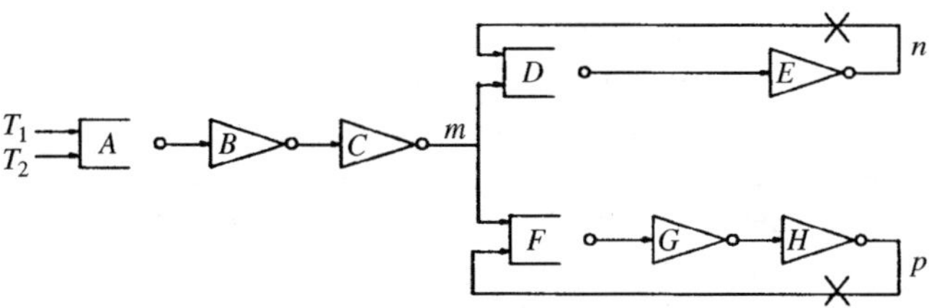

Figure 9.29 Digital circuit with cyclic and acyclic subcircuits.

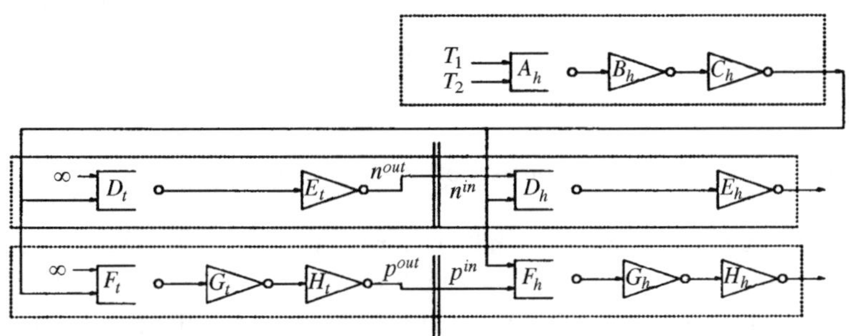

Figure 9.30 Event prediction network for Figure 9.29.

9.3.1.3 The P^2EDAS Algorithm. In P^2*EDAS*, the behavior models, corresponding to the architectural bodies of the VHDL entities, and event predictions network execute asynchronously and concurrently. The dependence between them is explained as follows. When external events are asserted at the primary input ports of the models, the event times are made available to the corresponding inputs of the pseudo components. In general, a model is permitted to execute appropriate events, limited by the value of the t_{win}. The event prediction network is responsible for computing updated values of t_{win}. It uses the T_i values at its primary inputs, the predicted event times at all other inputs, the event times associated with the input ports of the behavior models, and the propagation delay values to generate two quantities. The predicted event times at the outputs are computed by every pseudo component and propagated to the inputs of subsequent pseudo components. The head pseudo components, in addition, generate t_{win} values for the corresponding behavior models which are used by the simulation system to initiate execution of appropriate events. Upon execution of a behavior model, events may be generated at the output port(s). However, the events are not immediately asserted at the output ports. Only those output events are asserted at the respective outputs at appropriate times that are not preempted based on the rules established in [33, 46, 52, 105].

To preserve causality, and thereby guarantee correct simulation of the digital system, events must be executed in their causal order. This implies that, at every head pseudo component, the t_{win} value must increase monotonically.

Corresponding to every behavior model C^n is the "simulation clock value," $clock^n$. C^n has access to the t^n_{win} value associated with the corresponding head pseudo component. Assume that the symbol, S^n_o, represents the logical value of the most recent event asserted at the output "o" of C^n. Given that P^2*EDAS* employs preemption of inconsistent events, events generated as a consequence of execution of C^n are stored in an output queue and not immediately asserted at the output port of the behavior model. Assume that $t^n_{e_o}$ represents the time of the earliest event in the output event queue corresponding to the output "o" of C^n. Where the output event queue is empty, $t^n_{e_o}$ is set to ∞. If one or more input ports of C^n are primary, T_i defines the assertion time of their latest input event. Assume also that t_{e_i} represents the time of the earliest event at input "i" of C^n. P^2EDAS is explained for cyclic subcircuits, i.e. with feedback loops, since their simulation is most complex. Thus, assuming that C^n is included in a feedback loop, the event prediction network will consist of C^n_t and C^n_h, representing the tail and head pseudo components, respectively.

Corresponding to every head and tail pseudo component of C^n, the quantities W^n_i and W^n_o are associated with every input port "i" and output port "o" respectively. While W^n_i signifies the predicted event time at the input, W^n_o represents the predicted event time at the output.

INITIALIZATION. This consists of the following steps

- First, corresponding to each behavior model, C^n, $clock^n$ within the model and t^n_{win} in the head pseudo component within the event prediction network are set to 0.
- Second, for every pseudo component C^n_t and C^n_h, the predicted event times at all output ports are set to zero. Thus,

$$W^n_o = 0 \quad \forall n,\ \forall o \tag{9.8}$$

- Third, corresponding to those input ports of the tail pseudo components that are associated with the feedback arc set, the W^n_i values are set to ∞. The W^n_i values at all other input ports of pseudo components are initialized to zero.
- Fourth, the S^n_o values at every output of every head and tail pseudo component are user initialized.

SIMULATION PROCESS. The simulation system ensures the correct execution of the models triggered by appropriate events, the generation of output events, accurate determination of inconsistent events and their preemption, and propagation of correct output events to the subsequent behavior models.

- *Model execution:* For a given component, C^n, identify any and all events, t_{e_i}, at the input ports such that $t_{e_i} < t^n_{win}$. The behavior model is executed for all such events, starting with the earliest event. For every event executed, the $clock^n$ value is advanced, wherever possible. The value of $clock^n$ will always be less than t^n_{win}. The newly generated output events, if any, are included in the output event queue. They are not immediately asserted at the output. Utilizing the principles elaborated in Chapter 3, inconsistent events are identified and deleted.
- *Identification of inconsistent output events:* Due to the nature of the timing semantics of event driven simulation, one or more output events generated as a result of model execution may turn out to be inconsistent. Such inconsistent events are first identified and then preempted, utilizing the following principle that, in turn, is based on the principle of causality. The rules derived from these principles have already been explained in Chapter 3.

Consider two output assignments: e_1, which carries a logical value v_1 at $t = t_3$, and e_2, which carries a logical value v_2 at $t = t_4$, that are generated by input events i_1 at $t = t_1$ and i_2 at $t = t_2$ $(t_2 > t_1)$, respectively. If v_1 does not equal v_2 and $t_4 < t_3$, then e_1 and e_2 are inconsistent. In this case, e_2, the more recent effect of a later cause (input event), must take precedence over e_1, the less recent effect of an earlier cause. Therefore, e_1 is discarded, ensuring accuracy of results.

- *Propagation of correct output events:* For each output o of C^n, mark the earliest event, $t^n_{e_o}$, in the event queue, if the following relationship is true

$$t^n_{e_o} \leq clock^n \tag{9.9}$$

If affirmative, the marked event is noted as correct and is asserted at the output port of C^n. S^n_o is immediately updated. Other events in the output event queue, if any, that satisfy equation (9.9) are no longer subject to preemption and are also asserted at the output port "o". In P^2EDAS, whenever an input or output event queue is observed to become empty, a dummy event with assertion time ∞ is inserted in the queue. For improved communications efficiency, conceivably one may examine the output event queue with equation (9.9) after $clock^n$ advances to the maximum allowed value short of t^n_{win}, and propagate all correct events at once.

- *Updating event prediction network:* Whenever a new event is asserted at an input port of C^n, causing a new earliest event, the assertion time is propagated to the pseudo components. In addition, whenever the output event queue for an output of C^n is updated, the assertion time of the earliest event is propagated to the head and tail pseudo components. Following the assertion of a correct output event at the output port, the new logical value at the output of the behavior model is propagated to the pseudo components.
- The above three steps are continually executed until either all outstanding events are processed or $clock^n$ exceeds the maximum simulation time of interest.

EXECUTION OF THE EVENT PREDICTION NETWORK. Upon execution, a pseudo component generates a predicted event time, W^n_o, at the outputs. Where the output differs from its previous value, it is propagated right to trigger the execution of subsequent pseudo components in the form of a chain reaction. Should the output be identical to its previous value, it is not propagated. Whenever any of the arguments of a pseudo components' W^n_o value change, the pseudo component is executed. The underlying principle of execution of the event prediction network in P^2EDAS differs significantly from the data-flow network of YADDES. Also, unlike YADDES, P^2EDAS permits any number of propagation delays between every input to output path and ensures that inconsistent events are detected and preempted to guarantee correctness.

Corresponding to every head or tail pseudo component, a lookup table, represented through L^n, is constructed. The dimensions of the table are determined by the total number of input ports, total output ports, and total number of logical values at the outputs. For a given input port "i," output port "o," and current logical value "s," the entry in the table is the minimum of all possible transition times to other logical values at "o." P^2EDAS defines a function, *minimum_inertial_delay*( ) which accepts three arguments—input port number, output port number, and current logical value at the output—and generates the earliest transition time to any other legitimate logical value at the output. Although VHDL [33] proposes the use of inertial and transport delays in hardware descriptions, the semantics and implementation of transport delay are straightforward. The presentation of P^2EDAS in this section focuses on the more complex inertial delays.

EXECUTION OF THE HEAD AND TAIL PSEUDO COMPONENTS

- First, a pseudo component accesses the predicted event times, W_i^n, at every input port from the preceding pseudo components and assertion times of earliest events $t_{e_i}^n$ at every input port of the corresponding behavior model. It also receives the logical values S_o^n at every output port of the behavior model corresponding to the most recently asserted output events and the assertion times $t_{e_o}^n$ of the earliest events in every output event queue.
- Second, the pseudo component uses the accessed values to compute the predicted event time W_o^n at every output "o" using the following equation

$$W_o = minimum(t_{e_o}, (minimum(W_i, t_{e_i}) + minimum\_inertial\_delay(i, o, S_o))) \quad \forall i \tag{9.10}$$

It may be noted that, where there are nil events at an input port of the behavior model, t_{e_i} is set at ∞. Also, where there are nil events in the output event queue of a specific output, t_{e_o} is also set to ∞.

The head pseudo component also computes updated t_{win}^n values utilizing the accessed values and the following equation

$$t_{win}^n = minimum(W_i) \quad \forall i \tag{9.11}$$

Where a given input port, i, is primary, W_i is replaced by T_i in the computation of t_{win}^n. T_i represents the maximum simulation time of interest and its default value is infinity.
- Third, the newly computed W_o and t_{win}^n values are propagated to the subsequent pseudo components and behavior model, respectively when the values differ from their previous values.
- The above three steps are repeated until the simulation process terminates.

Figures 9.31 and 9.32 present the algorithms for the simulation model and pseudo components, in pseudo code.

The statement at label L1 and the two subsequent statements are executed only by the head pseudo component, X_h.

9.3.2 An Example to Illustrate P^2EDAS

To illustrate the working of P^2EDAS, consider a cyclic subcircuit, i.e. with feedback, whose execution is more complex relative to a combinational subcircuit. In Figure 9.33, each of the two NAND gates, A and B, have two inertial delay values given by $T_{plh} = 10$ ns and $T_{phl} = 5$ ns that correspond to low-to-high and high-to-low switching at the output, respectively. The function *minimum_inertial_delay*(i, o, S_0) therefore returns the value for T_{phl} corresponding to $s_0 = 0$ and the value of T_{phl} corresponding to $s_0 = 1$. To facilitate ease in representation, *minimum_inertial_delay*(i, o, S_0) is referred to as *delay*(i, o, s_0) in the remainder of this section. Assume that external inputs are asserted at the primary inputs, E_1 and E_2, as shown on Figure 9.33, and that the initial values at the output ports of A and B are "0" and "1" respectively. Initially, the window of temporal independence and the local

```
while (maximum simulation time is not exhausted)
do{
  check for messages;
    receive events at input ports from other components or external world;
    update event queue and order events according to assertion times;
  check for messages;
      receive updated window of temporal independence value from pseudo
      component;
  if (input event queue is modified){
       send time of earliest event in input event queue to pseudo components;
       }
  while (executable events at every input port or
          an executable event exists with time
            less than window of temporal independence)
      {
            advance simulation clock;
            assert events in output event queue that may not be preempted
              at output ports;
            execute simulation model;
            for each output port {
              if (new event is generated){
                 include it in the output event queue;
                 detect inconsistent events and preempt them;
              }
            }
            remove executed event(s) from input event queue(s).
      }
for each output{
  if (output event queue is modified){
        propagate assertion time of earliest event in output event queue
             to the pseudo components;
        propagate logical value of most recently asserted output event
             to the pseudo components;
      }
   }
}
```

Figure 9.31 Operations of a simulation model X.

clock of the models are set: $t_{win}^A = t_{win}^B = clock_A = clock_B = 0$. In addition, the event queues associated with the output ports of models A and B are empty, i.e. $t_o^A = t_o^B = \infty$.

For the circuit in Figure 9.33, the corresponding event prediction network is shown in Figure 9.34. For the pseudo components, the predicted event time values are set to 0, i.e. $W_{t_o}^A = W_{t_o}^B = W_{h_o}^A = W_{h_o}^B = 0$. Corresponding to a change in a predicted event time value in Figure 9.34, the value is propagated towards other pseudo components to the right, and a chain reaction is initiated, as described earlier in this section. In this section, the symbol "$\rightarrow$" represents the propagation of a predicted event time value. That is, $X \rightarrow Y$ indicates the

```
while(simulation model is executing){
     check for messages;
     receive assertion times of new earliest events at input ports of behavior
       model;
     receive logical value of most recently asserted output events of behavior
       model;
     receive assertion times of earliest events at output event queues;
     receive new W values at input ports from preceding pseudo component(s);

     for each output port {
       use the values received and delay values for every input output path
         to compute predicted event time, W, at output;
       if (W value for output is different) {
             propagate new W value to subsequent pseudo components;
         }
     }
     if (new W values received at input ports) {
          compute new window of temporal independence = minimum over all input
            W values;
         }
L1: if (window of temporal independence differs from its previous value) {
          propagate the new value to behavior model;
        }
}
```

Figure 9.32 Operations of a head (X_h) or tail (X_t) pseudo component.

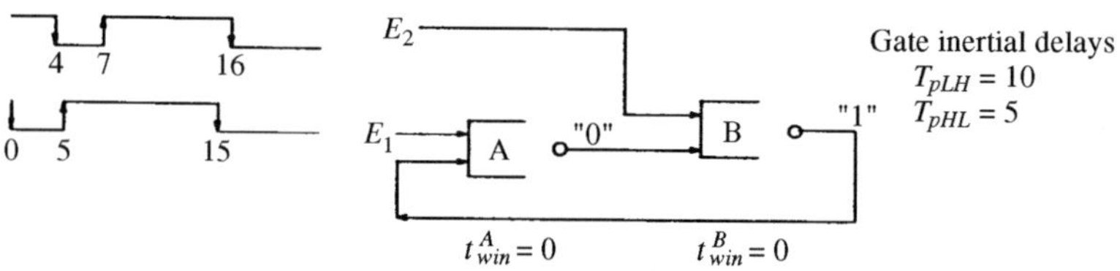

Figure 9.33 An example sequential subcircuit.

propagation of a predicted event time value from output X to input Y. Also, at any time instant, the event with the earliest assertion time at an input i of a model C is represented through t_i^C. Thus, for the event prediction network in Figure 9.34,

$$\infty \rightarrow W^A_{t_2} \qquad t^A_1 \rightarrow W^A_{t_1} \qquad W^A_{t_o} \rightarrow W^B_{t_2} \qquad t^B_1 \rightarrow W^B_{t_1}$$
$$W^B_{t_o} \rightarrow W^A_{h_2} \qquad t^A_1 \rightarrow W^A_{h_1} \qquad W^A_{h_o} \rightarrow W^B_{h_2} \qquad t^B_1 \rightarrow W^B_{h_1}$$

The execution of P^2EDAS is organized through the following steps.

Step 1: Following the assertion of the external transitions at the primary inputs E_1 and E_2, the pseudo components in the event prediction network are initiated. New predicted event times (W values) are computed, utilizing equation (9.10), and are shown in Figure 9.35.

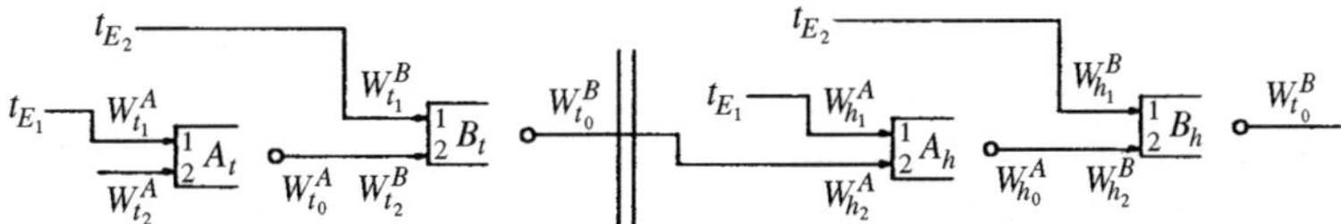

Figure 9.34 Event prediction network corresponding to Figure 9.33.

Corresponding to the newly generated predicted event time values, the window of temporal independence values for models A and B are computed utilizing equation (9.11) and shown in Figure 9.35.

$W_{t_0}^B = \min(0+\text{delay}(1,o,\text{"1"}),\ W_{t_2}^B+\text{delay}(2,o,\text{"1"}),\ t_0^B)$
$= \min(0+5, 5+10, \infty)$
$= 9(\text{changed}) \longrightarrow W_{h_2}^A$

$W_{t_0}^B = \min(0+\text{delay}(1,o,\text{"1"}),\ W_{t_2}^B+\text{delay}(2,o,\text{"1"}),\ t_0^B)$
$= \min(4+5, 5+10, \infty)$
$= 9(\text{changed}) \longrightarrow W_{h_2}^A$

$W_{t_0}^A = \min(4+\text{delay}(1,o,\text{"0"}),\ \infty+\text{delay}(2,o,\text{"0"}),\ t_0^A)$
$= \min(4+10, \infty+10, \infty)$
$= 10(\text{changed}) \longrightarrow W_{h_2}^B$

$W_{h_0}^A = \min(4+\text{delay}(1,o,\text{"0"}),\ W_{h_2}^A+\text{delay}(2,o,\text{"0"}),\ t_0^A)$
$= \min(0+10, 9+10, \infty)$
$= 10(\text{changed}) \longrightarrow W_{h_2}^B$

By equation (9.11) : $t_{win}^A = \min(\infty, 9) = 9 \longrightarrow A$
$t_{win}^B = \min(\infty, 10) = 10 \longrightarrow B$

Figure 9.35 Step 1: Execution of pseudo components.

Step 2: Input transitions to models A and B with assertion times less than 9 ns and 10 ns, respectively, may be executed. A and B may execute concurrently. Figure 9.36 describes the state of the subcircuit.

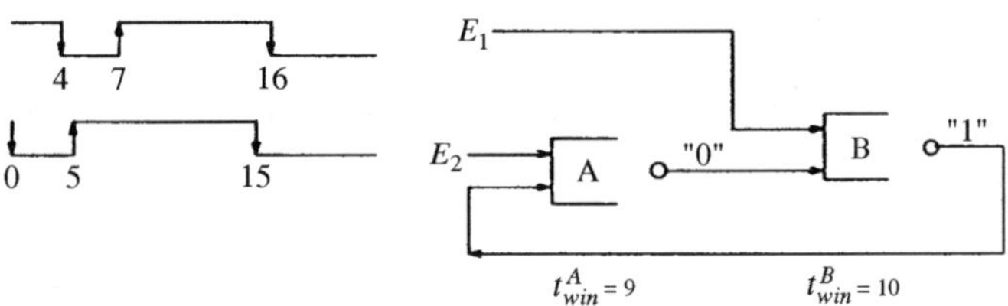

Figure 9.36 Step 2: Execution of models A and B utilizing a new window of temporal independence values.

First, the high to low transition at time 0 ns at the input of model A is executed. The execution of A generates an output event: $A_o = (0\ NAND\ 1) \Rightarrow 10 \uparrow$. Since the current logical value at A is 0, the output event queue of A will contain a low to high transition for time $t_o^A = 10$. Simultaneous with the execution of A, the high to low transition at the input of B is executed. The execution of B generates an output event: $B_o = (0\ NAND\ 0) \Rightarrow 14 \uparrow$. Thus, the newly generated event is a low to high transition at $t_o^B = 14$. Since the current logical value at the output of model B is already high, the generated event is deleted. The output event queue of B is empty and t_o^B is reset to ∞.

The processed input transitions are deleted. For the input event queue of A, $E_1 = 5 \uparrow$ defines the subsequent transition. For the input event queue of B, $E_2 = 7 \uparrow$ defines the subsequent transition. Both of these transitions may be executed, concurrently, since their times are defined within and less than the respective t_{win} values. The execution of A generates an output event: $A_o = (1\ NAND\ 1) \Rightarrow 10 \downarrow$. The previously generated event, stored in the output event queue of A, is thus rendered inconsistent. It is preempted utilizing the principles elaborated earlier in this book. The newly generated output event of A is also discarded since its logical value is indistinguishable from the current logical value at the output of A. Therefore, the output event queue of A is empty and t_o^A is set to ∞. The execution of B generates an output event: $B_o = (1\ NAND\ 0) \Rightarrow 17 \uparrow$ which is also indistinguishable from the current logical value at the output of B. This event is deleted and t_o^B remains at ∞.

Step 3: The processed input transitions are deleted. For the input event queue of A, $E_1 = 15 \downarrow$ defines the subsequent transition. For the input event queue of B, $E_2 = 16 \downarrow$ defines the subsequent transition. Neither of these transitions may be executed since they exceed the respective t_{win} values. For further execution of the models, the event prediction network must execute and update the t_{win} values. The state of the subcircuit is shown in Figure 9.37.

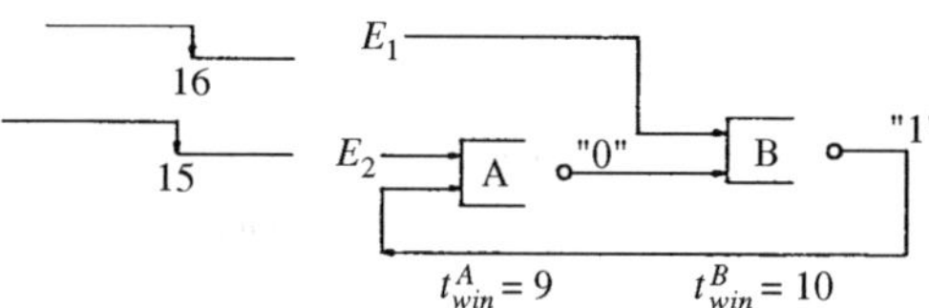

Figure 9.37 Step 3: State of subcircuit following execution of models A and B.

Step 4: The event prediction network is initiated which utilizes the fact that the transitions $E_1 = 15 \downarrow$ and $E_2 = 16 \downarrow$ are asserted at the respective input ports of the models, i.e. $t_1^A = 15 \rightarrow W_{t_1}^A,\ W_{h_1}^A,\ t_1^B = 16 \rightarrow W_{t_1}^B,\ W_{h_1}^B$. The execution of the event prediction network is shown in Figure 9.38.

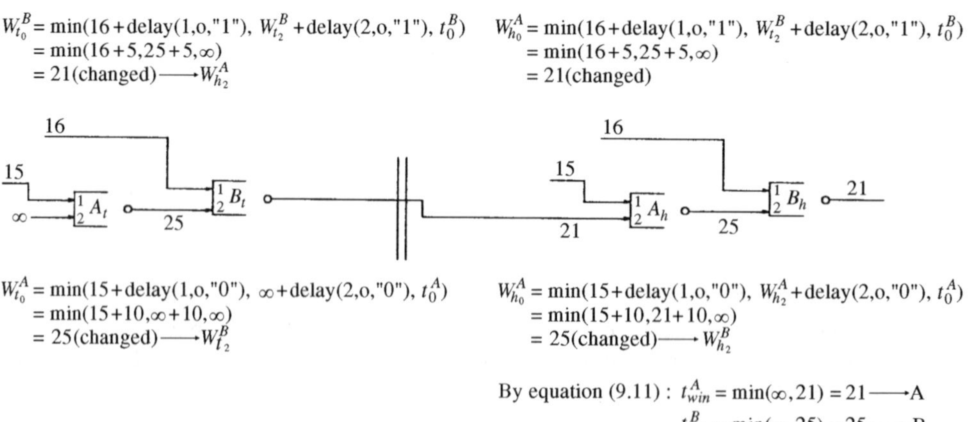

Figure 9.38 Step 4: Execution of the event prediction network.

In the event prediction network, the predicted event times, $W^A_{t_o}$, $W^B_{t_o}$, $W^A_{h_o}$, and $W^B_{h_o}$ are recomputed, as shown in Figure 9.38. As a result, the window of temporal independence values for models A and B are computed utilizing equation (9.11) and also shown in Figure 9.38. Thus, $t^A_{win} = 21 \rightarrow$ **A** and $t^B_{win} = 25 \rightarrow$ **B**. The state of the subcircuit is shown in Figure 9.39.

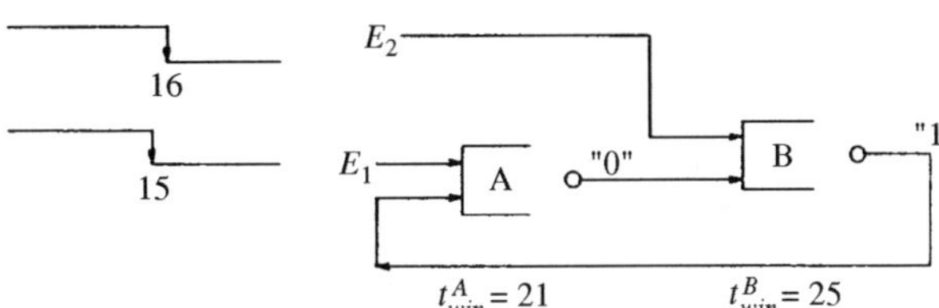

Figure 9.39 Step 4: State of subcircuit with new window of temporal independence values.

Step 5: Input transitions to models A and B with assertion times less than 21 ns and 25 ns, respectively, may be executed. First, the high to low transition at time 15 ns at the input of model A is executed. The execution of A generates an output event: $A_o = (0\ NAND\ 1) \Rightarrow 25 \uparrow$. Since the current logical value at A is 0, the output event queue of A will contain a low to high transition for time $t^A_o = 25$ ns. Simultaneous with the execution of A, the high to low transition at the input of B is executed. The execution of B generates an output event: $B_o = (1\ NAND\ 0) \Rightarrow 26 \uparrow$. Thus, the newly generated event is a low to high transition at $t^B_o = 14$ ns. Since the current logical value at the output of model B is already high, the generated event is deleted. The output event queue of B is empty and t^B_o is reset to ∞.

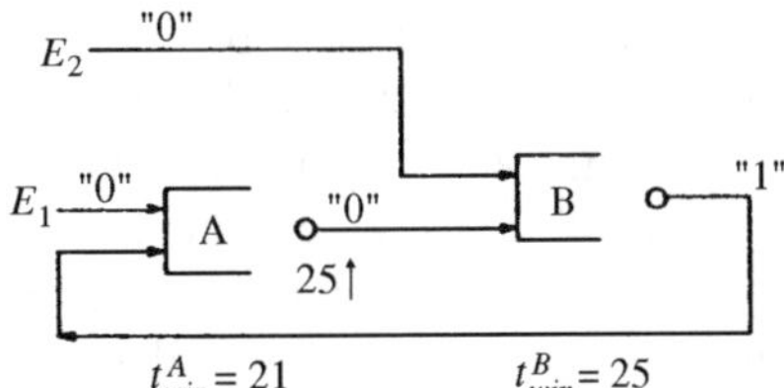

Figure 9.40 Step 5: State of subcircuit following execution of models A and B.

The processed input transitions are deleted. For the now empty input event queue of A, $t_{E_1} = \infty \rightarrow W^A_{t_1}$ and $t_{E_2} = \infty \rightarrow W^B_{t_1}$. The state of the subcircuit is shown in Figure 9.40.

Step 6: The event prediction network is initiated and it utilizes the fact that the transitions $E_1 = \infty$ and $E_2 = \infty$ are asserted at the respective input ports of the models.

In the event prediction network, the predicted event times, $W^A_{t_o}$, $W^B_{t_o}$, $W^A_{h_o}$, and $W^B_{h_o}$ are recomputed, as shown in Figure 9.41. As a result, the window of temporal independence values for models A and B are computed utilizing equation (9.11) and also shown in Figure 9.41. Thus, $t^A_{win} = 30 \rightarrow$ **A** and t^B_{win} remains at 25.

Step 7: Since $25 < 30$, the inequality $t^A_o < t^A_{win}$ is satisfied. That is, it is guaranteed that no new transitions will be asserted at any input port of A to preempt the previously generated output event, $t^A_o = 25$, at the output event queue of A. The output event is safely propagated to model B and t^A_o is reset to ∞. The state of the subcircuit is shown in Figure 9.42.

Step 8: Since $25 \not< 25$, the new transition at the input of model B cannot be executed with the current t_{win} value. The event prediction network is initiated and the pseudo components are executed.

$W_{t_0}^B = \min(\infty + delay(1,o,"1"),\ W_{t_2}^B + delay(2,o,"1"),\ t_0^B)$
$= \min(\infty + 5, 25 + 5, \infty)$
$= 30\,(changed) \longrightarrow W_{h_2}^A$

$W_{h_0}^B = \min(\infty + delay(1,o,"1"),\ W_{t_2}^B + delay(2,o,"1"),\ t_0^B))$
$= \min(\infty + 5, 25 + 5, \infty)$
$= 30\,(changed)$

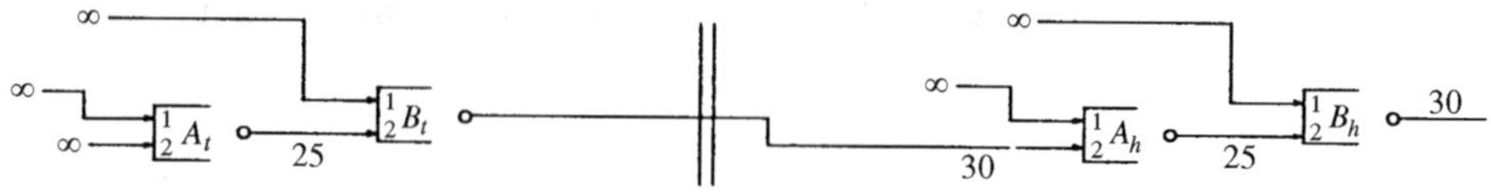

$W_{t_0}^A = \min(\infty + delay(1,o,"0"),\ \infty + delay(2,o,"0"),\ t_0^A)$
$= \min(\infty + 10, \infty + 10, 25)$
$= 25\,(nochange)$

$W_{h_0}^A = \min(\infty + delay(1,o,"0"),\ W_{h_2}^A + delay(2,o,"0"),\ t_0^A)$
$= \min(\infty + 10, 30 + 10, 25)$
$= 25\,(nochange)$

By equation (9.11) : $t_{win}^A = \min(\infty, 30) = 30 \longrightarrow A$
$t_{win}^B = \min(\infty, 25) = 25\,(nochange)$

Figure 9.41 Step 6: Execution of the event prediction network.

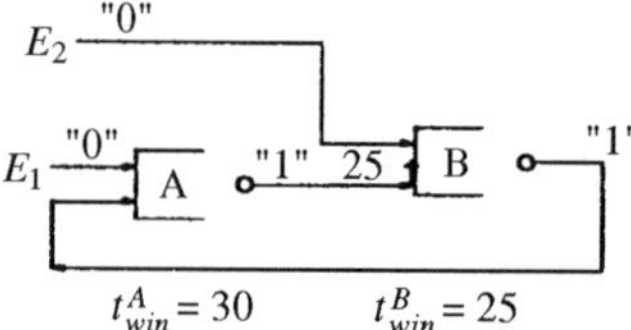

Figure 9.42 Step 7: State of the subcircuit following execution of models A and B.

In the event prediction network, the predicted event times $W_{t_o}^A$, $W_{t_o}^B$, $W_{h_o}^A$, and $W_{h_o}^B$ are recomputed, as shown in Figure 9.43. As a result, the window of temporal independence values for models A and B are computed utilizing equation (9.11) and also shown in Figure 9.43. Thus, $t_{win}^B = 35 \rightarrow$ **B** and t_{win}^A remains at 30.

Step 9: Since 25 < 35, the low to high input transition at the input of model B with assertion time of 25 ns is executed. The execution of B generates an output event: $B_o = (0\ NAND\ 1) \Rightarrow 35 \uparrow$. Since the current logical value at the output of model B is already high, the generated event is deleted. The output event queue of B is empty and t_o^B is

$W_{t_0}^B = \min(\infty + delay(1,o,"1"),\ W_{t_2}^B + delay(2,o,"1"),\ t_0^B)$
$= \min(\infty + 5, 25 + 5, \infty)$
$= 30\,(nochange)$

$W_{h_0}^B = \min(\infty + delay(1,o,"1"),\ W_{t_2}^B + delay(2,o,"1"),\ t_0^B))$
$= \min(\infty + 5, 35 + 5, \infty)$
$= 40\,(changed)$

$W_{t_0}^A = \min(\infty + delay(1,o,"0"),\ \infty + delay(2,o,"1"),\ t_0^A)$
$= \min(\infty + 5, \infty + 5, \infty)$
$= \infty\,(changed) \longrightarrow W_{t_2}^B$

$W_{h_0}^A = \min(\infty + delay(1,o,"0"),\ W_{h_2}^A + delay(2,o,"0"),\ t_0^A)$
$= \min(\infty + 5, 30 + 5, \infty)$
$= 35\,(changed) \longrightarrow W_{h_2}^B$

By equation (9.11) : $t_{win}^A = \min(\infty, 30) = 30$
$t_{win}^B = \min(\infty, 25) = 35 \longrightarrow B$

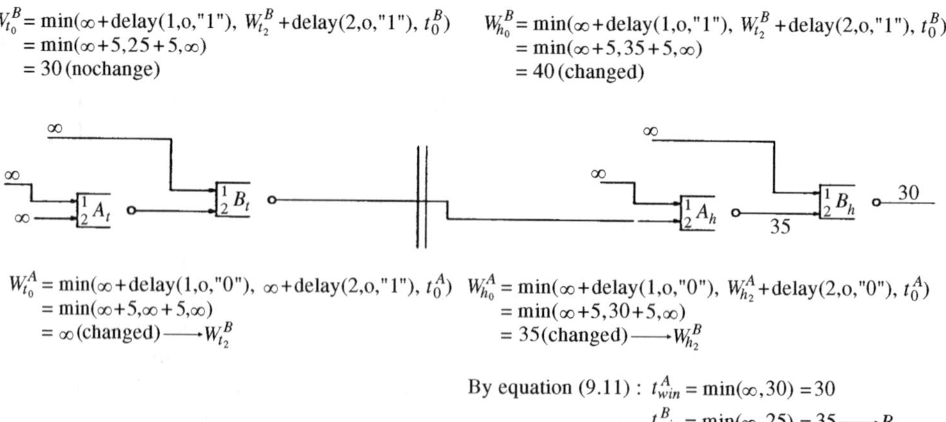

Figure 9.43 Step 8: Execution of the event prediction network.

reset to ∞. Also, the processed input transitions are deleted. The state of the subcircuit is shown in Figure 9.44.

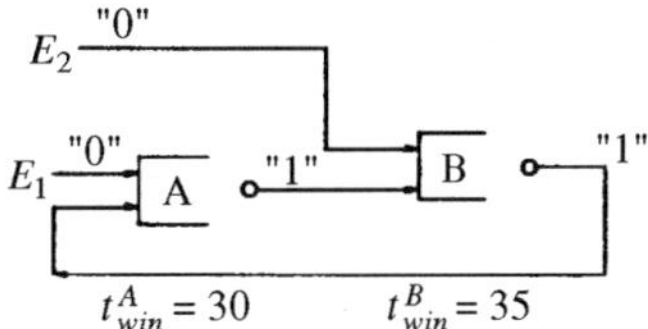

Figure 9.44 Step 9: State of the subcircuit following execution of model B.

Step 10: Corresponding to the update of the output event queue of B, namely t^B_o is reset to ∞, the event prediction network is initiated and the pseudo components are executed.

In the event prediction network, the predicted event times $W^B_{t_o}$, $W^A_{h_o}$, and $W^B_{h_o}$ are recomputed, as shown in Figure 9.45. The window of temporal independence values for models A and B are also computed utilizing equation (9.11) and shown in Figure 9.45. Thus, $t^B_{win} = \infty$ and $t^A_{win} = \infty$, indicating the termination of simulation.

$W^B_{t_0} = \min(\infty + delay(1,o,"1"), W^B_{t_2} + delay(2,o,"1"), t^B_0)$
$= \min(\infty+5, \infty+5, \infty)$
$= \infty \longrightarrow W^A_{h_2}$

$W^B_{h_0} = \min(\infty + delay(1,o,"1"), W^B_{t_2} + delay(2,o,"1"), t^B_0))$
$= \min(\infty+5, \infty+5, \infty)$
$= \infty$

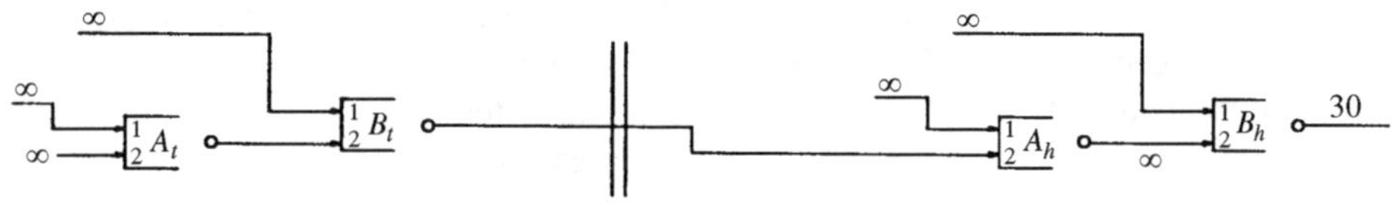

∞ ∞ ∞ A_t B_t ∞ ∞ A_h ∞ B_h 30

$W^A_{h_0} = \min(\infty + delay(1,o,"0"), W^A_{h_2} + delay(2,o,"1"), t^A_0)$
$= \min(\infty+5, \infty+5, \infty)$
$= \infty$ (changed) $\longrightarrow W^B_{h_2}$

By equation (9.11) : $t^A_{win} = \min(\infty,\infty) = \infty \longrightarrow A$
$t^B_{win} = \min(\infty,\infty) = \infty \longrightarrow B$

Figure 9.45 Step 10: Execution of the event prediction network.

Step 11 (Final): In the final step, the input and output event queues are empty and the number of outstanding transitions is nil. The final output waveform produced as a result of the execution of the subcircuit is shown in Figure 9.46.

9.3.3 Issues of Correctness, Freedom from Deadlock, and Termination of Simulation

9.3.3.1 Execution of Events in the Correct Order: Monotonicity in W_h Values. The conservative nature of P^2EDAS requires that events be executed in the correct order. That is, if an input event with assertion time t_z is executed by a model, the model may not execute any event with assertion time t_y in the future, where $t_y < t_z$. Similarly, if an output event with time t_z is asserted at the output, the model may not assert any output event with time t_y at the output in the future, where $t_y < t_z$. Since models are executed asynchronously and concurrently on

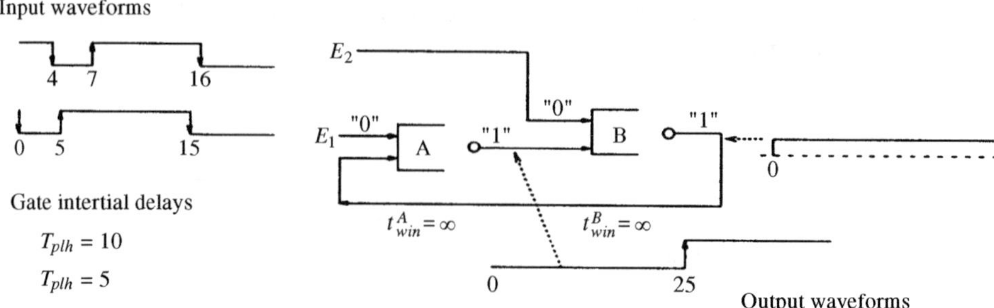

Figure 9.46 Step 11 (final): Output waveform generated by subcircuit execution.

different processors, the propagation of events from one model to another may require arbitrary time, and as output events may be rendered inconsistent requiring deletion, it is critical that P^2EDAS guarantees the execution of input and output events in the correct order. It is assumed, utilizing the guaranteed end-to-end message delivery in TCP/IP, that if two events with assertion times t_a and t_b ($t_a < t_b$) respectively, are propagated from model A to model B, the event with the smaller assertion time, t_a, will arrive at B prior to the event with assertion time t_b. Also, it is given that, at the primary inputs of the models, events are asserted in the order of increasing assertion time values.

In P^2EDAS, where multiple input events are available at a model, C^N, the $clock^N$ advances successively to the input event with the smallest assertion time, provided that $clock^N$ does not exceed the t^N_{win} value. Also, if the assertion time of an output event is less than or equal to the value of $clock^N$, the output event is asserted at the output. Therefore, if it can be shown that the t^N_{win} for every model only increases successively as the simulation progresses, i.e. it is a monotonically increasing function, then the execution of events in the correct order is guaranteed.

The t^N_{win} value for a model is given by equation (9.11) and is computed from the W_i values at the input ports of the corresponding head pseudo component in the event prediction network. Consider the circuit in Figure 9.47 and the corresponding event prediction network in Figure 9.48. In Figure 9.47, the input pin 1 of every model 1 through N is primary. While t^1_1 represent the event with the smallest assertion time associated with model 1, t^N_1 refer to the input with the smallest assertion time at model N. The event queues corresponding to the second input pin of each of the models 1 through N contain events generated as a result of execution of the models. For compactness in representation, the *minimum_inertial_delay*(i, o, S_0) is referred to as $d(i, o, s_0)$ in this section. Therefore,

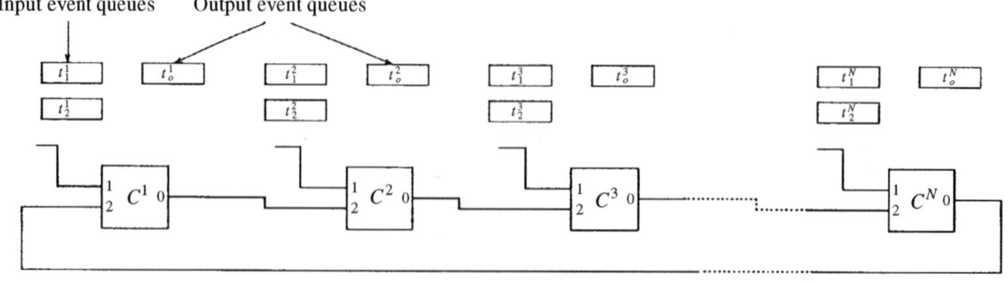

Figure 9.47 An example circuit to illustrate the monotonicity in the W_h values.

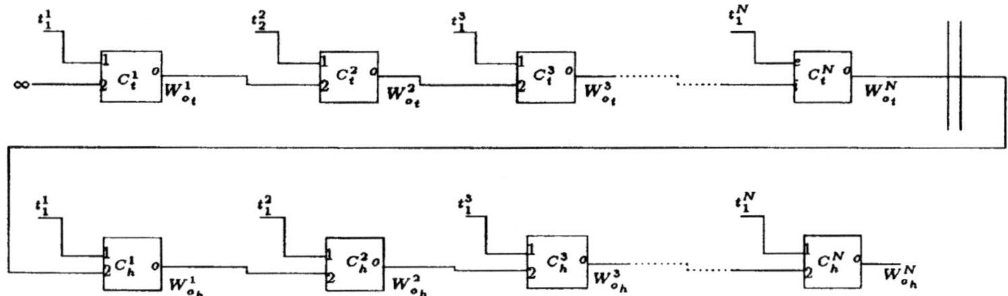

Figure 9.48 Event prediction network for the circuit in Figure 9.47.

to guarantee the execution of events in the correct order, it is adequate if it is proved that the W_i values at the inputs of head pseudo components and W_{oh} values at the output of the head pseudo components increase monotonically.

In Figure 9.48, the W value at the output of the pseudo component 1 in the tail event prediction network is computed as follows

$$\begin{aligned} W_{ot}^1 &= minimum(t_o^1,\ t_1^1 + d^1(1,\ o,\ S_o^1),\ d^1(2,\ o,\ S_o^1) + W_1^1) \\ &= minimum(t_o^1,\ t_1^1 + d^1(1,\ o,\ S_o^1),\ d^1(2,\ o,\ S_o^1) + \infty) \\ &= minimum(t_o^1,\ t_1^1 + d^1(1,\ o,\ S_o^1),\ \infty) \\ &= minimum(t_o^1,\ t_1^1 + d^1(1,\ o,\ S_o^1)) \end{aligned} \tag{9.12}$$

The W value at the output of the pseudo component 2 in the tail event prediction network is computed as follows

$$W_{ot}^2 = minimum(t_o^2,\ t_1^2 + d^2(1,\ o,\ S_o^2),\ t_2^2 + d^2(2,\ o,\ S_o^2),\ d^2(2,\ o,\ S_o^2) + W_{ot}^1) \tag{9.13}$$

Substituting equation (9.12) in equation (9.13), we get

$$\begin{aligned} W_{ot}^2 = minimum(&t_o^2,\ d^2(1,\ o,\ S_o^2) + t_1^2,\ d^2(2,\ o,\ S_o^2) + t_2^2,\ d^2(2,\ o,\ S_o^2) + t_o^1, \\ &d^2(2,\ o,\ S_o^2) + d^1(1,\ o,\ S_o^1) + t_1^1,\ d^2(2,\ o,\ S_o^2) + d^1(2,\ o,\ S_o^1) + t_2^1) \end{aligned} \tag{9.14}$$

The W value at the output of the pseudo component 3 in the tail event prediction network is computed as

$$W_{ot}^3 = minimum(t_o^3,\ t_1^3 + d^3(1,\ o,\ S_o^3),\ t_2^3 + d^3(2,\ o,\ S_o^3),\ d^3(2,\ o,\ S_o^3) + W_{ot}^2) \tag{9.15}$$

Substituting equation (9.14) in equation (9.15), we get

$$\begin{aligned} W_{ot}^3 = minimum(&t_o^3,\ d^3(1,\ o,\ S_o^3) + t_1^3,\ d^3(2,\ o,\ S_o^3) + t_2^3,\ d^3(2,\ o,\ S_o^3) + t_o^2, \\ &d^3(2,\ o,\ S_o^3) + d^2(1,\ o,\ S_o^2) + t_1^2,\ d^3(2,\ o,\ S_o^3) + d^2(2,\ o,\ S_o^2) + t_2^2, \\ &d^3(2,\ o,\ S_o^3) + d^2(2,\ o,\ S_o^2) + t_o^1,\ d^3(2,\ o,\ S_o^3) + d^2(2,\ o,\ S_o^2) + d^1(1,\ o,\ S_o^1) + t_1^1, \\ &d^3(2,\ o,\ S_o^3) + d^2(2,\ o,\ S_o^2) + d^1(2,\ o,\ S_o^1) + t_2^1) \end{aligned} \tag{9.16}$$

While equation (9.12) contains 3 terms, equation (9.14) contains 6 terms and equation (9.16) has 9 terms in it. The expression for W_{ot}^N will contain "3N" terms in it and may be derived through analogy as

$$\begin{aligned} W_{ot}^N = minimum(&t_o^N,\ d^N(1,\ o,\ S_o^N) + t_1^N,\ d^N(2,\ o,\ S_o^N) + t_2^N,\ d^N(2,\ o,\ S_o^N) + t_o^{N-1}, \\ &d^N(2,\ o,\ S_o^N) + d^{N-1}(1,\ o,\ S_o^{N-1}) + t_1^{N-1}, \ldots, \\ &d^N(2,\ o,\ S_o^N) + d^{N-1}(2,\ o,\ S_o^{N-1}) + \cdots + d^1(2,\ o,\ S_o^1) + t_2^1) \end{aligned} \quad (9.17)$$

The W_{ot}^N value is the input W value to the first pseudo component in the head section of the event prediction network. It influences the computation of t_{win}^1 for model 1 which, in turn, will determine the execution of the input and output events of model 1. It will also affect the expressions for W_{oh}^1 through W_{oh}^N. The expression in equation (9.17) takes into account the influences of every input transition and output event, if any, that may affect its value and, therefore, the execution of events by model 1. Thus, the terms in the expression in equation (9.17) represent the transitive closure over the t_1 values and t_o of all models. This section will show that W_{ot}^N is a monotonic function, i.e. its value only increases as a function of the progress of simulation.

The proof focuses on all worst-case scenarios that create the possible appearances of decrease in the value of W_{ot}^N between two successive computations and shows that they are false. The values of t_o^1 through t_o^N, t_1^1 through t_1^N, and t_2^1 through t_2^N changes as simulation progresses.

Scenario 1: Following the execution of an input transition at a primary input, say t_1^X, of model X, it is deleted and is replaced by the subsequent event at that input, if any, which has a higher value of assertion time. Since the models and pseudo component execute asynchronously and potentially on different processors, P^2EDAS first propagates the resulting t_o^X value, if different from the previous value, to the corresponding pseudo components, and then deletes the executed input event. This prevents a potential race condition that may momentarily cause a decrease in the W_{ot}^N value. When all transitions are consumed, the t_1^X value will be set to ∞. Also, when an output event is asserted at the output port, prior to deleting it, P^2EDAS ensures that the output event is propagated to the input event queues of the models that are connected to the output as well as the corresponding pseudo components in the event prediction network. Clearly, between any two successive values of W_{ot}^N, none of the t_1^X values will not decrease and will not contribute to lowering the value of W_{ot}^N.

Scenario 2: Next, assume an input event, t_{2a}^X, of model X. Assume that $clock^X = t_{2a}^X$ and t_{2a}^X is less than t_{win}^X. Therefore, t_{2a}^X is executed and say that an output event, t_{o2}^X, is generated at the output of model X. Clearly the assertion time of t_{o2}^X exceeds that of t_{2a}^X. The output event is included in the output event queue which already contains an event t_{o1}^X. Following its execution, the event t_{2a}^X is deleted and assume that the W_{ot}^N value is now determined by t_{o1}^X. Now assume that there is a second transition at the same input pin, t_{2b}^X, such that t_{2b}^X is less than t_{win}^X. Thus, the $clock^X$ is first advanced to t_{2b}^X. Assume that the entry t_{o1}^X in the output event queue satisfies the condition that t_{o1}^X is less than or equal to the new value of $clock^X$. Therefore, the output event t_{o1}^X is asserted at the output and the logical value, s_o^X, changes. The entry t_{o2}^X now occupies the head of the output event queue and assume that it controls the value of W_{ot}^N. Next, the input event t_{2b}^X is executed and an output event t_{o3}^X is generated. The generation of t_{o3}^X utilizes a different delay value as compared to the generation of t_{o2}^X and assume that t_{o3}^X is less than t_{o2}^X. Therefore, the output event t_{o2}^x will be preempted and event t_{o3}^X will become the head of the output event queue. If the value of W_{ot}^N has been controlled by the output event queue, the value of W_{ot}^N appears to suffer a decrease. This appearance is false for the following reason. Immediately preceding the execution of the input event t_{2b}^X, the value of W_{ot}^N was computed as the minimum of t_{2b}^X + delay( ), t_{o2}^X, and other factors. Clearly since t_{o3}^X is equal to t_{2b}^X + delay( ) and it is smaller than t_{o2}^X, the value of W_{ot}^N must have been determined by t_{2b}^X + delay( ) and not t_{o2}^X. The value of W_{ot}^N does not change when the output event t_{o2}^X is preempted. Therefore, the W_{ot}^N does not decrease.

Scenario 3: Consider the following situation that differs slightly from scenario 2. Consider that the two input events are associated with two different input pins of model X, i.e. t_{2a}^X and t_{1b}^X, where t_{2a}^X is executed prior to event t_{1b}^X. Also assume that the assertion time of the output event, t_{o3}^X, corresponding to t_{1b}^X is less than that of the event t_{o2}^X generated from the execution of t_{2a}^X. This is easily conceivable, since the generation of t_{o2}^X utilizes a delay value that is different from the one utilized for the generation of t_{o3}^X, regardless of whether there is a previously generated entry in the output event queue, t_{o1}^X, or not. Clearly, output event t_{o2}^X will be preempted. Under these circumstances, a proof may be constructed, similar to the one above, to show that the value of W_{ot}^N will be controlled by t_{1b}^X + delay( ) and not by t_{o2}^X. Therefore, W_{ot}^N will not decrease despite the incidence of preemption.

Scenario 4: Consider the following situation that also differs slightly from scenario 2. Consider that the two input events are associated with the same input pin 2 and that the output event queue is empty. Following the execution of the first input event, t_{2a}^X, the output event t_{o2}^X is generated. Since both input events are less than the current value of t_{win}^X, the comparison of the output event t_{o2}^X with the t_{win}^X is deferred until the second input event, t_{2b}^X, is executed. Following the execution of t_{2b}^X, output event t_{o3}^X is generated. Since the logical value at the output of model X, s_o^X, has not changed between the two consecutive executions of X, the same delay value is utilized in the generation of t_{o2}^X and t_{o3}^X. Since t_{2a}^X is less than t_{2b}^X, t_{o3}^X cannot be less than t_{o2}^X. Thus, there is neither preemption or any decrease in the value of W_{ot}^N.

It may be proved, analogously, that the values of W_{oh}^1 through W_{oh}^N, associated with the pseudo components 1 through N of the head section of the event prediction network in Figure 9.48, will also increase monotonically. The assertion that W_{oh}^X values increase monotonically may be proved for any digital circuit with any number of feedback loops.

9.3.3.2 *Proof of Freedom from Deadlock.* The principal characteristics of P^2EDAS that ensure freedom from deadlock include the following. A change in any input event queue of any of the models will trigger the initiation of the event prediction network. Changes in the event queue may arise as a result of new events asserted at the primary inputs or assertion of output events of a model at its output that subsequently alter the input event queues of other models that are connected to the output. During execution, the event prediction network propagates any and all changes to W_t and W_h values such that all necessary and sufficient W_t and W_h values are updated leading to the computation of correct values of t_{win}. Also, the execution of the pseudo components of the event prediction network are unidirectional, i.e. from left to right, and asynchronous and concurrent with respect to each other and the models.

Assume, on the contrary, that P^2EDAS deadlocks. First, consider the scenario where there are no outstanding input or output events. Thus, all external events are executed, implying that all input event queues are empty. Similarly, the output event queues are empty. As a result, all W_t, W_h, t_{win}, and clock values are set to ∞. Clearly, the simulation has terminated.

Next, consider the scenario wherein a model, C^X, contains an event in the output event queue, represented by t_o^X. The simulation is in deadlock and the output event t_o^X continues to remain unasserted at the output of X indefinitely. Therefore

$$t_{win}^X \leq t_o^X \tag{9.18}$$

In the event prediction network, t_o^X is the only event. Therefore, the W_h^{X-1} value at the output of the preceding model, C^{X-1}, must be given by

$$W_h^{X-1} = min(t_o^X + d^{X+1}(2,\ o,\ s_0) + \cdots + d^1(2,\ o,\ s_0) + \cdots + d^{X-1}(2,\ o,\ s_0)) \tag{9.19}$$

Now, t_{win}^X is computed as the minimum over all incoming W_h values, as per equation (9.11), and is given by

$$t_{win}^X = min(t_o^X + d^{X+1}(2,\ o,\ s_0) + \cdots + d^1(2,\ o,\ s_0) + \cdots + d^{X-1}(2,\ o,\ s_0)) \tag{9.20}$$

Since the delay values are non-negative, it follows from equation (9.20) that t_{win}^X is greater than t_o^X which contradicts equation (9.18). Thus, the assumption of deadlock leads to a contradiction which implies that the assumption is false. Therefore, deadlock may not occur relative to the event. In the above proof, if two or more events exist with the same assertion time, each event will independently lead to a contradiction.

The proof of freedom from deadlock is similar for the scenario where there are multiple outstanding input events and multiple events associated with the output event queues of models. The event—input or output, if any—with the smallest assertion time that must be executed first, in accordance with the correct order of execution of events, is identified. A proof is constructed similar to the one described earlier to prove that deadlock may not occur for this event. Then the event with the next larger assertion time is identified and it is shown that deadlock may not occur for this event. This process is continued successively to show that deadlock cannot occur for any input or output event. Thus, deadlock cannot occur in P^2EDAS.

9.3.3.3 Termination of Simulation. A simulation system terminates when the number of outstanding events is nil and consequently, every t_{win}, t_o, and W value is ∞ and t_{E_i} for every primary input of the circuit is ∞. Assume that a system with a finite number of externally applied transitions never terminates. Since it was proved that a system may not deadlock, it must therefore execute continuously. The execution of a model involves receiving from or transmitting messages to other models that require finite time, execution of the model description which must terminate in finite time, and the computation of the t_{win}, t_o, and W values in the event prediction network which must also terminate in finite time because the computations are unidirectional and limited by the finite number of pseudo components in the event prediction network. Consequently, models must continuously receive incoming messages as events. Since the t_{win} and W values increase monotonically, subsequent events must be associated with increasing assertion times. In the expression for any W or t_{win}, the external events appear along with events that may have been generated by the simulation models. Since the minimum operator is involved in the computation of W or t_{win}, the assertion times of externally asserted transitions must eventually approach ∞ which is contradictory. Consequently, the simulation system must terminate in finite time.

9.3.4 Implementation and Performance Analysis of P^2*EDAS*

P^2*EDAS* has been implemented in C for two parallel processing environments: (A) 4-node SUN Sparc 10 shared memory multiprocessor with 56 Mbytes of RAM and 200 Mbytes of virtual memory space supporting threads under Sun OS 5.5 operating system and (B) a network of twenty-five 90 MHz Pentium workstations with 16 Mbytes of RAM and 100 Mbytes of swap, under Linux operating system and connected through a 100 MHz Fast Ethernet and configured as a loosely-coupled parallel processor. The total length of the P^2*EDAS* implementation is 50,000 lines of C and it builds on top of the extensively modified Pittsburgh VHDL compiler [106].

9.3.4.1 Performance Results for Shared Memory Multiprocessor Implementation. For the shared memory multiprocessor supported by multiple threads, a total of 11 representative digital designs are simulated. They include BAR (a barrel shifter with 75 components), MULT1 (a multiplier with latches consisting of 442 components), MULT41 (a 4-bit multiplier with latches comprised of 1768 components), MULT2 (a multiplier with multiplexors consisting of 182 components), MULT42 (a 4-bit multiplier with multiplexors comprised of 728 components), RCNT8 (an 8-bit ripple counter with 118 components), RCNT32 (a 32-bit ripple counter with 454 components), RCNT432 (a set of four independent 32-bit ripple counters with 1816 components), RAM1 (a 32-bit RAM with 519 components), RAM2 (a 64-bit RAM with 3258 components), and SADD (a registered serial adder with 392 components). With the exception of BAR, all other circuits are cyclic in design.

Ghosh and Yu [107] had observed that the performance benefit of distributed simulation is best realized when the computation loads of the behavior entities are appreciable compared to the communications costs. This section also observes that the performance benefit of P^2*EDAS* is best realized when the computational loads of the models exceed the cost of executing the event prediction network. By design, P^2*EDAS* focuses on coarse-grain parallelism and targets large-scale system designs. The use of synthetic loads with the 11 representative digital designs serves to emulate a suite of large-scale designs with the topologies of the representative designs which, upon simulation, yield P^2*EDAS*'s performance. Thus, similar to the performance data presented in the earlier Section 9.2, the performance results reported in this section assume the presence of synthetic loads, i.e. within each component, the behavior model is designed to count to a large number, *Max_Count*. The value of *Max_Count* is assumed uniform across all models since the principal goal here is to study the nature of the algorithm's performance for different circuit topologies, not to focus on load balancing issues. While the value of *Max_Count* is set to 200,000 for smaller circuits, it is reduced to 50,000 for large circuits to limit the total simulation execution time. Figures 9.49, 9.50, 9.51 and 9.52 present the speedup graphs for the digital circuits, enumerated earlier. While the *X* axis represents the number of threads utilized for the simulations, the *Y* axis represents the speedup values. To obtain the speedup graphs, each of the designs are executed on a single processor, utilizing the Pittsburgh VHDL compiler/environment [106], to yield the corresponding uniprocessor execution time. In Figures 9.49 and 9.50, for each of the circuits, MULT41, MULT42, and RCNT432, two speedup graphs are reported corresponding to minor variations in the implementations that are not detailed here.

9.3.4.2 Performance Results for Loosely-coupled Parallel Processor Implementation. In general, messages in a loosely-coupled parallel processor communicate slower relative to data sharing in a shared memory multiprocessor. Thus, a set of two significantly larger

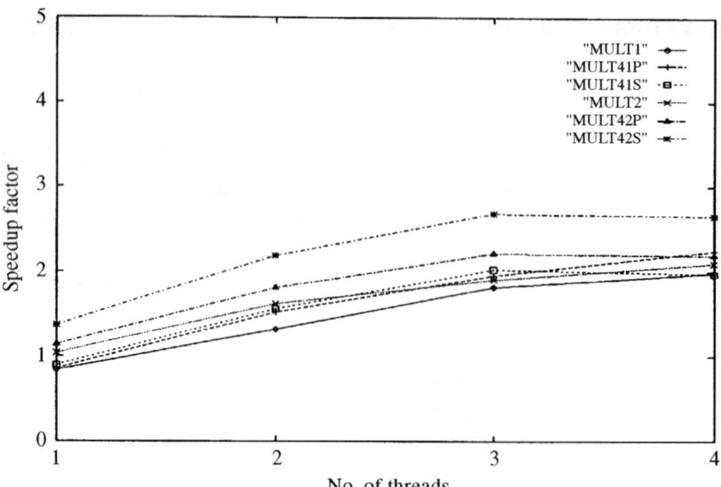

Figure 9.49 Speedup plot for multiplier circuits.

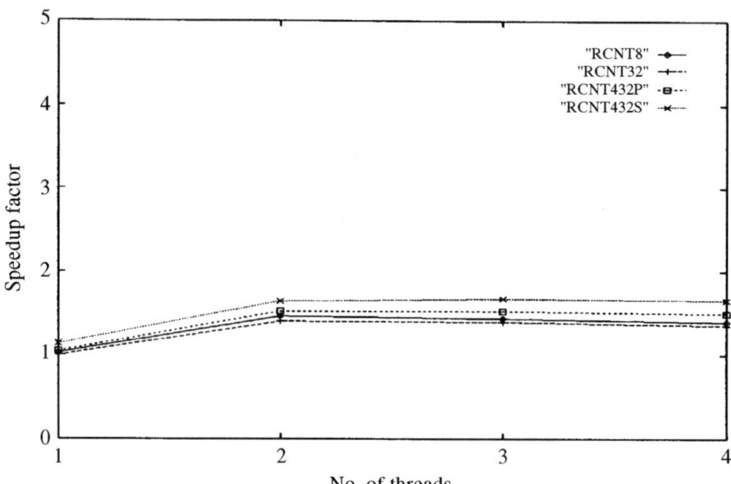

Figure 9.50 Speedup plot for ripple counters.

circuits, as opposed to the eleven designs utilized earlier, are chosen for performance analysis of P^2EDAS. As stated earlier, P^2EDAS is designed for coarse-grain parallelism and targets VHDL entities to constitute the models, especially those with significant computational complexities. The partitioning of a given system into models for P^2EDAS and their allocation to the available processors of a loosely-coupled parallel processor system are both determined by the user.

DLX COMPUTER ARCHITECTURE. Figure 9.53 presents the DLX computer system, a hypothetical architecture described by Hennessy and Patterson [108], that is described in VHDL and simulated in P^2EDAS. Table 9.1 presents a list of the eight major functional blocks of the system along with the constituent components and signals. These blocks also constitute the eight partitions utilized in the simulation.

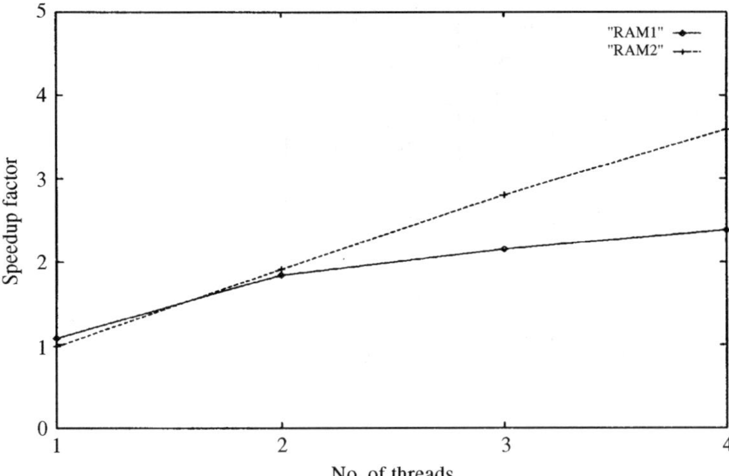

Figure 9.51 Speedup plots for RAM circuits.

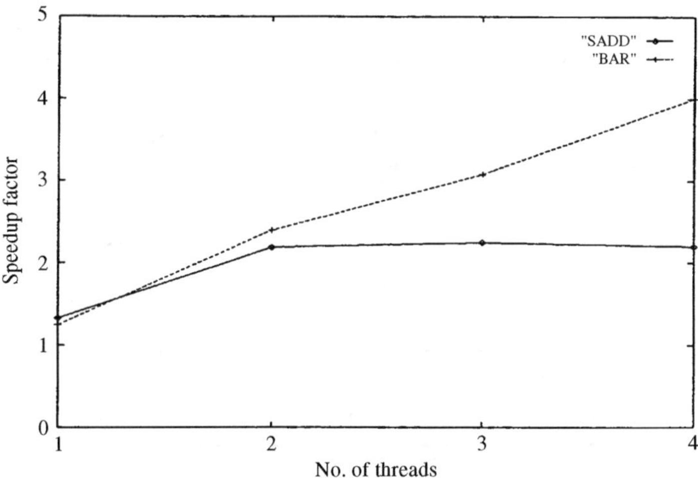

Figure 9.52 Speedup plots for SADD and BAR.

Table 9.2 presents the simulation execution time, in seconds, corresponding to the use of 3, 5, and 8 processors. The simulation time is identical to the longest of the execution times of all of the processors. The reduction in the simulation time from 1 processor to 3 processors is significant. It underscores the fact that the single processor with 16 Mbytes of RAM experiences significant swapping given that the circuit requires over 37 Mbytes of memory for representation. While the reduction in the simulation time from 3 to 5 processors clearly highlights the advantage of P^2EDAS, the slight increase in the simulation time from 5 to 8 processors simply reflects the dominance of communication for the increased number of processors. Observations from the execution of the simulation runs indicate that the distribution of the computational load is uneven among the processors. For instance, the processor that executes the DLX processor module performs the most computation and

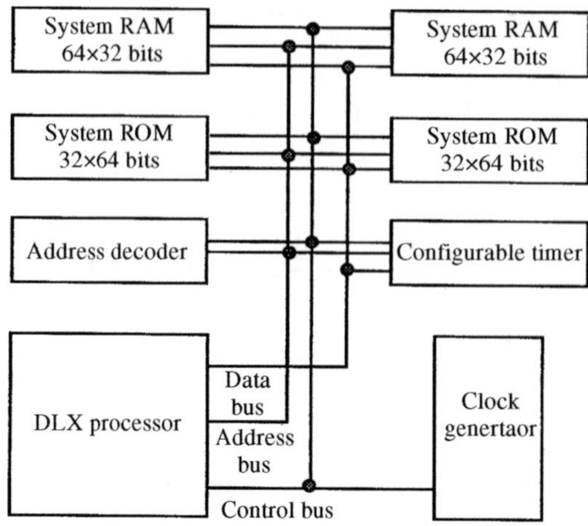

Figure 9.53 DLX computer architecture.

TABLE 9.1 Major blocks in the DLX architecture.

Code name	Circuit name	Components	Signals
CPU	DLX processor	19092	23749
TIMER	Timer	1063	1612
UROM	User ROM	852	984
SROM	System ROM	1502	1634
URAM	User RAM	12480	12581
SRAM	System RAM	12480	12581
Decoder	Address decoder	168	266
Resolver	Distributed resolver	32	259

TABLE 9.2 Performance results for DLX architecture in P^2EDAS.

No. of processors	1	3	5	8
Time (s)	$\gg$43200	4920	1800	2160

executes for the longest time. Conceivably, better performance may be achieved through further partitioning of the computationally intensive modules.

RANDOM STATE GENERATOR. The second circuit for the performance analysis study consists of a pseudorandom state generator that is selected because of the presence of the large number of feedback loops. To gain a better understanding of the benefits of P^2EDAS despite the presence of multiple feedback loops, substantial synthetic loads are assigned to each of the individual components of the generator. Table 9.3 presents the simulation execution time in seconds corresponding to different numbers of processors utilized. The simulation time decreases with increasing number of processors utilized, implying the advantages of P^2EDAS. However, when the number of processors utilized is 10, the increased cost of interprocessor communication outweighs the benefits from distributing the computational load among multiple processing engines.

TABLE 9.3 Performance results for a synthetically loaded random state generator.

No. of processors	1	2	4	6	8	10
Time (s)	659	411	306	302	320	424

The overall performance results indicate that, when the digital designs under simulation are compute intensive, P^2EDAS offers significant speedup benefits. In addition, for large circuits requiring significant memory for representation, partitioning them into multiple processors under P^2EDAS reduces the resource requirement on each processor, thereby implying better efficiency and higher performance. However, the continued increase in the number of processors allocated does not provide increasing speedup. Instead, beyond a certain number of processors, the interprocessor communication time starts to dominate and eclipses the potential advantages from sharing the computational load among multiple processing engines.

9.3.4.3 Performance Analysis. P^2EDAS is an asynchronous, distributed algorithm, and it is very difficult, if not virtually impossible, to develop an accurate analytic model of the asynchronous, concurrent execution of the behavior models on a parallel processor. That is, any effort to develop a complexity analysis to accurately predict the CPU time usage is impractical. Furthermore, an approach to classify digital designs into a finite number of classes, based on an accepted scientific definition, is still elusive. That is precisely why considerable effort has been expended to realize a practical implementation of P^2EDAS and test it against practical designs. However, to facilitate an understanding of the general nature of P^2EDAS's performance, an approximate analysis of the total computational requirement of P^2EDAS, termed cumulative computational burden (CCB), is developed and presented here. It is non-trivial, unlike the traditional sequential algorithms for which the computational complexities were successfully developed, to accurately convert the CCB into the CPU time requirement for asynchronous, distributed systems.

While the null message [107] and the virtual time [77] schemes are competitors of P^2EDAS, the virtual time technique suffers from a fundamental unpredictability, and is not considered in this comparative analysis. The analysis assumes a digital system with a single feedback loop and while it may be easily extended to multiple yet independent loops, the issue of analyzing system designs with multiple, interdependent loops is very complex.

In the event prediction network in P^2EDAS, the pseudo-primed and pseudo-unprimed components correspond to the models in the original system design but are merely mathematical units. Unlike the models that require, in general, significant computation, the mathematical units require only a minimum computation function. The propagation of the W and W' values and the execution of the primed and unprimed pseudo components are both event driven, i.e. triggered by the receipt of a new transition by a behavior model

$$CCB = (N.\alpha)(m + c) \tag{9.21}$$

Assume N behavior models rated by an average computational complexity, m. The average computational complexity of a pseudo component is p, where $p \ll m$. The communication cost per message is given by c. For a given input vector and assuming that, on average, a fraction, α, of the total number of models are activated as a result of the input vector, the ideal or minimum CCB is given by equation (9.21)

$$CCB = (N.\alpha)(m + c) + (N\alpha)f(N)(p + c) \tag{9.22}$$

For P^2EDAS, the CCB is given in equation (9.22), where $f(N)$ implies a function with argument N and whose exact nature is defined by the nature of the digital system, the nature of the input stimuli, and the asynchronous overlapping of the concurrent computations. At the present time, techniques to determine $f(N)$ accurately are elusive

$$CCB = (N.\alpha)(m + c) + (N\alpha)(k + c)\frac{T}{\sum \delta}. \tag{9.23}$$

For the null message technique, where the average time interval between two successive input vectors is $T$and the cumulative delay of all of the behavior models in the loop is given by $\sum \delta$, the CCB is given in equation (9.23). Where a model receives a null message, it first checks to determine whether any of the other inputs have received a new value (non-null message). If no new values have been received, it then updates its output tuple and, under these circumstances, the computational complexity is assumed to be "k". When a new value has been received, the model incurs an execution at a cost of "m," as described earlier.

From equations (9.22) and (9.23), the difference between P^2EDAS and the null message scheme reduces to a comparison between $(p + c)f(N)$ and $(k + c)T/\sum \delta$. Clearly, c plays a fundamental role in limiting the performance of all asynchronous, distributed algorithms but its adverse impact is exerted on both P^2EDAS and the null message scheme. Where the system design is an oscillator or the value of the fraction $T/\sum \delta$ is small, close to unity, which implies either a very high frequency input stimulus or a small to modest size system, the null message scheme may be preferred because of its simplicity. However, where the value of the fraction $T/\sum \delta$ is large, the null message scheme may be highly inefficient and P^2EDAS would be preferred. It is expected that future parallel processors will benefit from improved communications technology and network protocol design and witness a reduction in the communications time which, in turn, will boost the advantages of both P^2EDAS and the null message scheme.

10

On the Concept of Transport Delay in HDLs

10.1 INTRODUCTION

A key requirement of high-level hardware description languages is to accurately represent the timing behavior of hardware descriptions. One of the earliest high-level hardware description languages, ADLIB [32], proposed the notion of "delays" to facilitate the expression of timing behavior. ADLIB proposed to utilize delays to describe the timing behavior of both lumped components such as gates and LSI components as well as distributed components such as buses. However, the interpretation of "delays" and their consequent implementation in the underlying simulator, SABLE [38], for lumped components, led to erroneous results in scenarios that warranted the detection and descheduling of inconsistent events [46]. In its draft request for proposal, relative to VHDL [55], the U.S. DoD stressed the structure of timing constructs and accuracy of representing timing behavior as a key requirement in VHDL design. In response to the requirements, the VHDL language design [33] incorporates two timing constructs: "inertial" and "transport" delay. The semantics of inertial delay allows for both recognition and preemption of inconsistent events, as outlined in [46, 105], that may be generated during the simulation of the executable VHDL models. In contrast, the transport delay semantics assumes that a signal, asserted by a source at one end of a hardware device, will be propagated to the receiver and is not subject to preemption. In VHDL [33], the transport delay semantics extends to hardware devices such as buses and interconnect lines that exhibit infinite frequency response, i.e. any pulse is transmitted regardless of its duration. The design of transport delay suffers from two major flaws. First, there is an implicit assumption that, in an interconnect with multiple taps, only one tap is a driver and the signal reaches the other taps delayed only by the time necessary for the electromagnetic propagation. Thus, the perturbation due to reflection at the intermediate taps is ignored and this results in incorrect timing behavior. Second, for today's increasingly higher clock speeds and newer bus design techniques such as Intel's PCI [109, 110] bus, interconnects increasingly behave like transmission lines and effective simulation of such systems requires detailed transmission line analysis. As a result, VHDL fails to accurately model and simulate [111] hardware designs that incorporate subsystems such as Intel's PCI bus.

Steinberg and Wilson [112] note that in today's high-speed-circuit designs, high frequency signals cause PC-board traces to exhibit transmission line effects. Even on short traces, i.e. as small as 2 inches in an Advanced Shottky logic circuit operating at 50 MHz, reflective noises can cause signal voltages corresponding to "0" and "1" to wander into unknown logic state. Although techniques such as introducing series resistance, split resistance, AC coupling, selecting layout with daisy chaining over the use of stubs, and choosing logic families with clamping diodes, may reduce transmission line effects, their devastating effects warrant great care. Steinberg and Wilson [112] propose that suspect circuits must be thoroughly analyzed through PSpice simulation, each noise source identified and computed, and the total noise computed through a statistical root-sum-square formula. Since PSpice is analog simulation and very slow, they recommend simulating only the critical traces. Yu [113] provides guidelines to create optimal and efficient simulation models in VHDL and notes that, while wire or transport delay propagates all pulses, this issue has become increasingly important with today's shrinking geometries. Fazakerly and Torres [114] stress that, as physical elements get smaller and closer, in VHDL, the timing must be modeled more accurately, taking into account resistance, capacitance, and cross talk.

Rosenthal and Sartore [110] observe that the Peripheral Component Interconnect (PCI) bus, recently introduced by Intel Corporation, is fast gaining acceptance as the interface of choice for most desktop systems. Unlike its predecessor ISA and EISA buses, PCI can support from 132 Mbytes/s of data transfer for a 32-bit bus at 33 MHz to 528 Mbytes/s transfer rate for a 64-bit bus at 66 MHz. Telian [109] presents a detailed simulation study of the PCI bus. Utilizing transmission-line concepts and AC specifications in place of DC values, Telian permits designers to accurately define a buffer's drive capability. Historically, designers assumed simple capacitive loads and oversimplified the problem of designing buffers with adequate drive strength. As a result, although the analysis was tolerable for system clock rates up to 1 MHz in the past, Telian shows through HSpice simulation studies that they are clearly erroneous for today's 33 MHz and faster clock rates. In a simulation, Telian notes that the actual waveform crosses the switching threshold (1.5 V) 6 ns after the time predicted by the lumped-capacitance model—a significant error in systems driven by a 33 MHz clock. Telian strongly recommends adopting a transmission-line model for all high-speed system interfaces, particularly PCI which utilizes reflected-wave switching. Unlike the traditional, incident-wave switching scheme where a driver propagates a valid voltage level to enable switching of appropriate devices, in reflected-wave switching a driver asserts only half the voltage required for switching. The voltage and phase of the reflected wave are identical to that of the incident wave. As a result, following the reflection, the net voltage along the line is double that of the incident voltage. Device switching occurs following reflection and this results in lower surge current and reduced driver size, relative to the traditional scheme.

Chowdhury *et al.* [115] presents a transmission-line simulator, TLSIM, that is capable of computing signals at discrete time instants at user designated points on a multipoint interconnect. TLSIM is based on a method of iteratively updating the signal values at a node based on the previous signal values at neighboring nodes. It is computationally superior to SPICE3e1 and is a logical choice for clock and power lines that usually bear a large number of taps. In circuits, the capacitances, associated with the active devices, constitute the source of the differential equations. TLSIM first transforms these to a set of nonlinear algebraic equations utilizing the implicit Euler's method and then solves them through the Newton–Raphson iterative scheme. While TLSIMs solutions corroborate with the actual waveforms observed through a high-speed oscilloscope, it is also fast for the following reason. Transmission lines naturally partition a large circuit into smaller subcircuits which, in turn, can be processed independently and concurrently. TLSIM limits itself to lossless transmission lines.

The literature on the importance of modeling interconnect traces through transmission lines is rich. Palchesky [116] presents a new, two-port equivalent source model to simulate cable surges generated by nuclear environments in the form of a uniform, time invariant transmission line with independent distributed sources. Yee and Wu [117] presents the computation of the characteristic impedance of printed circuit transmission lines such as finlines, shielded microstrips, slotlines, etc., through transverse modal analysis. Liao *et al.* [118] present a method for analyzing the dynamic behavior of lossy electrical interconnects, with frequency-dependent parameters, in VLSI systems. Their method consists of deriving the circuit model for a transmission line from impulse response data and incorporating it into the UNATL program that performs time-domain analysis for coupled transmission lines with non-linear terminations. Kiziloglu *et al.* [119] report studies on measuring pertinent line parameters including resistance, capacitance per unit length, and characteristic impedance, of high-speed GaAs digital circuits up to 18 GHz. Their measurements confirm the quasi-TEM (Transverse ElectroMagnetic) properties of such interconnects. The measured distributed capacitance and inductance are insensitive to frequency. While the resistance increases by as much as 38%, use of DC resistance values for the line to compute the attenuation constant results in a relatively small error of only 8%. Hietala *et al.* [120] report efforts in fabricating MIS transmission lines on silicon that exhibit relatively low-loss propagation in the 1–12 GHz frequency range. Eisenstadt and Eo [121] report variation in interconnect impedance and capacitance and significant substrate loss for BiCMOS technology over a 45 MHz to 20 GHz frequency range. Brews [122] considers losses in VLSI interconnects a major feature and not merely a perturbation and provides a framework to construct equivalent circuits for lossy waveguide interconnections. Related studies on interconnect behavior are reported in [123–136].

A limitation of virtually every scheme enumerated earlier is that it resorts to analog simulation techniques that are slow and limited to a few interconnects. Furthermore, analog simulation techniques are incompatible with VHDL. VHDL is defined with digital systems as a priority [137] and, not surprisingly, analog systems are difficult to describe in VHDL. VHDL lacks both a scheme to naturally represent simultaneous equations as well as mechanisms to solve them. The proposed analog VHDL standard [138] utilizes continuous time simulation. Although TLSIM [115] is faster than SPICE3e1, it is limited in that it must solve differential equations. For large-scale systems that are likely to be addressed by VHDL, the large numbers of transmission lines, each with many taps, will cause the TLSIM tool to execute very slow. Furthermore, TLSIM's claim that transmission lines naturally partition a large circuit into smaller subcircuits which can be processed independently, is ambiguous for self-timed systems with multiple clocks and interconnects that are included in feedback loops.

To our knowledge, extensions to the timing semantics of VHDL grammar for enabling fast yet accurate digital simulation of high-speed buses and interconnects, coupled with simulation studies that experimentally validate the proposal, have not been proposed in the literature. This chapter recognizes the need for syntactic and semantic support at the level of the hardware description language so as to permit accurate modeling of the propagation of signals along interconnects in today's high-speed systems. This chapter is organized as follows. Section 10.2 reviews the theory of transmission lines and the requirements for modeling of digital buses, interconnects, and wire junctions or taps. Section 10.3 details the failure and inadequacy of the transport delay construct in VHDL. Section 10.4 presents grammar extensions and new semantics to enable simple yet accurate modeling of transmission line effects. Section 10.5 presents a simulation algorithm and develops a simulator that implements the proposed semantics, presents experimental results from modeling and simulating a number of representative bus schemes, and compares the accuracy of the simulation results against actual behavior. It is noted that, while the principle underlying the simulation algorithm is applicable to any event driven hardware simulation, this chapter focuses on the widely used VHDL language which encapsulates a strong effort to capture transmission line behavior.

10.2 CLASSICAL METHODS FOR TRANSMISSION LINE ANALYSIS

When an electromagnetic wave (wavelength λ) propagates along a conductor of length L and $\lambda \gg L$, traveling waves are not apparent on the conductor. By the time the polarity of the oscillator driver has changed appreciably, the energy fed into the line at the driver end has been delivered to the load. However, at very high frequencies where λ is very small, the potential difference along the line is logically described through traveling waves and the resulting analysis is termed transmission line analysis in the literature. The subject of transmission line analysis is a well established engineering science [139–141].

In high-speed digital systems, interconnects, micro-strips, and clock lines in VLSI, as well as interconnect wires that link circuit components on a printed circuit board or multi-chip module, exhibit strong transmission line characteristics. Experimental studies show that, in general, the line capacitance and inductance per unit length are independent of frequency while the resistance, though frequency dependent, tends to validate the quasi-TEM (Transverse ElectroMagnetic) assumptions. As a result, DC parameters that are typically used for switching circuits are not adequate and they must be superseded by parameters useful for transient analysis. In general, transient analysis requires a continuous time simulation system which is slow and inappropriate for VHDL. This chapter aims to develop suitable modeling assumptions and a scheme to accurately capture the transient behavior in an event driven simulation framework. The motivations are as follows. First, event driven simulation [46] is significantly faster and efficient, relative to continuous time simulation. Second, given that most digital simulation [97, 142] systems are event driven, this approach allows the transmission line effects to be modeled within a well understood environment without compromising the overall efficiency of the simulation environment.

For electromagnetic wave propagation in a uniform conductor, placed along the x axis, the voltage v_x, along the length of the conductor, is given by

$$\frac{\partial^2 v_x}{\partial x^2} = RGv_x + (RC + LG)\frac{\partial v_x}{\partial t} + LC\frac{\partial 2 v_x}{\partial t^2} \tag{10.1}$$

where R, L, G, and C are the series resistance, series inductance, shunt conductance, and shunt capacitance of the line, respectively, all measured per unit length [140]. Although solutions exist only for special cases, in general, equation (10.1) reveals the existence of two traveling waves along the conductor: a forward wave and a reflected wave. The interactions between the forward and reflected components give rise to the unique properties of signals on transmission lines. Although a few researchers report their efforts at modeling losses in transmission lines, a substantial body of literature suggests that the errors resulting from assuming lossless lines is relatively small. This chapter restricts itself to lossless transmission lines and assumes that the per unit length capacitance, C, and inductance, L, dominate the line characteristics. Lines with frequency dependent resistive losses do not lend themselves to convenient representations in a discrete, event driven simulation framework.

Unlike the approach in [115], this chapter focuses on the representation of the signals through combinations of one or more discrete events, i.e. at discrete times and at discrete locations along the line. The reason is that digital systems are discrete systems where only new information, i.e. change, is important. Thus, information relative to the state of a signal is required only at discrete locations in the system and at discrete times. This implies faster results with adequate accuracy.

Figure 10.1 presents a uniform lossless conductor of length L and characteristic impedance $Z_o = \sqrt{L/C}$. While a source $v_s(t)$ with series impedance Z_s drives one end of

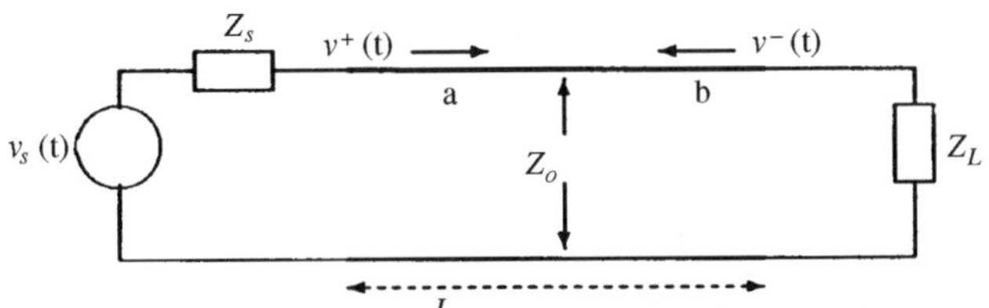

Figure 10.1 Transmission line analysis.

the conductor, a load Z_l is imposed on the other end of the conductor. For digital systems, $v_s(t)$ is usually a pulse generator. Assume that a and b are the locations along the conductor where the source and load are connected and that the voltages are $v_a(t)$ and $v_b(t)$ respectively. Assume that "+" represents the forward wave from a to b while "−" refers to the reflected wave traveling in the opposite direction. The propagation delay for the signal from a to b is given by δ. Define $r_s = (Z_s - Z_o)/(Z_s + Z_o)$ and $r_l = (Z_l - Z_o)/(Z_l + Z_o)$ as the reflection coefficients at the source and load respectively. Also define $\beta = Z_o/(Z_o + Z_s)$. The following equations may be written

$$v_b^+(t) = v_a^+(t - \delta) \tag{10.2}$$

$$v_a^-(t) = v_b^-(t - \delta) \tag{10.3}$$

$$v_b^-(t) = r_l v_b^+(t) \tag{10.4}$$

$$v_a^+(t) = \beta v_s(t) + r_s v_a^-(t) \tag{10.5}$$

Assume that changes to v_s occur only at discrete time instances as opposed to it being a continuous function. Thus, the transmission line may be viewed as an event driven system where an event refers to a change in the signal strength. The equations (10.2–10.5) may be used to derive

$$v_a^+(t) = \beta v_s(t) + r_s r_l v_a^+(t - 2\delta) \tag{10.6}$$

Equation (10.6) provides a convenient mechanism to compute the voltage at a point on a line. Thus, a voltage change at a will affect it again no earlier than 2δ units of time but will affect b after δ time units. No computation is necessary during the transmission interval. For a complex system with multiple drivers, the interactions of multiple traveling waves may be computed by superposition, i.e. by taking into consideration the contribution of each source and algebraically adding them to obtain the net effect. This approach is equivalent to the use of Bergeron diagrams [143, 144].

10.2.1 Modeling a Digital Bus

A typical digital bus consists of multiple drivers and loads placed at appropriate intervals along a conductor. Figure 10.2 presents a digital bus with drivers placed at locations "i," "j," and "k" along the length of the conductor. The source voltages are labeled $v_s^i(t)$, $v_s^j(t)$, and $v_s^k(t)$, while the source impedances are represented through Z_i, Z_j, and Z_{ki}. The characteristic impedances of the conductor segments between locations $\{i, j\}$ and $\{j, k\}$ are represented through $Z_{o_{ij}}$ and $Z_{o_{jk}}$.

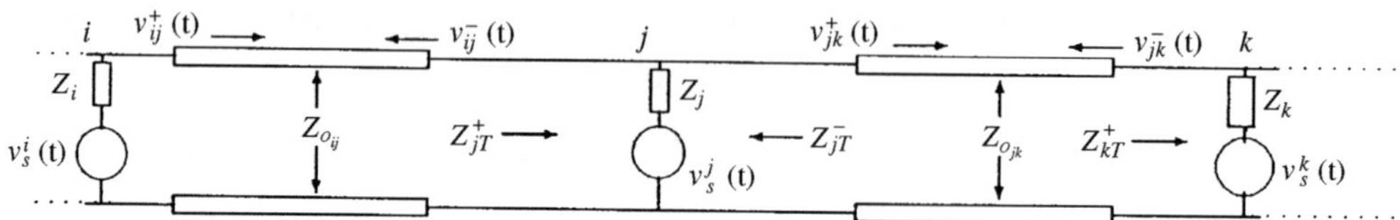

Figure 10.2 Modeling a digital bus.

In Figure 10.2, when a wave, $v_{ij}^+(t)$, traveling from i to j, reaches j, the contribution by the incident wave to the voltage already at j is given by $v_{ij}^+(t)(1 + r_{ij})$ where

$$r_{ij} = (Z_{j_T}^+ - Z_{o_{ij}})/(Z_{j_T}^+ + Z_{o_{ij}}) \tag{10.7}$$

is the reflection coefficient looking towards node j from node i, and $Z_{j_T}^+ = Z_j \| Z_{o_{jk}}$ is the total impedance looking into j from i. The voltage of the wave propagated from j to k is given by $v_{jk}^+(t)$ and it is equivalent to the total voltage change observed at j. This change is computed utilizing equation (10.6)

$$v_{jk}^+(t) = \beta_j v_s^j(t) + v_{ij}^+(t)(1 + r_{ij}) \tag{10.8}$$

where β_j is $(Z_{o_{ij}} \| Z_{o_{jk}})/(Z_j + Z_{o_{ij}} \| Z_{o_{jk}})$ and $\beta_j v_s^j(t)$ is the contribution of the source voltage, $v_s^j(t)$, on the line at j. The value $(1 + r_{ij}) = 2Z_{jT}/(Z_{j_T}^+ + Z_{o_{ij}})$ is the transmission coefficient [141], perceived by the incident wave $v_{ij}^+(t)$ at j. The wave transmitted towards i from j, at this instant, is the sum of the reflected component of $v_{ij}^+(t)$ and the change contributed by the source at j, and is given by $v_{ij}^-(t)$ computed as

$$v_{ij}^-(t) = \beta_j v_s^j(t) + v_{ij}^+(t) r_{ij} \tag{10.9}$$

It may be similarly shown that a wave traveling from $k \to j$ will contribute $v_{kj}^-(1 + r_{kj})$ to the voltage at j, where r_{kj} is the reflection coefficient looking into j from k. If the waves from i and k reach j simultaneously, the total voltage change at j is determined through superposition and is given by $v_j(t) = v_s^j(t) + v_{ij}^+(t)(1 + r_{ij}) + v_{kj}^-(1 + r_{kj})$. In general, where the incident waves at j have different arrival times, $v_j(t)$ is given by

$$v_j(t) = v_s^j(t) + v_{ij}^+(t - \delta)(1 + r_{ij}) + v_{kj}^-(t - \gamma)(1 + r_{kj}) \tag{10.10}$$

where δ, γ are the arrival times of the waves at j from i and k respectively. Clearly, the voltage at any location along a line may be computed by taking into consideration the contribution of the incident waves at that point, the source voltage at that point, if any, and the local impedances. There is no need to consider waves at other points along the line, if any, and, thus this computation is local. This characteristic, namely that the voltage computations are triggered and affected only when waves are incident at a point following reflection, etc., matches well with the general behavior of the event driven model underlying the VHDL environment.

10.2.2 Representation of a Wire Junction

A special case of the general bus model, and one that is widely used in digital systems, is the wire junction. In a wire junction, multiple conductors connect to the line at a point or junction. Figure 10.3 represents a wire junction in a digital system where m conductors, $1, 2, \ldots, k, \ldots, m$, meet at point p. Assume that the wave propagating towards a point is labeled with the superscript *in* while the wave propagating away from the point is represented through the superscript *out*.

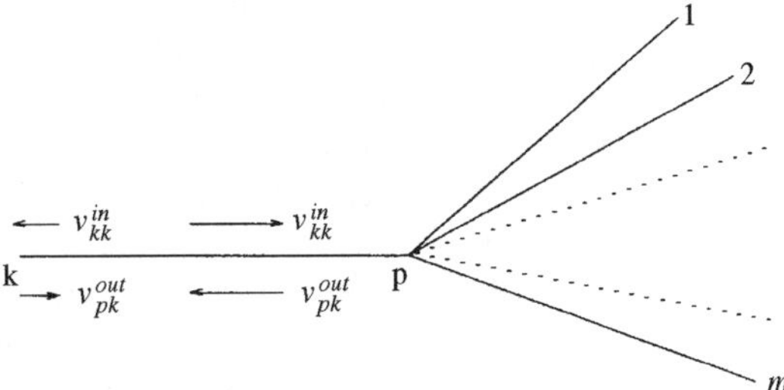

Figure 10.3 A wire junction in a digital system.

Analysis of wave interactions at a junction [115] reveals that the behavior is encapsulated through the equation

$$v_p^{out} = \mathbf{S} v_p^{in} \tag{10.11}$$

where the vectors v_p^{out} and v_p^{in} consist, respectively, of the components $v_{pk}^{out}\ \forall k \in \{1, \ldots, m\}$ and $v_{pk}^{in}\ \forall k \in \{1, \ldots, m\}$ and $\mathbf{S}$ represents a $m \times m$ scattering matrix,

$$\mathbf{S} = \begin{bmatrix} S_{11} & S_{12} & \cdots & S_{1m} \\ S_{s1} & S_{22} & \cdots & S_{2m} \\ \vdots & \vdots & \ddots & \vdots \\ S_{m1} & S_{m2} & \cdots & S_{mm} \end{bmatrix} \tag{10.12}$$

In the matrix (10.12), $S_{jk}(k \neq j)$ represents the fraction of v_{pk}^{in} that is refracted from conductor k to conductor j while S_{kk} implies the fraction of v_{pk}^{in} that is reflected back to the conductor k. Assuming that the characteristic impedances Z_ks of the conductors are known, Z_{kT} represents the parallel combination of the impedances, Z_ks, of all conductors except k. Therefore, $S_{kk} = (Z_{kT} - Z_k)/(Z_{kT} + Z_k)$ defines the reflection coefficient seen by the wave traveling from $k \to p$. Also, $S_{jk} = (1 + (Z_{kT} - Z_k)/(Z_{kT} + Z_k)) = 2Z_{kT}/(Z_{kT} + Z_k)$ represents the fraction of the wave voltage on conductor k that is transmitted to conductor j.

This chapter proposes to declare the topology of a wire junction and represent the static characteristic impedances of the conductors in a simulation language. The simulator may then compute the scattering matrix for any point along the line and utilize it in analyzing the line behavior. The proposal is analogous to network analyzers and timing verifiers [145] that extract comprehensive timing parameters and delays through back-annotation [113, 114], i.e. by first performing detailed, low-level simulation at the layout level and then translating the findings into parameters that are utilized efficiently at the higher levels.

10.2.3 Discretizing Voltage Computations along a Transmission Line through Line-Events

In a discrete event system, an event refers to a new input stimulus that may cause a change, in the future, in the state of the system. For digital systems [46, 142], an event refers to a change in the logical value of a signal at the input or output of a digital component. Where the behavior of the system is progressive, i.e. the state evolves as a function of time, a signal is additionally characterized by the time at which the change in the signal occurs. An event is a cause, corresponding to which there is an effect in the future. For a transmission line, a change in a signal is associated with a voltage value, the time of occurrence, and the direction of propagation of the wave. The knowledge of the direction is critical for computing the wave's refracted and reflected components and, therefore, its resulting value at any location along the line. Formally, an event on a transmission line, referred to as a line event in this chapter, is represented through $(e,\ t,\ dir)$, where e, t, and dir refer respectively to the signal voltage, the time of occurrence, and the direction of the traveling wave.

When a line-event starts to travel from a source, it will be subject to reflection and transmission when it encounters a line load or junction. The location of the line load nearest to the source of the wave defines the "distance of certainty" and it reflects the maximum distance that the wave may propagate with certainty, i.e. without any interference. Consider a wave initiated from i at t_1 towards j in Figure 10.2. While the wave is certain to reach j at $t_1 + \delta$, it is not guaranteed to reach k. For instance, assume that another line-event is scheduled at j for time $t_2 < (t_1 + \delta)$ that causes the impedance Z_j to become zero. Thus, when the wave from i arrives at j, it will be totally reflected and, under these conditions, it will fail to reach k.

In general, the interaction of waves along a transmission line may give rise to effects that are unique to specific locations along the conductor. For instance, consider a glitch that is caused by signal transitions in a wired-OR line [144]. The glitch will oscillate between the two junctions but is not visible to other points along the conductor. For accurate simulation of such behavior, this chapter utilizes the following scheme. When a source asserts a wave on a line, the wave is scheduled to propagate to the load point(s) immediately adjacent to the source. When the wave arrives at these points, the local impedance will determine whether the wave is refracted past the point, partially reflected, or reflected in its entirety.

While digital components function with logical values, transmission lines operate with analog voltage values. This requires the use of conversion routines to convert discrete logical values to voltages and vice versa. In actual transmission lines, the transients die out when the energy is dissipated through the resistances in the line and the line ultimately assumes the DC values. While the assumption of lossless lines simplifies the line behavior, it also fails to emulate the gradual dying out of the transients. Therefore, as an approximation, the voltages along the lines are set to their DC values at the end of the transient activity period of interest. The transient activity of interest consists in simulating the waves whose voltage magnitudes exceed that of the threshold.

10.2.4 Accuracy of Modeling Interconnects as Transmission Lines

Transmission line analysis of interconnects is indispensable where signal changes are very nearly instantaneous, e.g. step voltages, regardless of the lengths of the lines. In reality, however, line drivers exhibit exponential behavior and, conceivably, many interconnects do not warrant reflection analysis. There appears to be little consensus in the literature relative to when transmission line analysis must be extended to digital systems. Assuming a finite signal transition time (say, rise time), t_r, for a wave, Di Giacomo [143] proposes three domains of analysis based on the ratio t_r/T_{prop}: (1) lumped, where $\{t_r/T_{prop} \geq 4\}$, (2) short line, where $\{4 > t_r/T_{prop} > 2\}$, and (3) long line, where $\{t_r/T_{prop} \leq 2\}$. The T_{prop} is the propagation delay of the wave from the driver to the load.

While lumped interconnects permit analysis utilizing Kirchoff's laws, both the short and long lines require transmission line analysis. Short line analysis requires detailed, continuous time simulation since the incident wave may reflect at the load and return to the source before the transition time of the incident wave is complete. Evidently, this will bear a significant impact on the driver's behavior as well as the resultant signal throughout the line. This chapter focuses on long lines and the issue of short lines is beyond its scope. In many VLSI designs with submicron geometries, clock lines on the chip satisfy the inequality $\{t_r/T_{prop} \leq 2\}$ and may be classified as long transmission lines. While the future trend towards increasing densities implies shrinking line lengths and lower propagation delays, the drive towards higher clock frequencies may mandate newer gate designs with reduced rise and fall times. Thus, the $\{t_r/T_{prop} \leq 2\}$ may hold true for clock lines in future designs. For long lines, a rapidly changing signal at the driver may be viewed as an instantaneous signal for it will complete its transition before the reflected component returns to the driver. Under this scenario, a driver's output may be viewed as a step voltage, delayed only by the inertial, i.e. either rise or fall, delay. When the driver is in its saturation state, the load impedance it presents to the line is, in essence, resistive. Thus, for the subsequent reflection analysis, Z_s may be treated as purely resistive which, in turn, simplifies the analysis.

In digital systems, most loads, line drivers, input buffers, and tri-state buffers may be characterized by their resistive and capacitive components [143, 146]. The resistive components of the loads are typically much higher than the line impedance and are assumed to be concentrated at their respective locations along the transmission line. Where the transmission line and the loads are both treated as lumped, the rising edge of the signal is degraded, adding additional delay to the signal propagation. The extent of degradation, T_{deg}, is a function of Z_o and the lumped load capacitance, C_l, and the effective propagation delay is given by $T_{delay} = T_{prop} + T_{deg}$. For long lines, both the line and the loads are modeled using a distributed load model to represent reality more accurately. The distributed load capacitance, C_d, may be effectively absorbed into the transmission line characteristics by increasing the intrinsic capacitance C_o at those locations where the loads are attached. As a result the effective line impedance $Z_{o_{eff}}$ decreases and the effective unit propagation delay $T_{pd_{eff}}$ increases. The relationship between the Z_o and $Z_{o_{eff}}$ is given by $Z_{o_{eff}} = Z_o/(\sqrt{1 + C_d/C_o})$ and $T_{pd_{eff}}$ is given by $T_{pd_{eff}} = T_{pd}\sqrt{1 + C_d/C_o}$. The remainder of the distributed loads are essentially resistive. Thus, the preceding analysis allows one to isolate the input and output capacitances of the digital devices on a line and incorporate them into the line parameters. As a result, only the resistive components of the devices serve as loads at the interconnect nodes. Thus, the effective source, load, and line impedances are all resistive and this eliminates the need for the traditional, complex, analysis.

SUPERPOSITION LAW FOR COMPUTING TRANSIENTS. During transient analysis, when multiple waves are incident at a point on a line, the resultant wave may be computed utilizing the laws of superposition [124] wherein the analog voltages of the waves are algebraically added together. As a result, the sum of the incident, refracted, and reflected components at a point on a line may cause a resulting voltage which, in turn, may lead to circuit switching

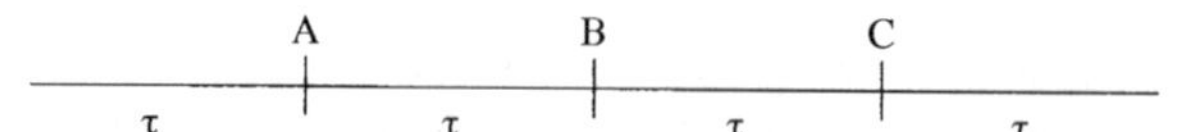

Figure 10.4 Circuit designed for reflective switching.

earlier or later than expected. The PCI bus [109], introduced by Intel Corporation, utilizes reflection and the notion of superposition to control the switching of devices on the bus.

For instance, consider the circuit in Figure 10.4 that is designed for reflective switching [109]. The line is left open circuit at both ends and assume that, while **A** and **C** represent drivers, **B** refers to a receiver. Also assume that the transmission delay between any two consecutive points is τ and the voltage required for switching is set at 2.6 V. If both **A** and **C** transmit 1.4 V signals at $t = t_1$, a valid switching voltage is established at **B** at $t = t_1 + \tau$, as per the superposition rules. However, if either **A** or **C** asserts a 1.4 V signal, a valid switching voltage is established at **B** at $t = t_1 + 3\tau$.

10.3 FAILURE OF VHDL TIMING SEMANTICS TO MODEL TRANSMISSION LINES

10.3.1 Transport Delays

The VHDL language reference manual [33] proposes the use of "transport delays" to model digital devices with infinite frequency response, i.e. where any new input signal, regardless of its duration, will initiate a change in the state of the component. Examples of such devices include bus and clock interconnect lines. When the reserved word "transport" is included in an assignment statement, the delay associated with the waveform element is construed as transport delay. The rules associated with transport delays [33] are

1. All old transactions that are projected to occur at or after the time at which the earliest new transaction is projected to occur are deleted from the projected output waveform.
2. The new transactions are appended to the projected output waveform in the order of their projected occurrence.

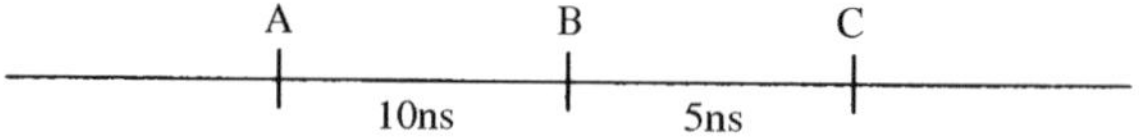

Figure 10.5 Failure of VHDL timing semantics to model reflective switching circuits.

Figure 10.5 presents a 1-bit digital bus to which three digital devices **A**, **B**, and **C** are connected at the locations shown. Assume that the propagation delay from **A** to **B** is 10 ns while that from **C** to **B** is 5 ns. Also, assume that at time $t = 0$, **A** asserts a pulse of width 2 ns on the line. According to the VHDL transport semantics, the leading edge of the pulse is scheduled to arrive at **B** at time $t = 10$ ns. Assume that at time $t = 2$ ns, **C** introduces a pulse also of width 2 ns on the line. The VHDL transport semantics will schedule the leading edge of the pulse at **B** at $t = 7$ ns. This most recent transaction will cause the scheduler to delete the earlier transaction, i.e. one that was executed at $t = 0$ ns, since $t = 10$ ns will occur after

$t = 7$ ns. This is erroneous since, in reality, both waves will reach **B**. VHDL permits a designer to define a signal of type bus and an associated resolution function [142,147], both of which are intended to define a common bus line to which multiple components of the system are connected. However, no resolution function can adequately address the error in Figure 10.5.

Clearly, the combination of the transport semantics, signal of type bus, and the notion of resolution function VHDL fails to accurately simulate buses and interconnects that demand rigorous transmission line analysis. The rules related to updating transactions on a signal representing a transmission line, as stated in [33], are incorrect. The principal reason for the failure lies in VHDL's erroneous assumption that the energy associated with a traveling wave along a transmission line may be arbitrarily deleted without any impact on the system.

10.3.2 Signals

The use of the assignment clause coupled with the transport delay in VHDL forces every component on a bus line to view an event, asserted on the bus line, at the same instant of time. In reality, different components on the bus will perceive the events at different time instants, depending on their distance from the driver, line and load characteristics, and the electrical design of the bus. Figure 10.6 presents a representative digital bus while Figure 10.7 describes VHDL's view of the bus wherein every receiver has been relocated to the same distance from the source. If a transceiver drives a bus, the receiver at that location must perceive the event at the same time the driver asserts the signal on the bus, assuming that the design does not rely on reflection effects. However, the timing semantics of VHDL delays the

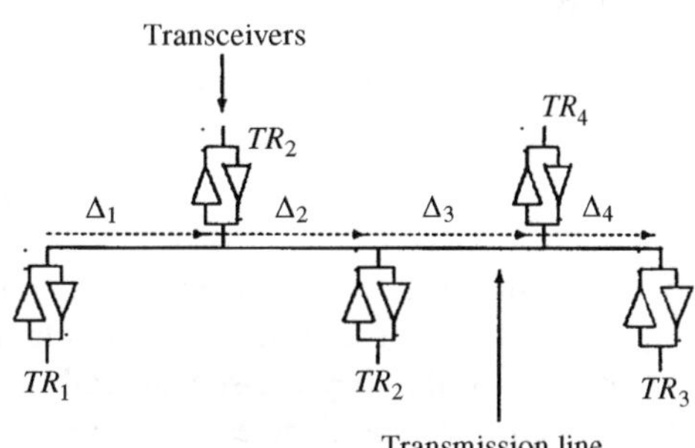

Figure 10.6 A representative digital bus.

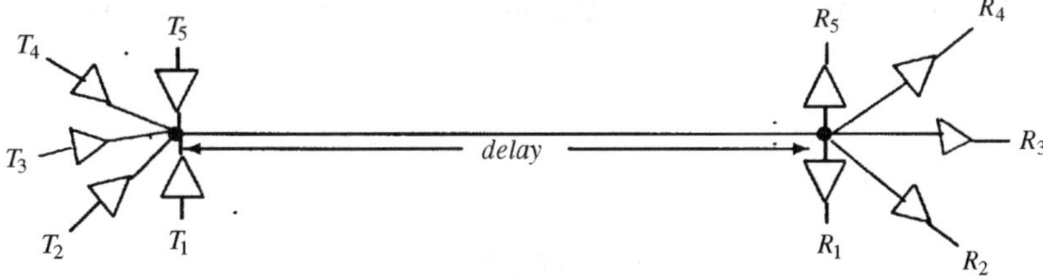

tline <= **transport** *t1_out* **after** *delay*

Figure 10.7 VHDLs view of the digital bus.

response of the receiver to a later time which, in turn, may fail to represent the subtle and critical behavior of the bus.

Furthermore, consider a signal, **S**, that is defined as a bus and assume that it is subject to the following assignments within a single process

```
S <= x0 transport after y0;
....
S <= x1 transport after y1;
```

Assuming $y0 \neq y1$, the sequence of the two assignments implies that first the data value $x0$ is assigned instantaneously to all devices on the bus and that they are all $y0$ time units away from the driver. In the second assignment, the data value $x1$ is assigned instantaneously to all device receivers on the bus and that they are all $y1$ time units away from the driver. The implication is unrealistic and inconsistent and it becomes even more ambiguous where the two assignments are executed in two different asynchronous processes. The resolving function fails to detect this inconsistency.

10.3.3 Semantic Inconsistencies

The VHDL language permits any arbitrary combination of assignment statements qualified by transport or inertial delay semantics. A signal is defined as a collection of objects and the delay semantics uniquely determines the update procedure. Thus, the permissive nature of VHDL may lead to indeterminate behavior specification. That is, the scheduler's behavior is undefined under such circumstances. Consider the following assignments to a signal, **S**

```
S <= x0 transport after y0;  - transport
....
S <= x1 after y1;            - inertial
```

While at one instant, the signal wire has infinite bandwidth, immediately thereafter it appears to possess a filtering characteristic. While the sequence of the two assignment statements are syntactically valid, they are semantically ambiguous. Clearly, the compiler will fail to detect such semantic inconsistencies and the modeling effort will result in error. Also, should a user inadvertently fail to specify the reserved word "transport" when referring to a bus, the compiler has no built-in mechanism to detect the mistake.

10.3.4 Issues of Logic "0" and Logic "1" on a Transmission Line

The concept of propagating pure logic 0 and logic 1 signals on a transmission line is valid only in very limited scenarios. In general, line losses, coupling effects, reflections, and the location of the driver must be taken into consideration to analyze every situation and it is conceivable that a signal may not be propagated to every receiver. VHDL lacks mechanisms to detect such scenarios and facilitate accurate modeling. While the development of analog VHDL is recommended [138] to model such behavior, it is likely to significantly reduce simulation performance.

10.4 PROPOSED EXTENSIONS TO TIMING SEMANTICS AND THE VHDL GRAMMAR

10.4.1 Extensions to the Timing Semantics

For accurate representation of interconnect lines in digital systems, this chapter proposes a new type of signal: transmission line. The definition of the signal includes the generation, propagation, combination, and removal of line-events from a line which, in turn, implies improved ability of the compiler to perform semantic checking and the capability of the system to model complex transmission line effects as discrete events. In addition, the environment associated with the new signal requires the following functions.

Line driving function: While the remainder of the digital system utilizes logical values, transmission line analysis requires analog voltages for detailed analysis. The conversion of the logical to analog voltage is achieved by the "line driving function" that constitutes a part of the behavior of the node where a driver connects to the line. When a driver asserts a logical value at the tap of the line, the "line driving function" computes the equivalent analog voltage, termed new voltage. Where the new voltage differs from the previous value, line events are generated and propagated to the adjacent nodes on the line. The "line driving function," ldf, assumes the form: `next_vnode = ldf(vnode:in}, Zo : in, logic : in, stateinout)`, where `next_vnode` represents the new voltage at the mode as a result of the logical value, `logic`, of the driver, `vnode` reflects the previous analog voltage at the node, and Z_o represents the impedance of the line. The refracted wave, `vtransmitted`, is determined as `vtransmitted = next_vnode-vnode`. The variable, `state`, is unique to each node, contains state information relative to the node, and may be both defined and utilized in the function. For instance, where the impedance at a node changes from 100 Ω to 75 Ω to 50 Ω during a transmission line analysis, `state` may be utilized to contain the subsequent impedance value.

Wave response function: When a waveform is asserted at a node of a line, both reflected and refracted waves may be generated. The magnitude and sign of the voltages are determined by the load impedance at the node at the time of incidence. From the perspective of transmission line analysis at a node, there are three scenarios. First, while a driver is connected at the node, a waveform generated elsewhere on the line may be incident at the node. The resulting analog voltage, `next_vnode`, at the node is computed through the wave response function which assumes the form: `next_vnode = wrf(vincident : in, vnode : in, Zo: in, logic : in, state : inout)`, where `vincident` is the incident waveform voltage, `vnode` is the previous voltage at the node, `logic` is the logical value of the driver, and `Zo` represents the impedance of the line. The variable `state` encapsulates the state information of the node. In the second scenario, only a receiver may be connected at the node. The corresponding wave response function is given by: `next_vnode = wrf(vincident : in, vnode : in, Zo: in)`. For both scenarios 1 and 3, the reflection at the node is computed by the formula, `vreflection = next_vnode - vnode - vincident`, while the refracted wave is computed through the equation, `vtransmitted = next_vnode - vnode`.

Line receiver function: For device receivers on a line, the line receiver function translates the resulting analog voltage at the node to a logical value. VHDL fails to model scenarios where different devices are characterized through unique input thresholds. In contrast, in this chapter, unique line receiver functions may be associated with the device receivers that return logic values that reflect the input threshold characteristics of the underlying devices. As a result, buffers with different input characteristics may be represented. The line receiver function assumes the form: `next_logicval = lrf(next_vnode : in, vnode : in, current_logicval : in)`, where `next_vnode` represents the voltage level that is returned by the wave response function, `vnode` is the previous voltage at the node, and `current_logicval` represents the logical value returned as a result of the previous execution of the line receiver function at the node. The type of arguments used in the function permits the inclusion of receivers that (1) respond to the differential voltage, (`next_vnode - vnode`), at the input node, and (2) those wherein the logic level is determined exclusively by the resulting analog voltage, `vnext_vnode`, at the node. For both types of receivers, where the change in the input voltage is less than adequate to cause a change in the logical value, the previous logic level, `current_logicval`, is returned as `next_logicval`.

DC resolution function: The objective of the DC resolution function is to return the final, steady-state DC voltage following the cessation of all transients on the line. The function accepts the logic levels of the device drivers on the line as its arguments. While the DC resolution function is similar in spirit to the resolution function in VHDL, it returns an analog voltage value while the resolution function in VHDL returns a logical value. The resolution function in VHDL bears no notion of transience.

10.4.2 Extensions to the VHDL Grammar

The proposed extensions to the VHDL grammar aim at providing designers with a scheme to systematically represent, model, and simulate transmission lines in digital systems. In addition, the VHDL compiler can detect inadvertent design and modeling errors at compile-time, thereby saving valuable simulation time. Conceivably, a user may model a transmission line of a digital design through an entity, describing the complex transmission line computations in the current VHDL language. The difficulty with this approach is that, upon encountering a transmission line, a user must bear the burden of writing a lengthy description carefully and debugging it for correctness. In contrast, the underlying philosophy of language design is to enhance user efficiency through a set of carefully synthesized generic constructs which are consistent, relatively simple to use, and, when utilized correctly, guarantee accuracy of description and execution results. Given the increasing frequency with which transmission lines are likely to occur in VLSI designs, the proposed extensions are timely.

10.4.2.1 Declaration of Transmission Line Signals. The grammar extension, presented in Figure 10.8, permits the user to declare a signal of type "tline" that includes specifications for the physical length of the line (L), the transport delay along the line (T_{prop}), impedance of the line (Z_o), the technology of the digital system which specifies the functions that define the device drivers, the default DC resolution function, and the user-defined type—voltage, current, etc.—that is returned by the resolution function. For two nodes along the line, x_1 and x_2, the transport delay between them is computed by the simulator as $(T_{prop}|x_1 - x_2|)/L$.

Thus, the following may define a specific instance of a line and is referred to as a declaration of type 1. In the declaration, the user-defined DC resolution function is "ttlresolve," which computes the final, steady-state, DC voltage value utilizing the logic values that the drivers assert on the line

```
signal wire tline 10 mm transport 5 ns impedance 75 ohms technology ttl
: ttlresolve;
```

```
signal_kind ::= bus | register
               | -trans_line_definition
_trans_line_definition ::= tline _line_value _line_parameter
_line_value ::= length_literal
               | name
_line_parameter ::= transport _time_value impedance _impedance_
  value_technology
               | impedance _impedance_value _technology
_time_value ::= time_literal
               | name
_impedance_value ::= impedance_literal
               | name
_technology ::= technology _node_function_package : _dc_resolution
  _function ;
_node_function_package  ::= name
_dc_resolution_function ::= name
               |NULL
```

Figure 10.8 Grammar for declaring transmission line signals.

A second instance of a line may consist of the following declaration and is referred to as a declaration of type 2. In it, "wiredelay" is a VHDL type that defines an array to store the actual transport delay values between the successive nodes along the line. The constant "connections" relates to a specific line. Also, the transport delays for three nodes along the line are specified the through the constant, "connections". The "null" implies the lack of a DC resolution function associated with the line

```
type wiredelays is array (natural range <> ) of time;

constant connections : wiredelays := (0 ns, 3 ns, 7 ns);

signal wire2 tline connections impedance 75ohms technology cmos :
null;
```

10.4.2.2 Assignments to Transmission Line Signals. Figure 10.9 defines grammar extensions to facilitate assignments to the transmission line signal. The "signal assignment statement" clause of the VHDL grammar is augmented to include assignment to lines. The use of the "waveform" clause in Figure 10.9 utilizes the definition in the VHDL grammar. The "`_line_value`" and "`_node_function_package`" literals in Figure 10.9 utilize the definition in Figure 10.8.

Any node along a transmission line is specified uniquely through the signal name that represents the line and one of two schemes to identify the exact location of the node. For instance, the construct "`s@x`" implies a node along signal line "`s`". Where "`s@`" uses the declaration of type 1, "`x`" refers to the distance from the origin, which is usually an end point of the line. However, if the declaration used is of type 2, "`x`" corresponds to the index into the array of transport delay values. The distance-related parameters are assumed available at

```
_signal_assignment_statement_instance ::= signal_assignment_statement
                                        | _line_signal_assignment_statement
_line_signal_assignment_statement ::=
        _connection_definition <= _connection_definition_option;
_connection_definition_option ::= waveform
                                | _connection_definition
_connection_definition ::= name @ _line_connect_distance
                                [: _node_function_package]
                                | name
_line_connect_distance ::= _line_value
                         | _tline_index
_tline_index ::= natural
```

Figure 10.9 Grammar for assigning to transmission line signals.

compile time. While a device driver is assumed to execute an assignment at a specific node of a line, a device receiver will read from a line at a specific node, and a device transceiver will both drive at and read off of a node along the line. Although a default package is specified through the technology clause in Figure 10.8, the presence of the clause `[: _node_function_package]` in Figure 10.9 permits the user to specify an alternate package that consists of three generic functions: line driving function, wave response function, and line receiver function. The valid DC resolution function continues to be that defined in the signal declaration. The alternate package is generally required to correctly represent the line behavior at the end points, i.e. whether the line is open, shorted, or is connected to a matching impedance.

The following constitute a partial set of valid language constructs that may be utilized to assign to transmission line signals

1. `s@x <= logic_value;`
   Semantics: The logical value is asserted at node *x* along the transmission line.
2. `s@x <= NULL;`
   Semantics: The connection at *x* is passive, i.e. at the node there are neither driving nor receiving functions. A resistive termination constitutes an example. Only a wave response function is associated with this node.
3. `s@x <= signal_name;`
   Semantics: The named signal, i.e. any VHDL signal, is connected to the transmission line at the specified node.
4. `s@x : my_drivers<= logic_value;`
   Semantics: The functions in the package *my_drivers* are used instead of those in the default package.
5. `signal_name <= s@x;`
   Semantics: The named signal reads the logical value off the line at node x. The events of a VHDL signal only refer to logical values.
6. `logical_value <= s@x : my_drivers;`
   Semantics: The functions in *my_drivers* package are used instead of those in the default package to read the logical value off the line at node x.

```
TYPE ohm IS RANGE 0.0 TO REAL'HIGH;
TYPE voltage IS RANGE REAL'LOW TO REAL'HIGH;
PACKAGE ethernet_driver IS
-- These are the default variables associated with every package;
-- Each node on a line gets its private copy of these variables
  VARIABLE state: INTEGER;              -- node state used/defined by driver &
                                           resp
  VARIABLE node_voltage: voltage;       -- current node voltage
  VARIABLE next_node_voltage: voltage; -- new node voltage computed in
                                           driver() or
  VARIABLE z_o: ohm;                    -- line impedance obtain from line
                                           signal d
  VARIABLE current_driving_logic: BIT; -- the last logic value to drive node;
                                           use
  VARIABLE next_logicval: BIT;          -- the new logic value at node computed
                                           by
  VARIABLE current_logicval: BIT;       -- the last logic value computed by
                                           receiv
  FUNCTION driver(new_driving_logic: BIT) RETURN voltage;
  FUNCTION response(incident_voltage: voltage) RETURN voltage;
  FUNCTION receiver(vnode: voltage) RETURN BIT;
END ethernet_driver;

PACKAGE ethernet_termination IS
  VARIABLE state: INTEGER;
  VARIABLE node_voltage: voltage;
  VARIABLE next_node_voltage: voltage;
  VARIABLE z_o: ohm;
  VARIABLE current_driving_logic: BIT;
  VARIABLE next_logicval; BIT;
  VARIABLE current_logicval: BIT;
                                        -- no driver or receiver function need
    FUNCTION response(incident_voltage: voltage) RETURN voltage;
END ethernet_termination;
ENTITY ethernet_wire IS
      PORT(node_1, node_2, node_3, node_4: IN BIT;
  onode_1, onode_2, onode_3, onode_4: OUT BIT ;);
END ethernet_wire;
ARCHITECTURE accurate_ethernet OF ethernet IS
   SIGNAL ethernet_bus TLINE 20m TRANSPORT 5us IMPEDANCE 75 ohm TECHNOLOGY
   ethernet_d
--define position of termination taps
  CONSTANT termination_tap0: distance := 0m;
  CONSTANT termination_tap1: distance := 20m;
```

Figure 10.10 Accurate model of ethernet utilizing proposed semantic and grammar extensions to VHDL. (*continued on next page*)

```
-- define position of connection connections;
  CONSTANT tap1: distance := 2m; CONSTANT tap2: distance := 8m;
  CONSTANT tap3: distance := 12m; CONSTANT tap4: distance := 19m;
BEGIN
  -- define passive terminations using the termination package
     ethernet_bus@termination_tap0:termination <= NULL;
     ethernet_bus@termination_tap1:termination <= NULL;
  -- define connections to wire
     onode_1 <= ethernet_bus@tap1;
     ethernet_bus@tap1 <= node_1;
     onode_2 <= ethernet_bus@tap2;
     ethernet_bus@tap2 <= node_2;
     onode_3 <= ethernet_bus@tap3;
```

Figure 10.10 *(continued)*

The following constitute a partial set of valid language constructs that may be utilized to specify the multiple wire interconnection. The topology is static and the corresponding scattering matrix is determined at compile-time

1. `s@x <= w@y;`
   Semantics: The intersection of transmission lines *s* and *w* occur at node *x* along *s* and node *y* along *w*, respectively. The scattering matrix is synthesized utilizing the line impedances of *s* and *w* and is used by the simulator.
2. `s@x <= w@y;`
   `s@x <= t@r;`
   Semantics: A three-wire junction where line *t* joins lines *s* and *w* at node *r*.
3. `s@x <= signal_name;`
   `s@x <= w@y;`
   Semantics: This is not allowed since, otherwise, the scattering matrix will be rendered dynamic by the active load impedance associated with the driver.

At a multi-wire junction, the DC resolution function is invoked only when all transient activities subside. Should the DC values returned by the resolution functions corresponding to the individual lines differ, an error message is generated. Figure 10.10 presents a model of an ethernet, utilizing the proposed semantics and grammar extensions, that accurately describe the ethernet behavior.

10.5 MODELING AND DISCRETE SIMULATION OF REPRESENTATIVE TRANSMISSION LINES

A total of three representative lines—ethernet wire, PCI bus, and a TTL wired-OR bus—are modeled in the proposed approach. The models are then subject to discrete simulation, and the results are compared against those obtained from detailed analog simulation utilizing PSpice [112]. The analog waveforms presented in this section other than those that bear the label "SPICE simulation" represent analog outputs from event driven simulation. SPICE waveforms are presented for the PCI and wired-OR glitch simulations. The purpose of the comparison is to evaluate the usefulness of the proposed approach.

10.5.1 Simulation Algorithm for Discrete Transmission Line Analysis

Clearly, one could start with a commercial VHDL compiler and environment and modify it subject to the semantic and grammar extensions proposed in this chapter. The augmented VHDL compiler would then synthesize a representation of the physical system and the simulation environment would operate on the representation, subject to the rules developed in the chapter. In contrast, in this effort, a simulator is designed and developed in C that utilizes the principles elaborated earlier in this chapter. The simulator is provided with, as its input, a description of the transmission line with the necessary physical details that would be specified in the extended VHDL grammar. While the digital drivers, connected to the line, assert the discrete events that they generate, logical values are extracted from the line by the digital receivers corresponding to the changes in the voltage values at the corresponding nodes along the line. Figure 10.11 presents the simulation algorithm, in pseudo-code, and defines the behavior of every node of a line.

The simulation algorithm is implemented in C and is approximately 1700 lines in length. It is organized through eight functions. They include "simulate node," "execute node," "propagate line event," "schedule line event," "update node event queue," "execute receiver," "DC voltage resolution," and "handle short." The `simulate_node` function corresponds to a node of a line and it is executed when either a driver at that node asserts a logical event or a line-event is asserted at the node, i.e. a wave is incident at the node. In turn, this function calls other functions. Although the simulator physically contains a single copy of this function, it maintains unique instances of the function corresponding to every node of the line. The `execute_node` function is responsible for emulating the behavior of the device driver and the wave response function. The `propagate_line_event` function represents the propagation of incident, reflected, and refracted waves along the line. The simulator maintains a queue for executing line-events in the correct time order. The `schedule_line_event` function manipulates this queue and schedules the execution of the line-events so as to preserve causality. The `update_node_event_queue` function also manipulates the queue and is responsible for removing those line-events that have been processed. Whenever the analog voltage at a node differs from its previous value, the `execute_receiver` function extracts the corresponding logical value as perceived by the receiver at that node. When the line-event queue becomes empty, the `DC_voltage_resolution` function asserts the steady-state DC voltage on the line. In the event of a short-circuit at a node, there is no change in the voltage despite reflections, and the `handle_short` function is invoked to accurately determine the resulting wave at the node.

The executable behavior of the functions are lengthy and are not presented here. For further details, the reader is requested to contact the authors.

10.5.2 Modeling and Simulation of an Ethernet Wire

Figure 10.12 presents an ethernet line with matching terminators at either endpoints and five transceivers TR1 through TR5 connected at nodes 1 through 5 respectively. The ethernet [148] utilizes the "carrier sense multiple access" protocol. While the end to end transport delay of the ethernet is 15 μs, the respective transport delays between the endpoints and nodes

```
if (logic event, i.e. logical value asserted by the driver) {
execute line driver function
        compute updated voltage at the node
determine change in node voltage from previous value
if (change in voltage at node is significant) {
determine transmission time to adjacent nodes
create and propagate line_events to adjacent nodes
        }
}
else if (line_event, i.e. wave is incident at the node) {
execute wave_response function at the node
        compute updated voltage at the node
compute reflected and refracted voltage waveforms
        if (reflected voltage waveform is significant) {
      generate and send reflected line_event to the source of incident wave
}
if (refracted voltage waveform is significant) {
          generate and propagate refracted line_event to adjacent nodes
        }
if (queue of events along the line is empty)
   if (DC resolution function is specified) {
execute DC resolution function corresponding to the node drivers
                update every node along the line with the steady-state, DC
                value
for (every node) {
                   execute receiver function at node
   if (logic value at receiver differs from previous value) {
              propagate logic value to the node circuit
                   }
                }
}
}
if (voltage at node differs from previous value) {
   execute receiver function at node
   if (logic value at receiver differs from previous value) {
       propagate logic value to the node circuit
   }
}
```

Figure 10.11 Simulation algorithm, in pseudo-code, corresponding to every node along a line.

are given by $\Delta_0 = 1.57$, $\Delta_1 = 2.70$, $\Delta_2 = 0.8$, $\Delta_3 = 2.3$, $\Delta_4 = 4.74$ and $\Delta_5 = 3.16\,\mu s$, as shown in Figure 10.12. By design, the transmitters are autonomous and asynchronous, except that in the event of collisions, they back off, wait for a random period of time, and then retry. That is, any of the transmitters may drive signals on the line at irregular time intervals, irrespective of other transceivers on the line. By definition, the nodes on the ethernet must be separated by a minimum physical distance and the maximum length of the ethernet wire is

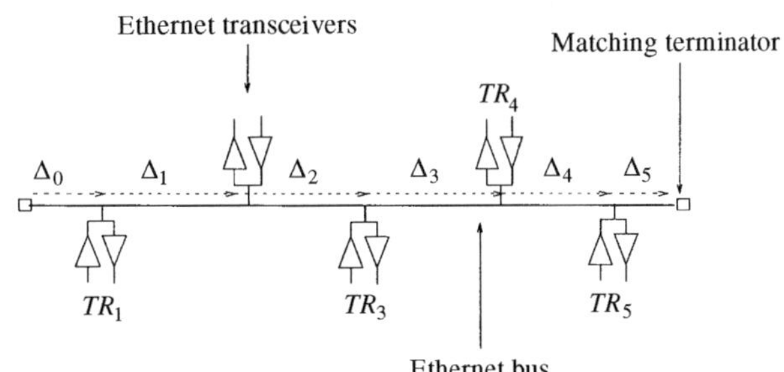

Figure 10.12 An ethernet line with five transceivers.

limited. When a transmitter is active, signals are propagated in both directions and the corresponding receiver listens on the line to detect for possible collisions. Accurate modeling of the ethernet in VHDL is difficult because of the following reasons. First, VHDL's view of a digital bus, as shown through Figure 10.7, lumps all receivers at a single location which contradicts the definition of the ethernet. Second, the timing aspect of collisions on the ethernet are extremely difficult to model in VHDL. In the event of a collision, the receivers corresponding to the transmitters that had been listening on the line must detect the collision. Then, the transmitters must abort the transmission, and reattempt later, unlike in VHDL where a resolution function merely resolves into a logical value.

10.5.2.1 Transmission Without Collisions. Where no two transmitters attempt to transmit on an ethernet line at the same time, collisions will not occur and a wave transmitted by a device driver will be incident at every node of the line. Consider that the transmitter at node 2 in Figure 10.12 transmits a pattern [1]01101000[$1\frac{1}{2}$], where [1] represents the start bit and [$1\frac{1}{2}$] reflects the one-and-a-half stop bit. Figure 10.13 presents a superposition of the logic pattern transmitted by the transmitter at node 2, the resulting analog line voltage at node 2, and the logic pattern that is read by the receiver at node 2. Given the absence of collisions, the identity between the logic pattern transmitted from and received at node 2 is expected. Figure 10.14 shows the analog line voltage and logical pattern that are received at node 4. Clearly, node 4 and all other nodes on the line receive the same logical pattern as transmitted by node 2 except that each waveform is delayed in time by a quantity that is equal to the transport delay from node 2 to the receiver in question.

10.5.2.2 Transmission Under Collisions. Since nodes along an ethernet are physically separated, the finite speed of electromagnetic transmission may give rise to a "blind spot" wherein multiple device drivers may initiate transmission without being aware of others' activities. In Figure 10.12, assume that node 4 transmits a signal "10110100" concurrently with the transmission from node 2. While Figure 10.15 presents a superposition of the logic pattern transmitted by the transmitter at node 2, the resulting analog line voltage at node 2, and the logic pattern that is read by the receiver at node 2, Figure 10.16 shows the corresponding waveforms at node 4. As expected, as a result of collision, the logic patterns transmitted at both nodes 2 and 4 differ from the corresponding logic patterns and analog voltages, received at nodes 2 and 4 respectively.

In Figure 10.15, although the analog voltage waveform experiences transitions from −8 V to −4 V and from −4 V to +1 V, the corresponding logical waveform only exhibits a

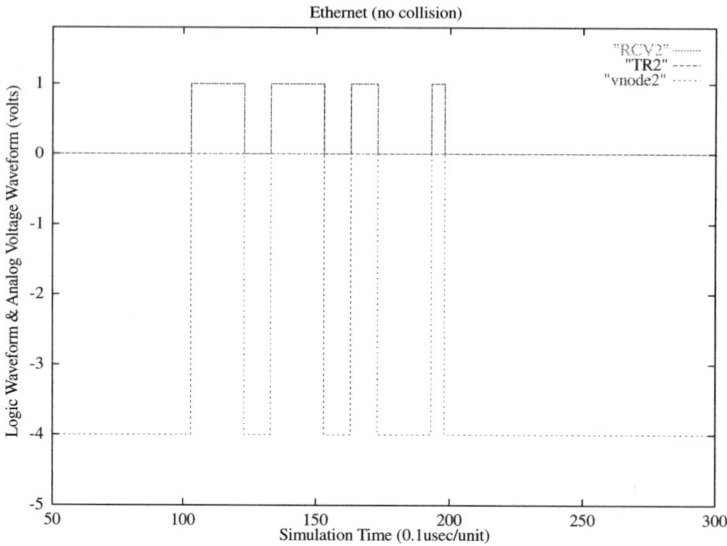

Figure 10.13 Logic pattern transmitted from node 2, the analog line voltage at node 2, and the logic pattern read by the receiver at node 2.

transition from 0 to 1. Also, corresponding to a transition from +1 V to −4 V, the logical waveform swings from 1 to 0, even though the analog waveform has not reached its low value of −8 V. The reason for this behavior lies in the assumption that the outputs of the receivers of the transceivers, connected to the ethernet, are determined by the difference between the current and previous analog voltage values, not the actual voltage values. This is the direct

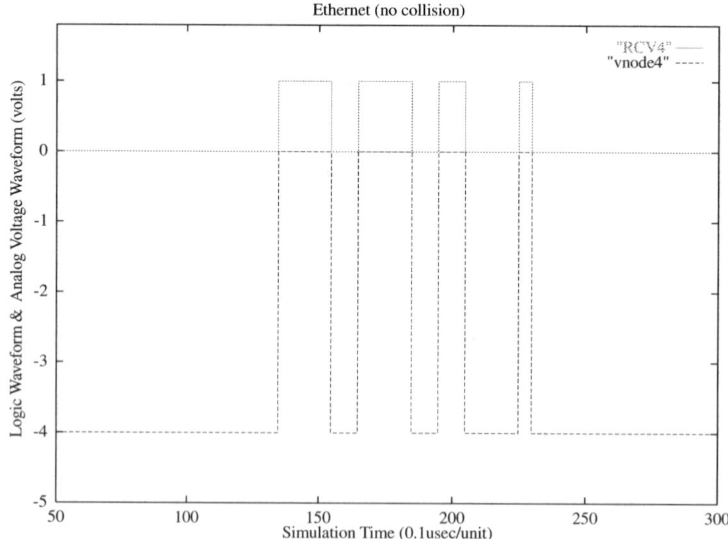

Figure 10.14 Analog line voltage received at node 4 and the logic pattern read by the receiver at node 4.

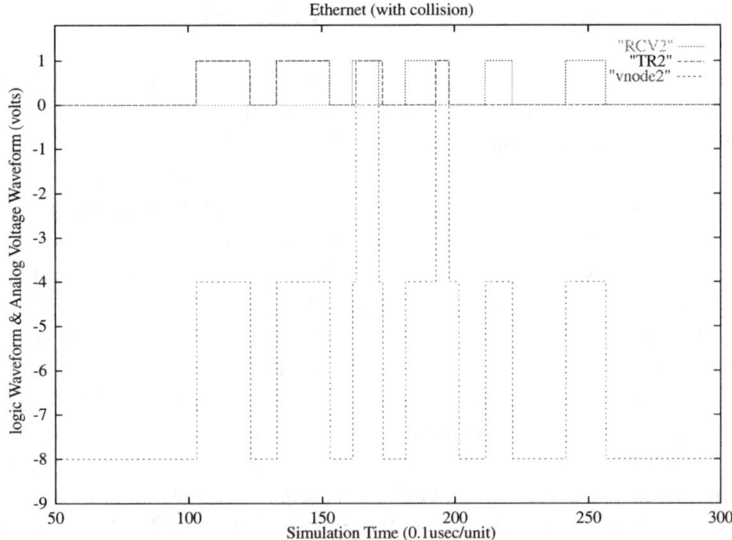

Figure 10.15 Logic pattern transmitted from node 2, the analog line voltage at node 2, and the logic pattern read by the receiver at node 2.

result of this chapter's effort in accurately modeling the AC behavior of the ethernet. Thus, corresponding to a change from -8 V to -4 V around 16 μs along the simulation time axis, the logical waveform jumps from 0 to 1. The logical waveform remains at 1 when the voltage changes again from -4 V to $+1$ V. Corresponding to the subsequent transition from $+1$ V to -4 V, the logical waveform changes from 1 to 0 and continues to be held at 0 even when the analog waveform dips from -4 V to -8 V. Similar behavior is exhibited through the graphs in Figure 10.16. For further clarity, the reader is referred to Section 10.4.1.

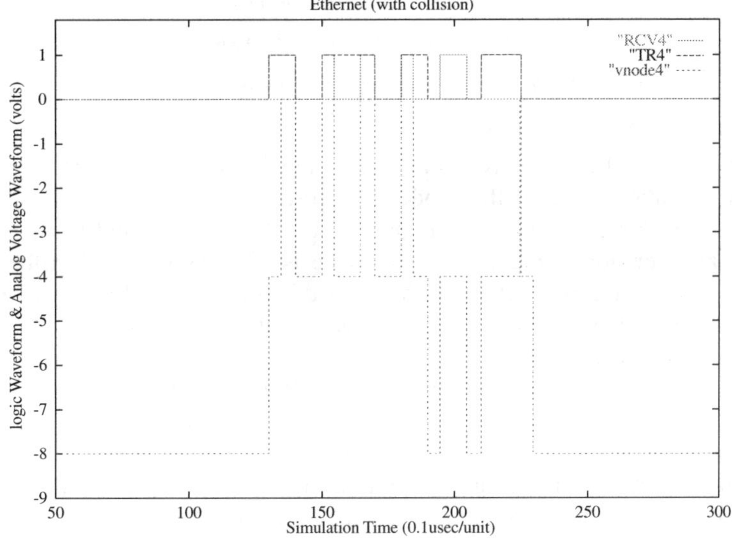

Figure 10.16 Logic pattern transmitted from node 2, the analog line voltage at node 2, and the logic pattern read by the receiver at node 4.

10.5.3 Modeling and Simulation of a PCI Speedway Bus

Figure 10.17 presents a representative PCI bus, introduced by Intel Corporation, to which five components are connected utilizing the PCI Speedway [149] scheme that is the recommended method of connecting PCI devices. PCI is a component-level standard bus [109, 149] and has attracted much attention due to its targeted speeds of 33 MHz and 66 MHz and bus widths ranging from 32 bits to 64 bits. The high speed operation of PCI requires transmission line analysis for accurate analysis. In addition, PCI utilizes reflective switching [109] to include drivers with minimal current driving capability. The driver asserts only half of the switching voltage on the line. The generated wave travels to the end of the line, which is deliberately left open, and is reflected back with the same amplitude and polarity. The reflected wave voltage is added on top of the incident wave which then pulls the line voltage down to the "low level." The device closest to the open circuit end at the right is the first one to be exposed to the switching voltage while other devices to its left are exposed successively. This is in contrast to the conventional buses where the receiver nearest to the driving node is the first one to be exposed to the switching voltage. In addition, unlike the ethernet, should a driver be located at the end opposite to the open end along a PCI bus, it will be the last one to be exposed to the full switching voltage.

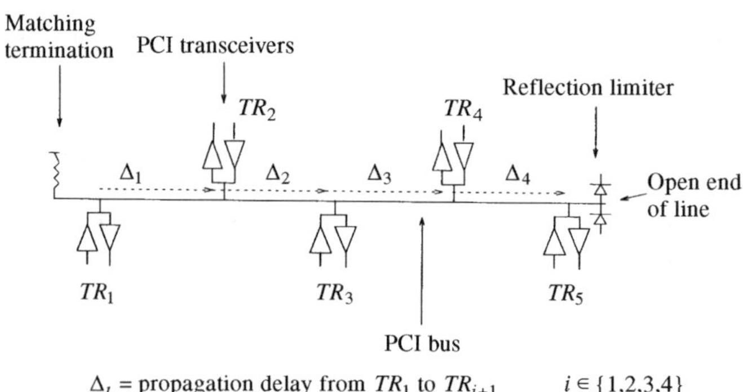

Figure 10.17 A peripheral component interconnect (PCI) bus with 5 nodes.

In Figure 10.17, the five components are located equidistant from each other. The transport delays between the nodes are given by, $\Delta_0 = \Delta_5 = 0.01$ and $\Delta_1 = \Delta_2 = \Delta_3 = \Delta_4 = 1.25$ ns, as shown in Figure 10.17. Thus, the round-trip delay along the PCI bus, for an operational frequency of 33 MHz, is 10 ns as included in the PCI specification [109]. The open end of the PCI bus is clamped by diodes to deliberately limit the magnitude of the reflected wave. Otherwise, this chapter's assumption of lossless line may lead the bus to exhibit unstable behavior.

The graphs in Figures 10.18 and 10.19 present the simulation results when the transmitter at node 1 is the driver. Both the logical and analog voltage waveforms reveal the (1) delay between the start of the driver-asserted logical value and the realization of the full switching voltage on the line, and that (2) the receiver at node 1 is the last one exposed to the full switching voltage. Similar behavior is exhibited through the graphs in Figures 10.21 and 10.22 that correspond to the transmitter at node 3 driving the bus. In Figure 10.18, the delay for node 1 to undergo switching equals approximately 10 ns which is the time for the

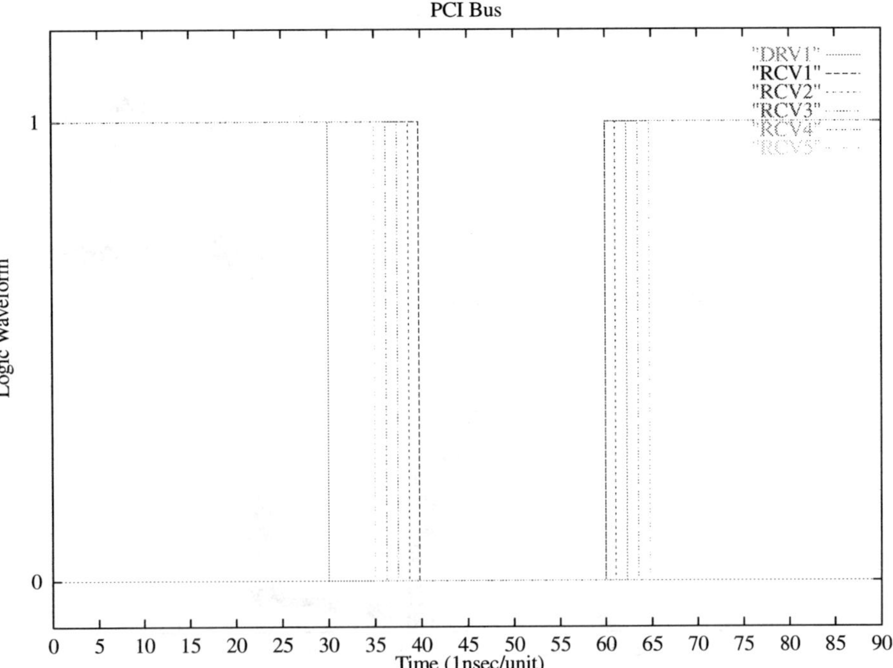

Figure 10.18 Logic waveform of driver at node 1 and resulting waveforms at the nodes of the PCI bus.

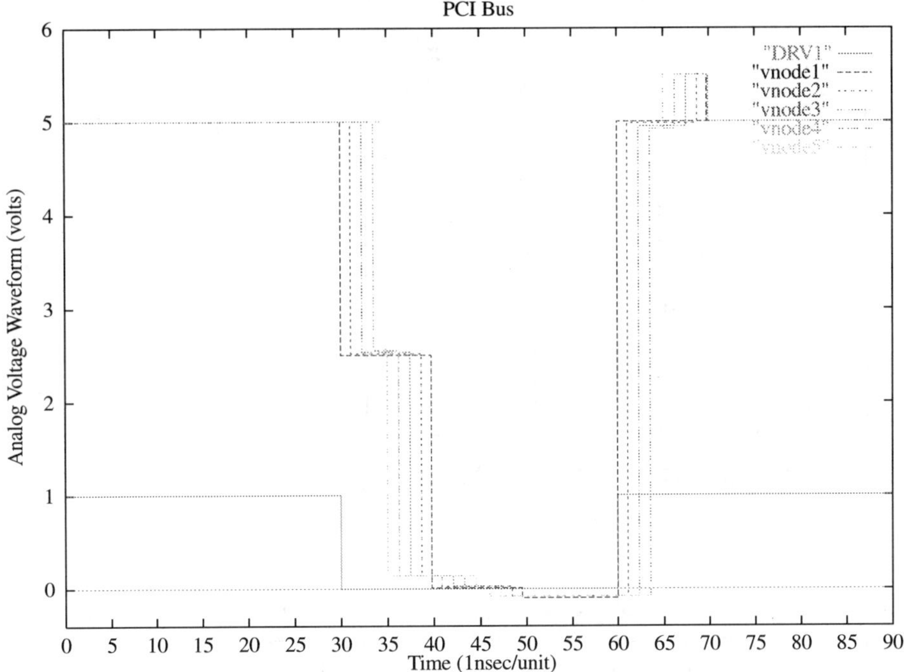

Figure 10.19 Voltage waveform of driver at node 1 and resulting waveforms at the nodes of the PCI bus.

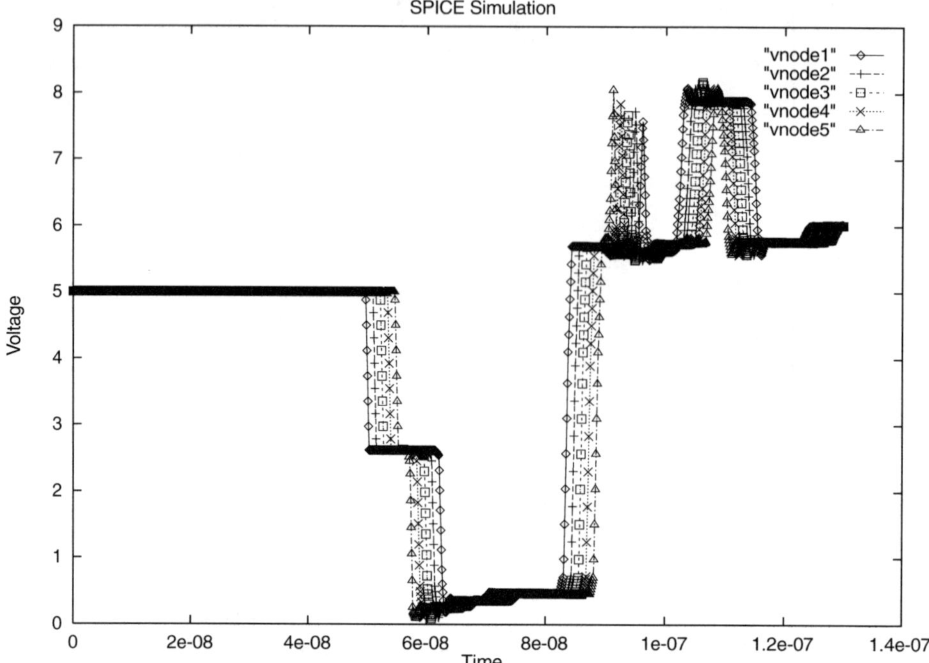

Figure 10.20 Voltage waveforms from PSpice simulation at node 1.

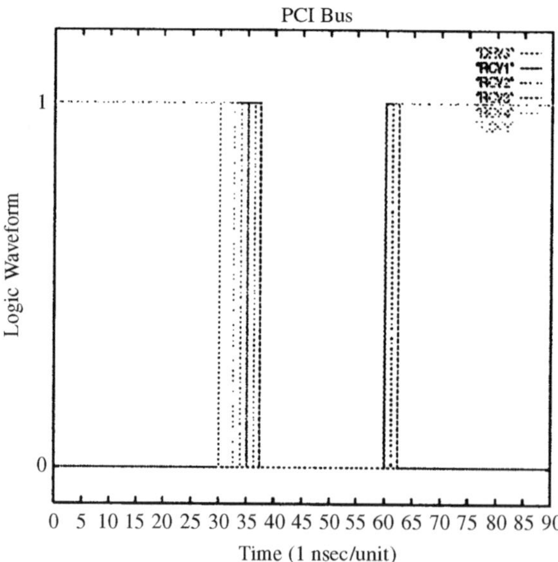

Figure 10.21 Logic waveform of driver at node 3 and resulting waveforms at the nodes of the PCI bus.

incident wave to travel towards the open circuit end of the bus and return to node 1. In Figure 10.21, this delay reduces to 5 ns for node 3 since the driver is located midway along the length of the bus. The simulation results are consistent with the corresponding HSPICE results reported in [109]. The graphs in Figure 10.20 represent results from PSpice simulation when the driver is at the end of the line and they contrast with the waveforms in Figure 10.19.

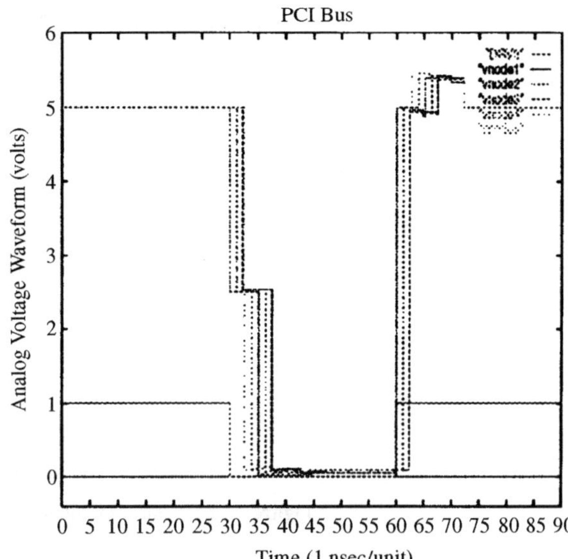

Figure 10.22 Voltage waveform of driver at node 3 and resulting waveforms at the nodes of the PCI bus.

10.5.4 Modeling and Simulation of a TTL Wired-OR Bus

Figure 10.23 presents a TTL wired-OR bus line that is often present in backplane designs [144]. Examples include the "bus busy" and "bus error" signals. In Figure 10.23, five devices are connected to the bus at five nodes and the transport delays between successive nodes are given by $\Delta_0 = 1.57$, $\Delta_1 = 2.7$, $\Delta_2 = 0.8$, $\Delta_3 = 2.3$, $\Delta_4 = 4.74$ and $\Delta_5 = 3.16$ ns respectively, resulting in the net transport delay along the line of $T_{prop} = 15$ ns. The purpose of the diodes is to clamp, i.e. limit, the negative voltage along the line. The behavior of the bus line is analyzed when it is driven by a single driver versus when there are two or more drivers.

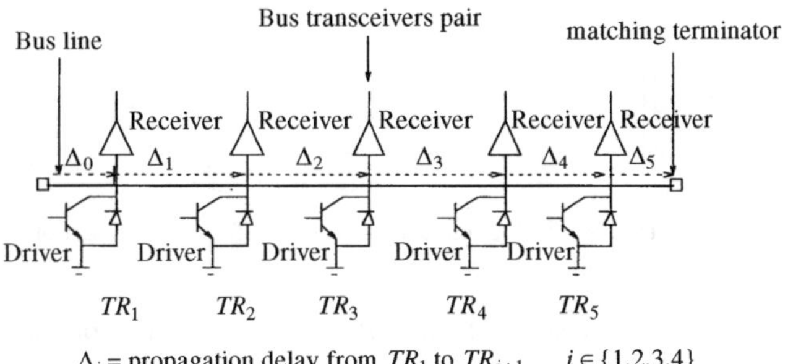

Figure 10.23 A TTL wired-OR bus line with five device drivers.

10.5.4.1 Behavior Under Single Driver. Under normal circumstances, i.e. when a single device driver asserts signals on the line, the receivers at the nodes view the exact transmitted signal, delayed only by the respective transport delays. This is revealed in the graph in Figure 10.25 that presents the logic and voltage waveforms at node 3, resulting from node 2 driving the bus with logical waveform as shown in Figure 10.24. Similar to node 3,

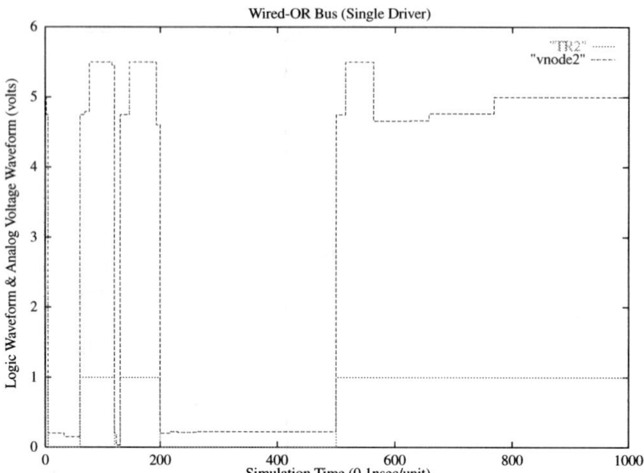

Figure 10.24 Logic and voltage waveforms of the driver at node 2.

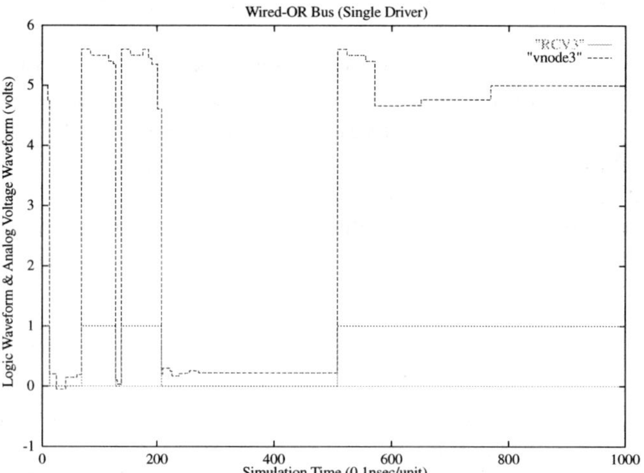

Figure 10.25 Logic and voltage waveforms at receiver at node 3.

each of the waveforms at nodes 1, 4, and 5 are observed to be delayed by the transmission time along the line.

10.5.4.2 Glitches in TTL Wired-OR Bus Line Under Multiple Drivers. Although the bus is wired-OR and two or more simultaneous drivers cannot assert conflicting logical values on the line, transmission line analysis reveals unique and important behavioral peculiarities [144] when two or more devices switch concurrently.

The integrating behavior exhibited by the analog voltage waveform between 15 ns and 45 ns along the simulation time axis in Figure 10.26 is an artifact of the proposed event driven

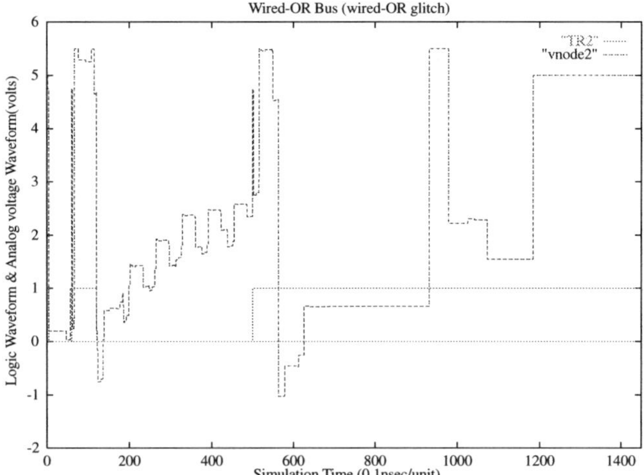

Figure 10.26 Logic and analog voltage waveforms asserted at node 2.

model and is explained as follows. When nodes 2 and 4 are pulled down from high to low, they behave as effective short circuits and the incident waveforms, originating at nodes 2 and 4 respectively, are successively reflected at the opposite ends. At each reflection, the polarity of the waveform changes nearly 180° while the magnitude following each reflection is determined by the corresponding reflection coefficient. The net voltages are obtained from algebraically adding the waves at each of the nodes 2 and 4. The TTL wired-OR bus model in this chapter assumes that the reflection coefficients at nodes 2 and 4 are voltage-dependent and, therefore, dynamic. In the proposed event driven approach, the voltage at a node between any two successive events affecting that node is assumed to remain constant. The net result is the appearance of an integrating behavior of the waveform. in reality, however, line losses will cause voltage decay. This clearly underlies a limitation of the proposed approach.

A serious problem is the generation of spurious pulses, referred to as the wired-OR glitch, that arise when two or more drivers concurrently switch from high to low. By design, a wired-OR bus is normally held at 1 and it switches to 0 when one or more devices drive the bus. The graphs in Figure 10.26 present the logical and analog waveforms that are asserted on the bus by node 2. The graphs in Figure 10.27 correspond to the assertions by the driver at node 4. Note that, around 10 ns, both drivers at nodes 2 and 4 simultaneously assert logical low values on the bus. As a result, high to low voltage traveling waves originate at nodes 2 and 4 and they begin to propagate towards both ends of the line. The wave propagating left from node 2 and the wave propagating right from node 4 are eventually absorbed by the matching terminators. However, since nodes 2 and 4 are both held at logic 0, they behave as short circuits, and the wave traveling right from the originating node 2 and that propagating left from its origin, node 4, are both reflected back as low to high voltage waveforms. The reflections continue repeatedly at the boundaries, nodes 2 and 4, and at every reflection the reflected wave is identical in amplitude to the incident wave but 180° out of phase. This results in an oscillation which manifests through a series of voltage pulses, as correctly exhibited by the logic and voltage waveforms at node 3, shown in Figure 10.28. The time period of the pulses is approximately equal to twice the round-trip delay between nodes 2 and 4, or $2 \times (0.8 + 2.3) = 6.2$ ns, that is correctly revealed in Figure 10.28. In contrast, the logic

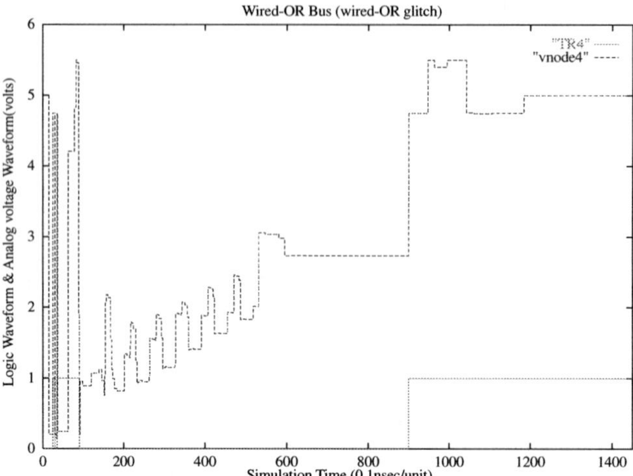

Figure 10.27 Logic and analog voltage waveforms asserted at node 4.

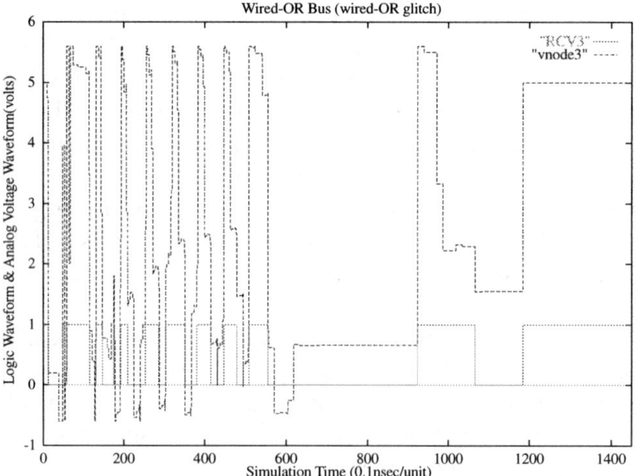

Figure 10.28 Logic and analog voltage waveforms observed at node 3.

and voltage waveforms at nodes 1 and 5 are not subject to the oscillations, as correctly revealed through Figures 10.30, 10.31, 10.32, and 10.33. Figure 10.29 presents the PSpice simulation results for node 3 and may be contrasted with those in Figure 10.28.

Table 10.1 presents the execution times required by the different circuits under event driven simulation and PSpice, both executing on a 133 MHz Pentium workstation under Linux operating system. The data reveals a significant speedup of the event driven algorithm over PSpice.

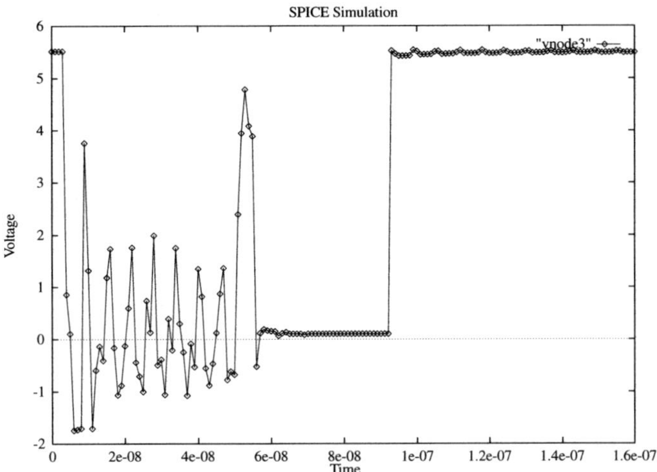

Figure 10.29 Voltage waveforms from PSpice simulation at node 3.

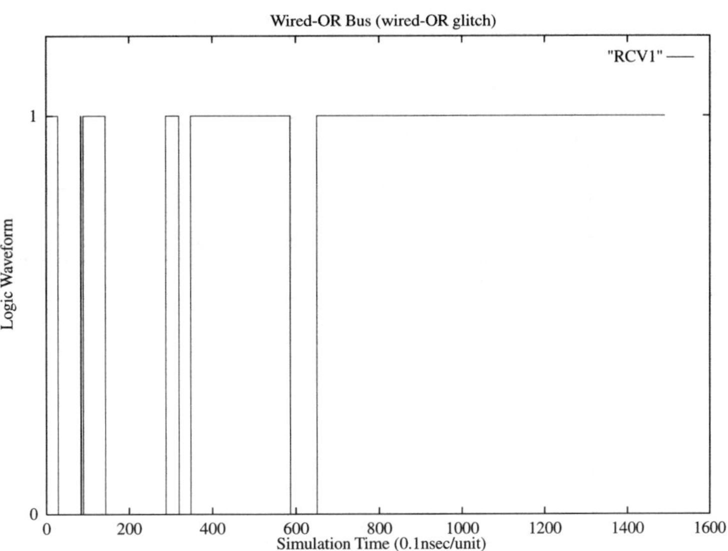

Figure 10.30 Logical waveform observed at node 1.

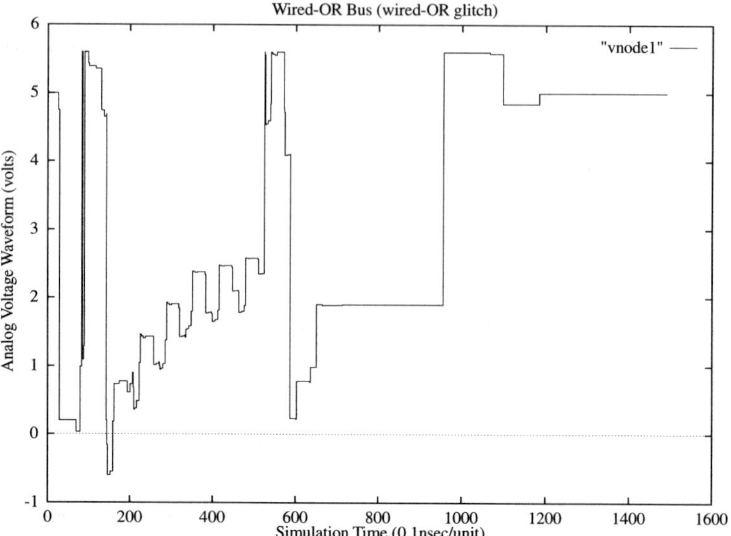

Figure 10.31 Analog voltage waveform observed at node 1.

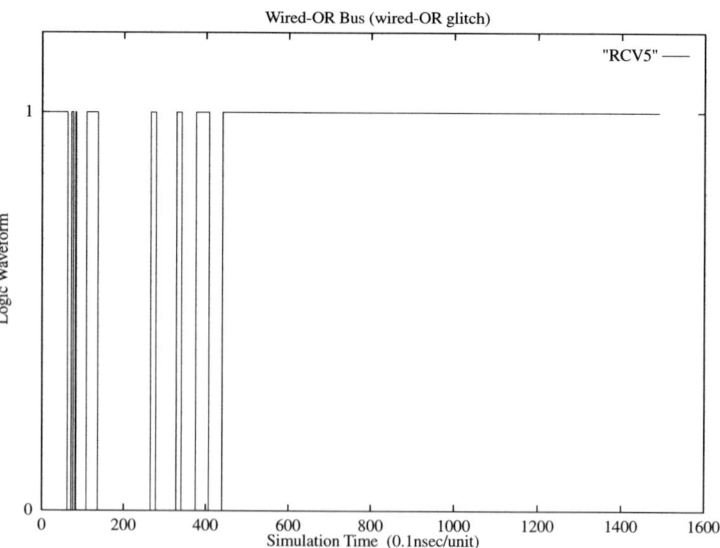

Figure 10.32 Logical waveform observed at node 5.

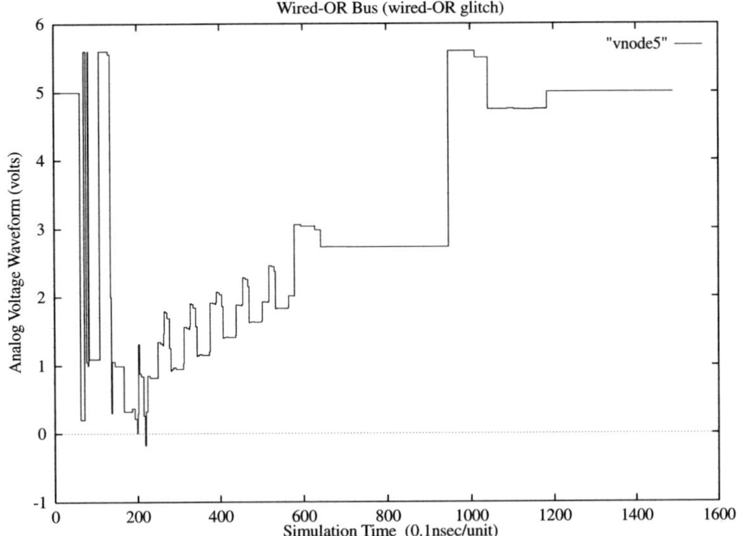

Figure 10.33 Analog voltage waveform observed at node 5.

TABLE 10.1 Simulation execution times on a Pentium workstation under Linux.

Circuit	Event driven simulation (s)	PSpice run (s)	Speedup over PSpice
Ethernet (collision)	0.030495	2.75	90
Ethernet (no-collision)	0.011462	2.43	212
PCI (driver at end)	0.007037	1.66	235
PCI (driver at midpoint)	0.002536	1.54	607
Wired-OR (glitch)	0.139689	4.03	29
Wired-OR (single driver)	0.035580	2.62	73

11

The Future of HDLs and Philosophical Reflections

The previous chapters were based on logical reasoning, mathematics, and sound science and engineering principles. In this chapter, I will take the liberty of being speculative, postulate a few key positions, reflect on the future possibilities, and share some intuitions that are yet to be proved scientifically.

The VHDL effort is unprecedented in terms of the number of individuals that it has involved since 1982, the net amount of work put in by academicians, industry personnel, and government and defense scientists, and its growing impact on electrical engineers and computer scientists worldwide. As an HDL standard, VHDL marks a significant achievement in the discipline of CAD. However, VHDL is not without its share of problems. It is bulky, slow to execute, and most important, it is perceived by most beginners and by many long time HDL veterans as difficult and confusing. The natural enthusiasm for VHDL appears to be stifled. Most of the problems with VHDL have been identified in the previous chapters and virtually all of them may be successfully addressed through a redesign by a team of individuals, skilled in language design, digital hardware, and distributed processing, through changes in the VHDL grammar and semantics. The resulting VHDL is expected to be precise, lean, and easy to learn and use. Not only will VHDL simulations execute faster, but designers will turn around VHDL models significantly faster, and with greater precision and enthusiasm.

One issue that I wish to evoke is as follows. Future hardware systems are expected to execute faster and, therefore, the slopes of the signal transitions may require greater attention. I would like to suggest that the possibility of introducing a "slope" into the signal assignment clause be explored. For example, "`sig1 <= '1' after 10 ns slope 2 ps,`" would imply that the signal, `sig1`, encounters a transition to 1 starting at 10.000 ns and completing at 10.002 ns.

It is also important to examine VHDL from the perspective of synthesis. At the present time, i.e. in 1999, to synthesize a hardware design, first the actual delays of the corresponding VHDL description are replaced with zero delays. The synthesis tool accepts the description which is, in essence, a functional specification, i.e. void of the timing, and determines a hardware design, built from standard gates and modules from the design library. There are several important limitations with this approach. First, the replacement of the actual delays with zero delays completely changes the original specification. Once lost, there is no scientific principle known to us through which the original specification, i.e. logic plus timing, can be recovered. Thus, the synthesizer cannot generate a hardware with the exact same characteristics as the original specification. Second, upon comparing the input–output results of the synthesized design against those of the original specification, if there is a discrepancy, there is no known scientific method to resolve it. Third, for increasingly higher IC densities and faster clock speeds in the future, the relative timing between the sub-components will only become more critical. The precise timing cannot be ignored without incurring a stiff penalty. Fourth, it is predicted [150] that the future will witness asynchronous processor designs with superior performance relative to the current, synchronous, designs. Recently, during its effort to design the 1.0 GHz Pentium processor, utilizing the synchronous design philosophy, Intel [151] faced severe difficulty with excessive power requirements by the modules at the clock transitions. Intel solved the problem by differentially delaying the clock signal to the different modules through arbitrary, small delays. Finally, the chief reason cited for ignoring the actual delays in the synthesis tools is the complexity of the VHDL descriptions. This is incorrect and unfair since the weakness lies in the lack of superior principles and techniques to perform synthesis, not in the hardware descriptions. Ideally, a description of a hardware system that is natural and consistent with the underlying hardware should be suitable for synthesis. However, other issues may be pertinent and the overall issue deserves a comprehensive analysis. The issue of synthesis from VHDL descriptions is discussed in [67, 152].

In the conclusion of the article [17], Baudet, Cutler, Davio, Peskin, and Rammig state a revealing prophecy: "In contrast, the 1990s are likely to be an era of inertia, as the FORTRAN/COBOL syndrome takes effect. The large body of existing 'code,' or machine descriptions will stifle innovations because of the cost to rewrite these descriptions. This will be unfortunate, as the tremendous increase in the experience due to the writing of those descriptions will expose the weakness . . . ". Apart from its obvious implications relative to HDLs, this commentary reflects the need to guard against complacency and to foster humbleness in our work to ensure the continued growth of science. The commentary also corroborates Nobel Laureate Leo Esaki's [153] advice to scientists, "do not avoid confrontation," as well as the sayings of the ancient philosophers, "For the continued progress of knowledge and science, old ideas must withdraw themselves on their own and make room for the new ideas which alone determine their own future."

My chief concern with the future of HDLs is as follows. In my vision, the future will witness very large asynchronous systems over immense geographical distances, dwarfing even today's largest system, with the individual sub-systems executing at speeds beyond our current level of comprehension. Two key issues will then surface: (1) our ability to comprehend and describe such systems effectively, and (2) the speed at which these descriptions may be executed or simulated accurately. Our present understanding and modeling of digital system behavior through changes in the input voltage values may not adequately address issue (1). A higher-level, more compact and crystallized representation of systems must be developed. Systems will probably be defined through some energy function perhaps. The basic characteristics of HDL including the notion of entity, outlined in Chapter 3, will continue to remain in effect. However, the expressive power of the HDL may need to be enhanced to accommodate the new representations. Parallel with this development, new simulation principles must be developed to address issue (2), i.e. to quickly and accurately execute the new descriptions. We are currently investigating both issues (1) and (2) but do not have any concrete answers at the moment. In this context, I wish to share two deeply personal thoughts with the reader.

First, the power of philosophy is never to be underestimated for it is a rich source of intuition and imagination. Although engineering is a hard science and the rule of reproducibility reigns supreme, many of the ideas that are accepted as scientific concepts today originated in philosophy. The concept of inconsistent event detection and preemption was motivated by the philosophical discussions on the principle of causality. Second, one's intuitions are to be trusted and developed further with precision, patiently and methodically. Intuition coupled with precision constitute a powerful research tool. They are never to be abandoned or ignored, despite criticisms even from the experts. The truth is, no one knows for sure the exact image of true knowledge. Errors and incompleteness in understanding is only natural and can happen to anyone. As nature unfolds with the progress of time, we gain better and deeper understanding of things. So, to ensure the continued growth of science, everything must be questioned and nothing is to be considered sacrosanct.

Bibliography

[1] William Morris (Ed.). *The American Heritage Dictionary of the English Language*. Houghton Mifflin Company, Boston, MA, 1981.

[2] E.I. Organick, A.I. Forsythe, and R.P. Plummer. *Programming Language Structures*. Academic Press, New York, 1978.

[3] T.W. Pratt. *Programming Languages Design and Implementation*. Prentice Hall, New Jersey, 1984.

[4] S. Vivekananda. *Karma-Yoga*. Advaita Ashrama, Calcutta, 1974.

[5] R.P. Feynman. *The Character of Physical Law*. The Messenger Lectures. The MIT Press, Cambridge, 1990.

[6] S. Lubkin. Asynchronous Signals in Digital Computers. *Mathematical Tables and other Aids in Computation*, 6(40):238–241, Oct 1952.

[7] T.J. Chaney and C.E. Molnar. Anomalous Behavior of Synchronizer and Arbiter Circuits. *IEEE Transactions on Computers*, C-22(4):421–422, 1972.

[8] M.J. Stucki and J.R. Cox, Synchronization Strategies. In *Proceedings of the Caltech Conference on VLSI*, pages 375–386, Jan 1979.

[9] R. Penrose. *The Emperor's New Mind Concerning Computers, Minds, and the Laws of Physics*. Oxford University Press, New York, 1989.

[10] Fredrick J. Hill and Gerald R. Peterson. *Digital Systems: Hardware Organization and Design*, Chapter Introduction, pages 5–6. John Wiley and Sons, New York, 2 edition, 1978.

[11] Sumit Ghosh and P.A. Subrahmanyam. On the Notion of Control Signals in Digital Designs. *IEEE Circuits and Devices*, 8(4):47–52, July 1992.

[12] P.W. Case, H.H. Graff, and M. Kloomok. The Recording, Checking, and Printing of Logic Diagrams. In *Proceedings of the Eastern Joint Computer Conference*, pages 108–118, 1958.

[13] C.G. Bell and A. Newell. The PMS and ISP Descriptive Systems for Computer Structures. In *Proceedings of the AFIPS Conference, SJCC*, Vol. 36, pages 351–374, Reston, Virginia, 1970.

[14] S.Y.H. Su. A Survey of Computer Hardware Description Languages in the U.S.A. *IEEE Computer*, Dec 1974.

[15] W.M. vanCleemput. Computer Hardware Languages and their Applications. In *Proceedings of the 16th Design Automation Conference*, pages 554–560, ACM/IEEE San Diego, CA, June 1979.

[16] W.M. vanCleemput and H. Ofek. Design Automation for Digital Systems. *IEEE Computer*, pages 114–122, Oct 1984.

[17] G.M. Baudet, M. Cutler, M. Davio, A.M. Peskin, and F.J. Rammig. The Relationship between HDLs and Programming Languages. In *VLSI and Software Engineering Workshop*, pages 64–69, Port Chester, NY, June 1982.

[18] M.R. Barbacci and T. Uehara. Computer Hardware Description Languages: The Bridge Between Software and Hardware. *IEEE Computer*, pages 6–8, Feb 1985.

[19] S.G. Chappel and P.R. Menon. Functional Simulation in the LAMP System. *Journal of Design Automation and Fault Tolerant Computing*, pages 203–215, May 1977.

[20] G.R. Case and J.D. Stauffer. SALOGS-IV: A Program to Perform Logic Simulation and Fault Diagnosis. In *Proceedings of the 15th Design Automation Conference*, pages 392–397, ACM/IEEE June 1978.

[21] S.A. Szygenda and A.A. Lekkos. Integrated Techniques for Functional and Gate Level Digital Logic Simulation. In *Proceedings of the 10th Design Automation Conference*, pages 159–172, 1973.

[22] Y. Chu. An ALGOL-like Computer Design language. *Communications of the ACM*, pages 607–615, October 1965.

[23] Y. Chu, D.L. Dietmeyer, F. Hill, and D. Siewiorek. Introducing Computer Hardware Description Languages. *IEEE Computer*, 7(12):27–44, December 1974.

[24] D.L. Dietmeyer. Introducing DDL. *IEEE Computer*, 7(12):34–38, Dec 1974.

[25] M. Barbacci, D. Siewiorek, R. Gordon, R. Howbrigg, and S. Zuckerman. An Architectural research facility—ISP Descriptions, simulation, data collection. In *Proceedings of the AFIPS National Computer Conference*, 1977.

[26] W.M. vanCleemput. A Hierarchical Language for the Structural Description of Digital Systems. In *Proceedings of the 14th Design Automation Conference*, pages 378–385, ACM/IEEE New Orleans, June 1977.

[27] S.A. Szygenda and E.W. Thompson. Digital Logic Simulation in a Time-Based Table-Driven Environment, Part 1: Design Verification, *IEEE Computer*, 8(3):24–40, March 1975.

[28] E.W. Thompson and S.A. Szygenda. Digital Logic Simulation in a Time-Based Table-Driven Environment, Part 2: Parallel Fault Simulation. *IEEE Computer*, 8(3):24–40, March 1975.

[29] H.Y. Chang, G.W. Smith, and R.B. Walford. LAMP: System Description. *The Bell System Technical Journal*, 53(8):1431–1449, October 1974.

[30] Swami Venkatasananda. *The Concise Yoga Vasistha*. State University of New York Press, Albany, New York, 1984.

[31] S. Sivananda. *What becomes of the Soul after Death*. The Divine Life Society, Tehri-Garhwal, U.P., India, 1979.

[32] D. Hill. Adlib Users Manual. *Technical Report 177, Computer Systems Lab., Stanford University*, 1979.

[33] The Institute of Electrical and Electronic Engineers. IEEE Standard VHDL Language Reference Manual. ANSI/IEEE Std 1076-1993, IEEE, Institute of Electrical and Electronics Engineers, Inc., 345 East 47th Street, New York, NY 10017, USA, April 14 1994.

[34] M.R. Barbacci. A Comparison of Register Transfer Languages for Describing Computers and Digital Systems. *IEEE Transactions on Computers*, C-24(2):137–150, Feb 1975.

[35] R. Piloty and D. Borrione. The Conlan Project: Concepts, Implementations, and Applications. *IEEE Computer*, C-24(2):81–92, Feb 1985.

[36] M. Shahdad, R. Lipsett, E. Marschner, K. Sheehan, and H. Cohen. VHSIC Hardware Description Language. *IEEE Computer*. Feb. 1985.

[37] The Engineering Staff of TI Inc. *The TTL Databook for Design Engineers*. Texas Instruments Incorporated, Dallas, Texas, 1976.

[38] D. Hill. Language and Environment for Multi-level Simulation. *Technical Report 185, Computer Systems Lab., Stanford University*, 1980.

[39] Bengt Magnhagen. Probability Based Verification of Time Margins in Digital Designs. Ph.D. thesis, Department of Electrical Engineering, Linkoping University, S-581 83 Linkoping, Sweden, September 1977.

[40] Daisy Systems Corporation. Daisy Behavioral Language. Technical Report, Daisy Systems Corporation, Sunnyvale, California 94086, December 1983.

[41] S.G. Chappell, C.H. Elmendorf, and L.D. Schmidt. LAMP: Logic Circuit Simulators. *The Bell System Technical Journal*, 53(8):1451–1476, October 1974.

[42] M.A. Breuer and A.D. Friedman. *Diagnosis and Reliable Design of Digital Systems*. Computer Science Press, Maryland, 1976.

[43] Sumit Ghosh. A Rule-Based Design Verifier. Ph.D. thesis, Computer Systems Laboratory, Department of Electrical Engineering, Stanford University, Palo Alto, CA 94305, August 1984.

[44] D.C. Luckham, S. Ghosh, Y. Huh, and A. Stanculescu. Analysis of the VHSIC Hardware Description Language. Draft, Computer Systems Laboratory, Stanford University, Palo Alto, CA 94305, May 7 1985.

[45] D. Luckham, A. Stanculescu, Y. Huh, and S. Ghosh. The Semantics of Timing Constructs in Hardware Description Languages. In *Proceedings of the International Conference on Computer Design (ICCD'86)*, pages 156–164, IEEE Rye Town Hilton, Port Chester, New York, October 6–9 1986.

[46] Sumit Ghosh and Meng-Lin Yu. A Preemptive Scheduling Mechanism for Accurate Behavioral Simulation of Digital Designs. *IEEE Computer*, 38(11):1595–1600, November 1989.

[47] M.A. Breuer and A.C. Parker. Digital System Simulation: Current Status and Future Trends. In *Proceedings of the 18th Design Automation Conference*, pages 269–275, ACM/IEEE 1981.

[48] A. Pawlak and W. Wrona. Modern Object Oriented Programming Language as HDL. In *Computer Hardware Description Languages and their Applications*, pages 343–362, Elsevier Science Publishers, Edited M.R. Barbacci and C.J. Koomen, 1987.

[49] D.D. Hill and W.M. vanCleemput. SABLE: A Tool for Generating Structured Multi-Level Simulations. In *Proceedings of the 16th Design Automation Conference*, pages 272–278, ACM/IEEE June 1979.

[50] D.D. Hill and D. Coehlo. *Multi-Level Simulation for VLSI Design*. Kluwer Academic Publishers, Boston, MA, 1987.

[51] Sumit Ghosh. A Distributed Approach to Timing Verification of Synchronous and Asynchronous Digital Designs. *IEEE Transactions on Computer Aided Design of ICs and Systems*, CAD-6(4):666–677, July 1987.

[52] Sumit Ghosh. Behavioral-Level Fault Simulation. *IEEE Design and Test of Computers*, pages 31–42, June 1988.

[53] W.E. Cory and W.M. vanCleemput. Developments in Verification of Design Correctness—A Tutorial. In *Proceedings of the 17th Design Automation Conference*, pages 156–164, ACM/IEEE June 1980.

[54] D. Hill. Beginnings of ADLIB/SABLE. *IEEE Design and Test of Computers*, pages 58–60, September 1992.

[55] Department of the U.S. Air Force. Draft Request for Proposal F33615-83-R-1003, VHSIC Hardware Description Language (VHDL). *Sources Sought Synopsis PMRE 82-116*, September 1982.

[56] E. Sternheim, R. Singh, and Y. Trivedi. *Digital Design with Verilog HDL*. Automata Publishing Company, Cupertino, CA, 1990.

[57] D.E. Thomas and P.R. Moorby. *The Verilog Hardware Description Language*. Kluwer Academic Publishers, Boston, 1991.

[58] U. Golze. *VLSI Chip Design with the Hardware Description Language Verilog*. Springer, Berlin, 1996.

[59] Open Verilog International. *Proceedings of the 1996 International Verilog HDL Conference*, Santa Clara, California, February 26–28 1996.

[60] C. Burns. An Architecture for a Verilog Hardware Accelerator. In *Proceedings of the 1996 International Verilog HDL Conference*, pages 2–11, Santa Clara, California, February 26–28 1996.

[61] M. Becker. Faster Verilog Simulations Using a Cycle Based Programming Methodology. In *Proceedings of the 1996 International Verilog HDL Conference*, pages 24–31, Santa Clara, California, February 26–28 1996.

[62] G. Arnout and H. DeMan. The Use of Threshold Function and Boolean-Controlled Network Elements for Macromodelling of LSI Circuits. *IEEE Journal of Solid State Circuits*, June 1978.

[63] R.A. Newton. The Simulation of LSI Circuits. Memo # ucb/erl m78/52, Department of Electrical Engineering, University of California, Berkeley, July 1978.

[64] T.W. McWilliams. The SCALD Timing Verifier: A New Approach to Timing Constraints in Large Digital Systems. In *Proceedings of the International Symposium on Circuits and Systems*, pages 415–423, May 1980.

[65] J. Hartmanis and R. Stearns. *Algebraic Structure Theory of Sequential Machines*. Prentice Hall, Inc., NJ, 1966.

[66] US Department of Defense. *Reference Manual for the Ada Programming Language*. Government Printing Office, Washington, DC 20402, 1980.

[67] D.L. Perry. *VHDL*. Computer Engineering. McGraw Hill, Inc., New York, 1993.

[68] S. Mazor and P. Longstraat. *A Guide to VHDL*. Kluwer Publishers, Boston, MA, 1993.

[69] R. Goering. VHDL Shared Variables Create Dissent—Spec Divides EDA. *EETimes*, (937). January 20 1997.

[70] Advanced Micro Devices Inc. *Bipolar Microprocessor Logic and Interface—AM2900 Family, 1985 Data Book*. Advanced Micro Devices, Inc., Sunnyvale, CA, 1985.

[71] Z.G. Vranesic and S.G. Zaky. *Microcomputer Structures*. Holt, Reinhart and Winston, New York, 1989.

[72] N. Ishiura, H. Yasuura, and S. Yajima. Time First Evaluation Algorithm for High-Speed Logic Simulation. In *Proceedings of the ICCAD*, pages 197–199, IEEE Santa Clara, CA, 1984.

[73] R.M. Fujimoto. Parallel Discrete Event Simulation. *Communications of the ACM*, 33(10):30–53, October 1990.

[74] J. Misra. Distributed Discrete-Event Simulation. *Computing Surveys*, 18(1):39–65, March 1986.

[75] R.D. Chamberlain and M.A. Franklin. Hierarchical Discrete-event Simulation on Hypercube Architectures. *IEEE Micro*, 10(4):10–20, 1990.

[76] L. Soule and T. Blank. Parallel Logic Simulation on General Purpose Machines. In *Proceedings of the 25th Design Automation Conference*, ACM/IEEE June 1988.

[77] D. Jefferson. Virtual Time. *ACM Transactions on Programming Languages*, 7(3):404–425, July 1985.

[78] P.J. Ashden, H. Detmold, and W.S. McKeen. Parallel Execution of VHDL Models. Technical Report, Department of Computer Science, University of Adelaide, Australia, 1993.

[79] J.V. Briner, J.L. Ellis, and G. Kedem. Breaking the Barrier of Parallel Simulation of Digital Systems. In *Proceedings of the 28th Design Automation Conference*, ACM/IEEE San Francisco, June 1991.

[80] J. Sissler. Assessing the Potential of Multi-threaded VHDL Simulation. In *VHDL Boot Camp Proceedings, Fall '93 Conference*, pages 131–136, Red Lion Inn, San Jose, CA, October 10–13 1993.

[81] P. Chawla, H. Carter, D. Barker, and S. Bilik. Preliminary Design and Analysis of a Hardware Accelerator for VHDL Simulation. In *VHDL Boot Camp Proceedings, Fall '93 Conference*, pages 267–279, Red Lion Inn, San Jose, CA, October 10–13 1993.

[82] F. Mattern. Efficient Algorithms for Distributed Snapshots and Global Virtual Time Approximation. *Journal of Parallel and Distributed Computing*, 18(4):423–434, 1993.

[83] I.F. Akyildiz, L. Chen, S.R. Das, R. Fujimoto, and R. Serfozo. The Effect of Memory Capacity on Time Warp Performance. *Journal of Parallel and Distributed Computing* 18(4):411–422, 1993.

[84] D.M. Nicol and P. Heidelberger. Optimistic Parallel Simulation of Continuous Time Markov Chains Using Uniformizatio. *Journal of Parallel and Distributed Computing*, 18(4):395–410, 1993.

[85] T.K. Som and R.G. Sargent. A New Process to Processor Assignment Criterion for Reducing Rollbacks in Optimistic Simulation. *Journal of Parallel and Distributed Computing*, 18(4):509–515, 1993.

[86] C.P. Wen and K.A. Yelick. Parallel Timing Simulation on a Distributed Memory Multiprocessor. In *Proceedings of the ICCAD*, pages 131–136, IEEE Santa Clara, November 1993.

[87] J.S. Steinman. SPEEDES: Synchronous Parallel Environment for Emulation and Discrete Event Simulation. In *Fifth Workshop on Parallel and Distributed Simulation (PADS91)*, pages 95–103, San Diego, CA, 1991. The Society for Computer Simulation.

[88] D. West. Lazy Rollback and Lazy Reevaluation. Master's thesis, University of Calgary, January 1988.

[89] V. Krishnaswamy and P. Banerjee. Actor Based Parallel VHDL Simulation Using Time Warp. In *Proceedings of the Tenth Workshop on Parallel and Distributed Simulation (PADS96)*, Philadelphia, PA, pages 135–142, May 22–24 1996.

[90] M.L. Bailey. A Time-Based Model for Investigating Parallel Logic-Level Simulation. *IEEE Transactions on CAD of Integrated Circuits and Systems*, 11(7):816–824, 1992.

[91] K.M. Chandy and J. Misra. Asynchronous Distributed Simulation via a Sequence of Parallel Computations. *Communications of the ACM*, 24(4):198–206, April 1981.

[92] K.M. Chandy and J. Misra and L. Haas. Distributed Deadlock Detection. *ACM Transactions on Computer Systems*, 1(2):144–156, May 1983.

[93] J.K. Peacock and Wang and Manning. Distributed Simulation Using a Network of Processors. *Computer Networks*, 3(1):44–56, 1979.

[94] Ronald DeVries. Reducing Null Messages in Misra's Distributed Discrete Event Simulation Method. *IEEE Transactions on Software Engineering*, 16(1):82–91, January 1990.

[95] D.A. Reed and Allen Malony. Parallel Discrete Event Simulation: The Chandy–Misra Approach. In *Proceedings of the SCS Multiconference on Distributed Simulation*, pages 8–13, San Diego, CA, February 1988.

[96] K.M. Chandy and R. Sherman. The Conditional Event Approach to Distributed Simulation. In *SCS Multiconference on Distributed Simulation*, pages 93–99, Tampa, FL, March 1989.

[97] Eric Debenedictis, Sumit Ghosh, and Meng-Lin Yu. An Asynchronous Distributed Discrete Event Simulation Algorithm for Cyclic Circuits using Data-flow Network. *IEEE Computer*, 24(6):21–33, June 1991.

[98] L. Lamport. Time, Clocks, and the Ordering of Events in a Distributed System. *Communications of the ACM*, 21(7):559–565, 1978.

[99] E.G. Ulrich. Exclusive Simulation of Activity in Digital Networks. *Communications of the ACM*, 12(2):102–110, 1969.

[100] Narsingh Deo. *Graph Theory with Applications to Engineering and Computer Science*. Prentice Hall Inc, NJ, 1974.

[101] E. Debenedictis. Multiprocessor Programming with Distributed Variables. In *Proceedings of the Conference on Hypercube Multiprocessor*, August 1985.

[102] J.T. Rayfield and H.F. Silverman. Operating System and Applications of the Armstrong Multiprocessor. *IEEE Computer*, 21(6):38–52, 1988.

[103] H-C. Shih, P.G. Kovijanic, and R. Razdan. A Global Feedback Detection Algorithm for VLSI Circuits. In *Proceedings of the IEEE ICCD*, pages 37–40, IEEE September 1990.

[104] S. Davidson and J. Lewandowski. ESIM/AFS—A Concurrent Architectural Level Fault Simulator. In *Proceedings of the IEEE International Test Conference*, pages 375–383, IEEE October 1986.

[105] S. Ghosh, D. Luckham, Y. Huh, and A. Stanculescu. The Semantics of Non Anticipatory Timing Constructs in Behavior-Level Hardware Description Languages. Technical Report 11354-860321-02tm, Bell Laboratories Research, Holmdel, NJ 07733, USA, March 1986.

[106] Steven Levitan. VHDL Compiler (Vcomp) and VHDL Simulator (Vsim). Technical Report, University of Pittsburgh Department of Electrical Engineering, June 1993.

[107] S. Ghosh and M.-L. Yu. An Asynchronous Distributed Approach for the Simulation and Verification of Behavior-Level Models on Parallel Processors. *IEEE Transactions on Parallel and Distributed Systems*, 6(6):639–652, June 1995.

[108] John L. Hennessy and David A. Patterson. *Computer Architecture: A Quantitative Approach*. Morgan Kaufmann Publishers Inc., 1990.

[109] Donal Telian. Treat PC-board Traces as Transmission Lines to Specify Drive Buffers. EDN, A Cahners Publication, Newton, Mass, 38(18):129–136, 2 September 1993.

[110] B. Rosenthal and R. Sartore. Designing PCI-Compliant Master/Slave Interfaces for Add-On Cards. EDN, A Cahners Publication, Newton, Mass, 40(7):97–104, 30 March 1995.

[111] Cadence Design Systems, Chelmsford, MA 01824. *Private communications with C. Kumar*, February 1994.

[112] C.J. Steinberg and I.M. Wilson. Simulation Programs Iron Out Transmission-line Effects. EDN, A Cahners Publication, Newton, Mass, 39(5):115–118, 3 March 1994.

[113] Tonny Yu. Create Optimal Simulation Libraries Using VHDL. EDN, A Cahners Publication, Newton, Mass, 38(10):133–140, 13 May 1993.

[114] William Fazakerly and Jose A. Torres. Face-Off: Traditional Gate-level vs. VHDL Simulation for ASIC Sign-off. *Computer Design's ASIC Design*, pages A14–A19, February 1994.

[115] S. Chowdhury, J.S. Barkatullah, D. Zhou, Er-Wei Bai, and K.E. Lonngren. A Transmission line Simulator for High-Speed Interconnects. *IEEE Transactions on Circuits and Systems—II: Analog and Digital Signal Processing*, 39(4):201–211, April 1992.

[116] J.W. Palchefsky. A Two-Port Model for a Transmission Line with Distributed Sources. *IEEE Transaction on Nuclear Science*, NS-34(6):1488–1491, December 1987.

[117] H.-Y. Yee and K. Wu. Printed Circuit Transmission-Line Characteristic Impedance by Transverse Modal Analysis. *IEEE Transaction on Microwave Theory and Techniques*, MTT-34(11):51–74, November 1986.

[118] J.C. Liao, O.A. Palunsky, and J.L. Prince. Computation of Transients in Lossy VLSI Packaging Interconnections. *IEEE Transactions on Components, Hybrids and Manufacturing Technology*, 13(4):833–838, December 1990.

[119] Kürsad Kiziloglu, Nadir Dagli, George L. Matthaei, and Stephen I. Long. Experimental Analysis of Transmission Line Parameters in High-Speed GaAs Digital Circuit Interconnects. *IEEE Transactions on Microwave Theory and Techniques*, 39(8):1361–1367, August 1991.

[120] V.M. Hietala, Y.R. Kwon, and K.S. Champlin. Low-Loss Slow-Wave Propagation along a Microstructure Transmission Line on a Silicon Surface. *Electronics Letters*, 22(14):755–756, July 1986.

[121] W.R. Eisenstadt and Y. Eo. S-Parameter-Based IC Interconnect Transmission Line Characterization. *IEEE Transactions on Components, Hybrids and Manufacturing Technology*, 15(4):483–489, August 1992.

[122] J.R. Brews. Transmission Line Models for Lossy Waveguide Interconnections in VLSI. *IEEE Transactions on Electron Devices*, ED-33(9):1356–1365, 1986.

[123] Giovanni Ghione, Ivan Maio, and Giuseppe Vecchi. Modeling of Multiconductor Buses and Analysis of Crosstalk, Propagation Delay, and Pulse Distortion in High-Speed GaAs Logic Circuits. *IEEE Transaction on Microwave Theory and Techniques*, 37(3):445–456, March 1989.

[124] H. Nakanishi and A. Ametani. Transient calculation of transmission line using superposition law. *IEEE Proceedings*, 133 Part C(5):263–269, July 1986.

[125] T. Sakurai. Approximation of Wiring Delay in MOSFET LSI. *IEEE Journal of Solid State Circuits*, SC-18(4):418–426, 1983.

[126] H. You and M. Soma. Analysis and Simulation of Multiconductor Transmission Lines for High-Speed Interconnect and Package Design. *IEEE Transactions on Components, Hybrids, and Manufacturing Technology*, 13(4):839–846, 1990.

[127] M. Cases and D.M. Quinn. Transient Response of Uniformly Distributed RLC Transmission Lines. *IEEE Transactions on Circuits and Systems*, CAS-27(3):200–207, 1980.

[128] G.R. Deily. Closed-Form Solutions for Voltage-Step Response of Open and Shorted Distributed RC Lines. *IEEE Transactions on Circuits and Systems*, CAS-22(6):534–541, June 1975.

[129] R.J. Antinone and G.W. Brown. The Modeling of Resistive Interconnects for Integrated Circuits. *IEEE Journal of Solid State Circuits*, SC-18(2):200–203, April 1983.

[130] T.A. Fjeldly, A. Paulsen, and O. Jensen. A GaAs MESFET Small-Signal Equivalent Circuit Including Transmission Line Effects. *IEEE Transactions on Electron Devices*, 36(9):1557–1563, September 1989.

[131] M.E. Mokari-Bolhassan and S.M. Kang. Analysis and Correction of VLSI Delay Measurement Errors Due to Transmission-Line Effects. *IEEE Transactions on Circuits and Systems*, 35(1):19–25, January 1988.

[132] R.R. Troutman and M.J. Hargrove. Transmission Line Modeling of Substrate Resistances and CMOS Latchup. *IEEE Transactions on Electron Devices*, ED-33(7):945–954, July 1986.

[133] K.M. El-Shennawy, G.P. Finani, and M.B. Tayel. Closed-Form Solutions for Voltage Pulse Response of Open, Shorted, and Loaded Distributed RC Thin Film Structure. *IEEE Transactions on Circuits and Systems*, 38(12):1567–1571, December 1991.

[134] W.C. Elmore. The Transient Response of Damped Linear Networks with Particular Regard to Wideband Amplifiers. *Journal of Applied Physics*, 19:55–63, Jan 1948.

[135] J. Rubenstein, Paul Penfield, and M.A. Horowitz. Signal Delay in RC Tree Networks. *IEEE Transactions on Computer-Aided Design*, CAD-2(3):202–211, July 1983.

[136] L.T. Pillage and R.A. Rohrer. Asymptotic Waveform Evaluation for Timing Analysis. *IEEE Transactions on Computer-Aided Design*, CAD-9(4):152–166, April 1990.

[137] Lisa Gunn. A Plethora of VHDL Products Parade at DAC. *Electronic Design*, A Penton Publication, Cleveland, Ohio, pages 51–74, 14 June 1990.

[138] David W. Smith and Brian S. Cohen. The Communication Gap between VHDL and the Analog Designer. *Electronic Design*, A Penton Publication, Cleveland, Ohio, pages 58–58, 14 June 1990.

[139] David Halliday and Robert Resnick. *Physics Part II*. John Wiley and Sons Inc., New York, 1962.

[140] Philip C. Magnusson. *Transmission Lines and Wave Propagation*. Allyn and Bacon, Inc., Boston, 1965.

[141] Harold S. Stone. *Microcomputer Interfacing*. Addison-Wesley Publishing Company, Inc., Reading, MA, 1982.

[142] Design Automation Standards Committee and IEEE Standards Coordinating Committee 20. IEEE Standard VHDL Language Reference Manual. ANSI/IEEE Std 1076-1993, IEEE, Institute of Electrical and Electronics Engineers, Inc., 345 East 47th Street, New York, NY 10017, USA, June 1994.

[143] Joseph Di Giacomo. *Digital Bus Handbook*. McGraw-Hill, New York, 1990.

[144] Arthur L. Dexter. *Microcomputer Bus Structures and Bus Interface Design*. Electrical Engineering and Electronics. Marcel Dekker, Inc, 1986.

[145] Ravender Goyal. Managing Signal Integrity. *IEEE Spectrum*, 31(3):54–58, March 1994.

[146] Paul Horowitz and Winfield Hill. *The Art of Electronics, Second Edition*. Cambridge University Press, 1989.

[147] Ronald Waxman and Larry Saunders. VHDL Language Reference Manual: IEEE Standard 1076. Technical Report, The Institute of Electrical and Electronic Engineers, New York, May 1987.

[148] Andrew S. TanenBaum. *Computer Networks*, Chapter The Medium Access Sublayer, pages 141–165. Prentice-Hall International, Inc., New Jersey, 2 edition, 1988.

[149] HSPICE Application Note. *PCI Modeling Using HSPICE*. Meta-Software, Inc., 1300 White Oaks Road, Campbell, CA 95008, October 1992.

[150] Tony Werner and Venkatesh Akella. Asynchronous Processor Survey. *IEEE Computer*, 30(11):67–76, November 1997.

[151] Intel Corporation, Santa Clara, CA. Private communications with R. Wirt, July 1998.

[152] K. Skahill. *VHDL for Programmable Logic*. Addison-Wesley Publishing Company, Inc., Reading, MA, 1996.

[153] Leo Esaki. The Secret to Winning a Nobel Prize. *India Tribune Weekly, 100 West 32nd Street, 4th Floor, New York, NY 10001*, 19(2), January 14 1995.

[154] L. Soule and A. Gupta. Characterization of Parallelism and Deadlocks in Distributed Digital Logic Simulation. In *Proceedings of the 26th Design Automation Conference*, ACM/IEEE June 1989.

Index

H

I

L

M

N

P

R

S

T

U

V

Z

About the Author

Sumit Ghosh received his B.Tech degree from IIT Kanpur, India and his M.S. and Ph.D. degrees from Stanford University. He serves as a tenured associate professor and the associate chair for research and graduate programs in the Computer Science and Engineering Department at Arizona State University.

Previously Dr. Ghosh had been on the faculty at Brown University and a member of the technical staff (principal investigator) at Bell Laboratories Research in Holmdel, NJ. His industrial experience includes Silvar-Lisco in Menlo Park, CA, Fairchild Advanced Research and Development, and Schlumbergber Palo Alto Research Center. Dr. Ghosh's research interests include fundamental problems from the disciplines of asynchronous distributed algorithms, stability of complex asynchronous algorithms, networking, impact of topology on network performance, network security modeling security attacks in ATM networks, mobile wireless networks, modeling and distributed simulation of complex systems, algorithms for real-time banking, distributed resources allocation, hardware description languages, continuity of care in medicine, intelligent transportation, distributed visualization of deep space networks, synthetic creativity, qualitative metrics for evaluating advanced graduate courses, and issues in the Ph.D. process.

Dr. Ghosh's past and present grant support has been from the National Security Agency, U.S. Air Force Labs in Rome, New York through Motorola, DARPA, U.S. Army Research Office, U.S. Army Research Lab, Intel, NYNEX, Air Force Office of Scientific Research, IEEE, Bellcore, Ballistic Missile Defense Organization (Pentagon), and the National Library of Medicine (NIH). He has written over 65 Journal/Transactions papers and 65 refereed conference papers. Dr. Ghosh serves on the editorial board of the IEEE Press Book Series on Microelectronic Systems.

About the Author